TRENDS IN BIOPHYSICS

From Cell Dynamics Toward Multicellular Growth Phenomena

TRENDS IN BIOPHYSICS

From Cell Dynamics Toward Multicellular Growth Phenomena

Edited by
Pavel Kraikivski, PhD

Apple Academic Press
TORONTO NEW JERSEY

Apple Academic Press Inc.
3333 Mistwell Crescent
Oakville, ON L6L 0A2
Canada

Apple Academic Press Inc.
9 Spinnaker Way
Waretown, NJ 08758
USA

Exclusive worldwide distribution by CRC Press, a member of Taylor & Francis Group

Printed in the United States of America on acid-free paper

International Standard Book Number-13: 978-1-926895-36-9 (Hardcover)

Library of Congress Control Number: 2012951946

Library and Archives Canada Cataloguing in Publication

Trends in biophysics: from cell dynamics toward multicellular growth phenomena/edited by Pavel Kraikivski.

Includes bibliographical references and index.
ISBN 978-1-926895-36-9
1. Cell physiology. 2. Cells--Growth. 3. Cell cycle--Regulation.
4. Biophysics. I. Kraikivski, Pavel

QH631.T74 2013 571.6 C2012-906414-9

Apple Academic Press also publishes its books in a variety of electronic formats. Some content that appears in print may not be available in electronic format. For information about Apple Academic Press products, visit our website at **www.appleacademicpress.com** and the CRC Press website at **www.crcpress.com**

About the Editor

Pavel Kraikivski, PhD

Pavel Kraikivski, PhD, is a research faculty member in the Department of Biological Sciences at Virginia Polytechnic Institute and State University, Blacksburg, Virginia, USA. He has acted as a reviewer for several journals and has several published articles to his name as well as several presentations at professional meetings. His current research interests include biophysics, system biology, molecular biophysics and structural biology, cell cycles, intracellular systems, and intracellular transport.

Dr. Kraikivski has a unique academic background from his education at Russian, German, and American universities and an extensive scientific research education. He has conducted research at the Max-Planck Institute for Colloids and Interfaces in Germany, where his research focused on active bio-polymers, and has also worked at the Center for Cell Analysis and Modeling at the University of Connecticut Health Center. He has made significant contributions in computational cell biology and biophysics, particularly in the areas of intracellular transport and cytoskeleton dynamics.

Contents

List of Contributors .... *ix*
List of Abbreviations .... *xiii*
Introduction .... *xv*

1. **Mammalian Cell Cycle Regulation** **1**
Rajat Singhania, R. Michael Sramkoski, James W. Jacobberger, and John J. Tyson
2. **Spindles and Active Vortices** **25**
David A. Head, W. J. Briels, and Gerhard Gompper
3. **Actomyosin-Dependent Cortical Dynamics** **49**
Patrizia Sommi, Dhanya Cheerambathur, Ingrid Brust-Mascher, and Alex Mogilner
4. **Filament Depolymerization During Bacterial Mitosis** **71**
Edward J. Banigan, Michael A. Gelbart, Zemer Gitai, Ned S. Wingreen, and Andrea J. Liu
5. **Actin Ring Constriction** **95**
Alexander Zumdieck, Karsten Kruse, Henrik Bringmann, Anthony A. Hyman, and Frank Jülicher
6. **Curved Activators and Cell-Membrane Waves** **109**
Barak Peleg, Andrea Disanza, Giorgio Scita, and Nir Gov
7. **Actin Waves and Phagocytic Cup Structures** **133**
Günther Gerisch
8. **Yeast and Scaling** **145**
Daniel Riveline
9. **Cellular Shapes from Protrusive and Adhesive Forces** **157**
Roie Shlomovitz, Kathrin Schloen, Theresia Stradal, and Nir S. Gov
10. **Amoeboid Cells' Protrusions** **193**
Peter J. M. Van Haastert
11. **Collective Cell Migration** **205**
Masataka Yamao, Honda Naoki, and Shin Ishii

12. Differential Adhesion and Cell Sorting 237

Ying Zhang, Gilberto L. Thomas, Maciej Swat, Abbas Shirinifard, and James A. Glazier

Author Notes 263

Index 269

List of Contributors

Edward J. Banigan
Department of Physics and Astronomy, University of Pennsylvania, Philadelphia, Pennsylvania, United States of America

W. J. Briels
Computational Biophysics, University of Twente, 7500 AE Enschede, The Netherlands

Henrik Bringmann
Max Planck Institute for Molecular Cell Biology and Genetics, Dresden, Germany

Ingrid Brust-Mascher
Department of Molecular and Cellular Biology, University of California Davis, Davis, California, United States of America

Dhanya Cheerambathur
Department of Cellular and Molecular Medicine, Ludwig Institute for Cancer Research, University of California San Diego, La Jolla, California, United States of America

Andrea Disanza
IFOM, the FIRC Institute for Molecular Oncology Foundation, Milan, Italy

Michael A. Gelbart
Graduate Program in Biophysics, Harvard University, Boston, Massachusetts, United States of America, Department of Physics, Princeton University, Princeton, New Jersey, United States of America

Günther Gerisch
Max-Planck-Institut für Biochemie, D-82152 Martinsried, Germany

Zemer Gitai
Department of Molecular Biology, Princeton University, Princeton, New Jersey, United States of America

James A. Glazier
Biocomplexity Institute and Department of Physics, Indiana University, Bloomington, Indiana, United States of America

Gerhard Gompper
Theoretical Soft Matter and Biophysics, Institute of Complex Systems, Forschungszentrum Jülich, 52425 Jülich, Germany

Nir S. Gov
Department of Chemical Physics, The Weizmann Institute of Science, Rehovot, Israel

David A. Head
Theoretical Soft Matter and Biophysics, Institute of Complex Systems, Forschungszentrum Jülich, 52425 Jülich, Germany; Computational Biophysics, University of Twente, 7500 AE Enschede, The Netherlands; School of Computing, Leeds University, Leeds LS2 9JT, UK

Anthony A. Hyman
Max Planck Institute for Molecular Cell Biology and Genetics, Dresden, Germany

Shin Ishii
Graduate School of Informatics, Kyoto University, Uji, Kyoto, Japan; RIKEN Computational Science Research Program, Wako, Saitama, Japan

James W. Jacobberger
Case Comprehensive Cancer Center, Case Western Reserve University, Cleveland, Ohio, United States of America

Frank Jülicher
Max Planck Institute for the Physics of Complex Systems, Dresden, Germany

Doron Kabaso
Department of Chemical Physics, The Weizmann Institute of Science, Rehovot, Israel

Karsten Kruse
Max Planck Institute for the Physics of Complex Systems, Dresden, Germany

Andrea J. Liu
Department of Physics and Astronomy, University of Pennsylvania, Philadelphia, Pennsylvania, United States of America

Alex Mogilner
Department of Neurobiology, Physiology and Behavior, University of California Davis, Davis, California, United States of America; Department of Mathematics, University of California Davis, Davis, California, United States of America

Honda Naoki
Graduate School of Informatics, Kyoto University, Uji, Kyoto, Japan

Barak Peleg
Department of Chemical Physics, the Weizmann Institute of Science, Rehovot, Israel

Daniel Riveline
Laboratory of Yeast Genetics and Cell Biology, The Rockefeller University, New York, New York, United States of America; Laboratoire de Spectrométrie Physique (CNRS), UMR 5588, Université Joseph Fourier, Saint-Martin d'Hères, France

Kathrin Schloen
Helmholtz Center for Infection Research, Braunschweig, Germany

Giorgio Scita
IFOM, the FIRC Institute for Molecular Oncology Foundation, Milan, Italy, Department of Medicine, Surgery and Dentistry, Università degli Studi di Milano, Milan, Italy

Abbas Shirinifard
Biocomplexity Institute and Department of Physics, Indiana University, Bloomington, Indiana, United States of America

Roie Shlomovitz
Department of Chemical Physics, The Weizmann Institute of Science, Rehovot, Israel

Rajat Singhania
Department of Biological Sciences, Virginia Polytechnic Institute and State University, Blacksburg, Virginia, United States of America

Patrizia Sommi
Human Physiology Section, Department of Physiology, University of Pavia, Pavia, Italy

R. Michael Sramkoski
Case Comprehensive Cancer Center, Case Western Reserve University, Cleveland, Ohio, United States of America

Maciej Swat
Biocomplexity Institute and Department of Physics, Indiana University, Bloomington, Indiana, United States of America

Theresia Stradal
Helmholtz Center for Infection Research, Braunschweig, Germany

Gilberto L. Thomas
Instituto de Física, Universidade Federal do Rio Grande do Sul, Porto Alegre, Brazil

John J. Tyson
Department of Biological Sciences, Virginia Polytechnic Institute and State University, Blacksburg, Virginia, United States of America

Ned S. Wingreen
Department of Molecular Biology, Princeton University, Princeton, New Jersey, United States of America

Masataka Yamao
Graduate School of Information Science, Nara Institute of Science and Technology, Ikoma, Nara, Japan

Ying Zhang
Cancer and Developmental Biology Laboratory, National Cancer Institute, Frederick, Maryland, United States of America

Alexander Zumdieck
Max Planck Institute for the Physics of Complex Systems, Dresden, Germany

List of Abbreviations

ADF	Actin depolymerizing factor
AFM	Atomic force microscopy
APC-C	Anaphase Promoting Complex-Cyclosome
ATPase	Adenosine triphosphate
BAR	Bin/Amphiphysin/Rvs
BNs	Boolean networks
BSA	Bovine serum albumin
cAMP	Cyclic adenosine monophosphate
CAM	Cell adhesion molecule
CDM	Cis-dimer model
CDR	Circular Dorsal Ruffles
DAH	Differential Adhesion Hypothesis
DAPI	4′,6-diamidino-2-phenylindole
DMEM	Dulbecco's modified minimal essential media
DMSO	Dimethyl sulfoxide
DNA	Deoxyribonucleic acid
DROK	Drosophila Rho-associated kinase
EB1	End binding protein
ECM	Extra-cellular matrix
EGM-2	Endothelial Cell Growth Medium-2
EST	Expressed sequence tag
FBS	Fetal bovine serum
FCS	Fetal calf serum
FITC	Fluorescein isothiocyanate
GEF	Guanine nucleotide exchange factors
GFP	Green fluorescent protein
GGH	Glazier-Graner-Hogeweg
GTP	Guanosine 5'-triphosphate
HBL	Heterotypic boundary length
HUVEC	Human Umbilical Vein Endothelial Cell
IMD	IRSp53-Missing-In-Metastasis domains
IRSp53	Insulin receptor substrate protein of 53 kDa
LZM	Linear-zipper model
MCS	Monte Carlo Step
MEF	Mouse embryonic fibroblasts
MIM	Missing–in–metastasis
MLC	Myosin light chain
mRFP	Monomeric red fluorescent protein
MT	Microtubule
NEB	Nuclear envelope breakdown
NETO	New End Take-Off

N-WASP	Neural Wiskott-Aldrich syndrome protein
NWHBL	Normalized weighted heterotypic boundary length
ODEs	Ordinary differential equations
PB	Phosphate buffer
PBS	Phosphate buffered saline
PDGF	Platelet-derived growth factor
PFA	Paraformaldehyde
PIP3	Phosphatidylinositol (3,4,5)-triphosphate
PV	Pressure-volume
RKO	Human colorectal carcinoma
RNAi	Ribonucleic acid interference
Sced	Staphylococcus epidermidis
SCF	Skp, Cullin, F-box
SD	Standard deviation
SM	Saturation model
SMC	Structural maintenance of chromosomes
TF	Transcription factors
THBM	Trans-homophilic-bond model
TIRF	Total internal reflection fluorescence
WGA	Wheat germ agglutinin
WHBL	Weighted heterotypic boundary length
WT	Wild-type

Introduction

The physiological behaviors of cells (growth and division, differentiation, movement, death, etc.) are controlled by complex networks of interacting genes and proteins, and a fundamental goal of computational cell biology is to develop dynamical models of these regulatory networks that are realistic, accurate and predictive. Historically, these models have divided along two basic lines: deterministic or stochastic, and continuous or discrete, with scattered efforts to develop hybrid approaches that bridge these divides.

In chapter 1 of this volume, using the cell cycle control system in eukaryotes as an example, Singhania and colleagues propose a hybrid approach that combines a continuous representation of slowly changing protein concentrations with a discrete representation of components that switch rapidly between "on" and "off" states, combining the deterministic causality of network interactions with the stochastic uncertainty of random events. The hybrid approach can be easily tailored to the available knowledge of control systems, and it provides both qualitative and quantitative results that can be compared to experimental data to test the accuracy and predictive power of the model.

In chapter 2, Head, Briels, and Gompper present the results of numerical simulations of a discrete filament-motor protein model confined to a pressurized cylindrical box. Stable spindles, nematic configurations, asters, and high-density semi-asters spontaneously emerge. State diagrams are presented delineating each stationary state as the pressure, motor speed and motor density are varied. The authors further highlight a parameter regime where vortices form exhibiting collective rotation of all filaments, but have a finite lifetime before contracting to a semi-aster. They demonstrate that discrete filament-motor protein models provide new insights into the stationary and dynamical behavior of active gels and subcellular structures, because many phenomena occur on the length-scale of single filaments.

In yet another scenario, the assembly of the Drosophila embryo mitotic spindle during prophase depends upon a balance of outward forces generated by cortical dynein and inward forces generated by kinesin-14 and nuclear elasticity. Myosin II is known to contribute to the dynamics of the cell cortex but how this influences the prophase force-balance is unclear. Sommi and her colleagues investigate this question in chapter 3; they did so by injecting the myosin II inhibitor, Y27632, into early Drosophila embryos. They observed a significant increase in both the area of the dense cortical actin caps and in the spacing of the spindle poles. Their results suggest that two complementary outward forces are exerted on the prophase spindle by the overlying cortex. Specifically, dynein localized on the mechanically firm actin caps and the actomyosin-driven contraction of the deformable soft patches of the actin cortex, cooperate to pull astral microtubules outward. Thus, myosin II controls the size and dynamic properties of the actin-based cortex to influence the spacing of the poles of the underlying spindle during prophase.

Reliable chromosome segregation is crucial to all dividing cells. In some bacteria, segregation has been found to occur in a rather counterintuitive way: the chromosome attaches to a filament bundle and erodes it by causing depolymerization of the filaments. Moreover, unlike eukaryotic cells, bacteria do not use molecular motors and/or macromolecular tethers to position their chromosomes. This raises the general question of how depolymerizing filaments alone can continuously and robustly pull cargo as the filaments themselves are falling apart. In chapter 4, Banigan and his colleagues introduce the first quantitative physical model for depolymerization-driven translocation in a many-filament system. Their simulations of this model suggest a novel underlying mechanism for robust translocation, namely self-diffusiophoresis, motion of an object in a self-generated concentration gradient in a viscous environment. In this case, the cargo generates and sustains a concentration gradient of filaments by inducing them to depolymerize. The authors demonstrate that their model agrees well with existing experimental observations such as segregation failure, filament-length-dependent translocation velocity, and chromosomal compaction. In addition, they make several predictions–including predictions for the specific modes by which the chromosome binds to the filament structure and triggers its disassembly–that can be tested experimentally.

Next, in chapter 5, Zumdieck and his coauthors present a physical analysis of the dynamics and mechanics of contractile actin rings. In particular, they analyze the dynamics of ring contraction during cytokinesis in the Caenorhabditis elegans embryo. They present a general analysis of force balances and material exchange and estimate the relevant parameter values. The authors show that on a microscopic level contractile stresses can result from both the action of motor proteins, which cross-link filaments, and from the polymerization and depolymerization of filaments in the presence of end-tracking cross-linkers.

In chapter 6 we turn our attention to cells that exhibit propagating membrane waves which involve the actin cytoskeleton. One type of such membranal waves are Circular Dorsal Ruffles (CDR), which are related to endocytosis and receptor internalization. Experimentally, CDRs have been associated with membrane bound activators of actin polymerization of concave shape. Peleg and colleagues present experimental evidence for the localization of convex membrane proteins in these structures, and their insensitivity to inhibition of myosin II contractility in immortalized mouse embryo fibroblasts cell cultures. These observations lead the authors to propose a theoretical model that explains the formation of these waves due to the interplay between complexes that contain activators of actin polymerization and membrane-bound curved proteins of both types of curvature (concave and convex). Their model predicts that the activity of both types of curved proteins is essential for sustaining propagating waves, which are abolished when one type of curved activator is removed. Within this model waves are initiated when the level of actin polymerization induced by the curved activators is higher than some threshold value, which allows the cell to control CDR formation. The authors demonstrate that the model can explain many features of CDRs, and give several testable predictions. This chapter demonstrates the importance of curved membrane proteins in organizing the actin cytoskeleton and cell shape.

Chapter 7 deals with actin waves that are spontaneously generated on the planar, substrate-attached surface of Dictyostelium cells. Gerisch reveals that the waves have the following characteristics:

1. They are circular structures of varying shape, capable of changing the direction of propagation.
2. The waves propagate by treadmilling with a recovery of actin incorporation after photobleaching of less than 10 seconds.
3. The waves are associated with actin-binding proteins in an ordered 3-dimensional organization: with myosin-IB at the front and close to the membrane, the Arp2/3 complex throughout the wave, and coronin at the cytoplasmic face and back of the wave. Coronin is a marker of disassembling actin structures.
4. The waves separate two areas of the cell cortex that differ in actin structure and phosphoinositide composition of the membrane. The waves arise at the border of membrane areas rich in phosphatidylinositol (3,4,5) trisphosphate (PIP3). The inhibition of PIP3 synthesis reversibly inhibits wave formation.
5. The actin wave and PIP3 patterns resemble 2-dimensional projections of phagocytic cups, suggesting that they are involved in the scanning of surfaces for particles to be taken up.

Lengths and shapes are approached in different ways in different fields: they serve as a read-out for classifying genes or proteins in cell biology, whereas they result from scaling arguments in condensed matter physics. In chapter 8, Riveline proposes a combined approach with examples illustrated for the fission yeast Schizosaccharomyces pombe.

Cells have highly varied and dynamic shapes, which are determined by internal forces generated by the cytoskeleton. These forces include protrusive forces due to the formation of new internal fibers and forces produced due to attachment of the cell to an external substrate. A longstanding challenge is to explain how the myriad components of the cytoskeleton self-organize to form the observed shapes of cells. In chapter 9, Kabaso and coauthors present a theoretical study of the shapes of cells that are driven only by protrusive forces of two types; one is the force due to polymerization of actin filaments, which acts as an internal pressure on the membrane, and the second is the force due to adhesion between the membrane and external substrate. The key property is that both forces are localized on the cell membrane by protein complexes that have convex spontaneous curvature. This leads to a positive feedback that destabilizes the uniform cell shape and induces the spontaneous formation of patterns. The authors compare the resulting patterns to observed cellular shapes and find good agreement, which allows them to explain some of the puzzling dependencies of cell shapes on the properties of the surrounding matrix.

Chapter 10 deals with amoeboid cells, which crawl using pseudopods, convex extensions of the cell surface. In many laboratory experiments, cells move on a smooth substrate, but in the wild cells may experience obstacles of other cells or dead material, or may even move in liquid. To understand how cells cope with heterogeneous environments, Van Haastert has investigated the pseudopod life cycle of wild type and mutant cells moving on a substrate and when suspended in liquid. He shows that the same pseudopod cycle can provide three types of movement that he addresses as walking, gliding and swimming. In walking, the extending pseudopod will adhere firmly to the substrate, which allows cells to generate forces to bypass obstacles. Mutant cells with compromised adhesion can move much faster than wild type cells on a smooth substrate (gliding), but cannot move effectively against obstacles that provide resistance. In a liquid, when swimming, the extending pseudopods convert to side-bumps that move rapidly to the rear of the cells. Calculations suggest that these bumps provide sufficient drag force to mediate the observed forward swimming of the cell.

During development, the formation of biological networks (such as organs and neuronal networks) is controlled by multicellular transportation phenomena based on cell migration. In multi-cellular systems, cellular locomotion is restricted by physical interactions with other cells in a crowded space, similar to passengers pushing others out of their way on a packed train. The motion of individual cells is intrinsically stochastic and may be viewed as a type of random walk. However, this walk takes place in a noisy environment because the cell interacts with its randomly moving neighbors. Despite this randomness and complexity, development is highly orchestrated and precisely regulated, following genetic (and even epigenetic) blueprints. Although individual cell migration has long been studied, the manner in which stochasticity affects multi-cellular transportation within the precisely controlled process of development remains largely unknown. To explore the general principles underlying multicellular migration, in chapter 11, the authors focus on the migration of neural crest cells, which migrate collectively and form streams. Yamoa, Naoki, and Ishii introduce a mechanical model of multi-cellular migration. Simulations based on the model show that the migration mode depends on the relative strengths of the noise from migratory and non-migratory cells. Strong noise from migratory cells and weak noise from surrounding cells causes "collective migration," whereas strong noise from non-migratory cells causes "dispersive migration." Moreover, the authors' theoretical analyses reveal that migratory cells attract each other over long distances, even without direct mechanical contacts. This effective interaction depends on the stochasticity of the migratory and non-migratory cells. On the basis of these findings, the authors propose that stochastic behavior at the single-cell level works effectively and precisely to achieve collective migration in multi-cellular systems.

The actions of cell adhesion molecules, in particular, cadherins during embryonic development and morphogenesis more generally, regulate many aspects of cellular interactions, regulation and signaling. Often, a gradient of cadherin expression levels drives collective and relative cell motions generating macroscopic cell sorting. Computer simulations of cell sorting have focused on the interactions of cells with only a few discrete adhesion levels between cells, ignoring biologically observed continuous variations in expression levels and possible nonlinearities in molecular binding. In the final chapter, the authors present three models relating the surface density of cadherins to the net intercellular adhesion and interfacial tension for both discrete and continuous levels of cadherin expression. Zhang and colleagues then use then the Glazier-Graner-Hogeweg (GGH) model to investigate how variations in the distribution of the number of cadherins per cell and in the choice of binding model affect cell sorting. They find that an aggregate with a continuous variation in the level of a single type of cadherin molecule sorts more slowly than one with two levels. The rate of sorting increases strongly with the interfacial tension, which depends both on the maximum difference in number of cadherins per cell and on the binding model. The authors' approach helps connect signaling at the molecular level to tissue-level morphogenesis, thus adding to our understanding of how biophysics relates to yet another realm of investigation.

— Pavel Kraikivski, PhD

1 Mammalian Cell Cycle Regulation

Rajat Singhania, R. Michael Sramkoski, James W. Jacobberger, and John J. Tyson

CONTENTS

1.1 Introduction 1
1.2 Methods 4
  1.2.1 Simulations 4
  1.2.2 Cells, culture, and fixation 7
  1.2.3 Immunofluorescence staining, antibodies, flow cytometry 7
  1.2.4 Data pre-processing 8
1.3 Discussion 8
1.4 Results 14
  1.4.1 Hybrid modeling approach 14
  1.4.2 Cyclin distributions in an asynchronous culture 18
  1.4.3 Contact inhibition of cultured cells 19
Keywords 21
Supporting Information 21
Acknowledgments 22
Author Contributions 22
References 22

1.1 INTRODUCTION

The timing of DNA synthesis, mitosis, and cell division is regulated by a complex network of biochemical reactions that control the activities of a family of cyclin-dependent kinases. The temporal dynamics of this reaction network is typically modeled by nonlinear differential equations describing the rates of the component reactions. This approach provides exquisite details about molecular regulatory processes but is hampered by the need to estimate realistic values for the many kinetic constants that determine the reaction rates. It is difficult to estimate these kinetic constants from available experimental data. To avoid this problem, modelers often resort to 'qualitative' modeling strategies, such as Boolean switching networks, but these models describe only the coarsest features of cell cycle regulation. In this chapter it describes a hybrid approach that combines the best features of continuous differential equations and discrete Boolean networks. Cyclin abundances are tracked by piecewise linear differential equations for cyclin synthesis and degradation. The cyclin synthesis is regulated by transcription factors whose activities are represented by discrete variables (0 or 1) and likewise for the activities of the ubiquitin-ligating enzyme complexes that govern cyclin degradation. The discrete variables change according to a predetermined sequence, with the times between transitions determined in part by cyclin accumulation and degradation and as well by exponentially distributed random variables. The model is evaluated in terms of flow cytometry measurements of cyclin proteins in asynchronous populations of human cell lines. The few kinetic constants in the model are easily estimated from the experimental data. Using this hybrid approach, modelers can quickly create quantitatively accurate, computational models of protein regulatory networks in cells.

The physiological behaviors of cells (growth and division, differentiation, movement, death, etc.) are controlled by complex networks of interacting genes and proteins, and a fundamental goal of computational cell biology is to develop dynamical models of these regulatory networks that are realistic, accurate and predictive. Historically, these models have divided along two basic lines: deterministic or stochastic, and continuous or discrete; with scattered efforts to develop hybrid approaches that bridge these divides. Using the cell cycle control system in eukaryotes as an example, we propose a hybrid approach that combines a continuous representation of slowly changing protein concentrations with a discrete representation of components that switch rapidly between 'on' and 'off' states, and that combines the deterministic causality of network interactions with the stochastic uncertainty of random events. The hybrid approach can be easily tailored to the available knowledge of control systems, and it provides both qualitative and quantitative results that can be compared to experimental data to test the accuracy and predictive power of the model.

The cell division cycle is the fundamental physiological process by which cells grow, replicate, and divide into two daughter cells that receive all the information (genes) and machinery (proteins, organelles, etc.) necessary to repeat the process under suitable conditions [1]. This cycle of growth and division underlies all biological expansion, development, and reproduction. It is highly regulated to promote genetic fidelity and meet the demands of an organism for new cells. Altered systems of cell cycle control are root causes of many severe health problems, such as cancer and birth defects.

In eukaryotic cells, the processes of DNA replication and nuclear/cell division occur sequentially in distinct phases (S and M) separated by two gaps (G1 and G2). The mitosis (M phase) is further subdivided into stagesprophase (chromatin condensation, spindle formation, and nuclear envelope breakdown), prometaphase (chromosome attachment and congression), metaphase (chromosome residence at the mid-plane of the spindle), anaphase (sister chromatid separation and movement to opposite poles of the spindle), telophase (re-formation of the nuclear envelopes), and cytokinesis (cell division). The G1 phase is subdivided into uncommitted and committed sub-phases, often referred to as G1-pm (postmitotic interval) and G1-ps (pre S phase interval), separated by the 'restriction point' [2]. In this chapter, it is refered as the sub-phases G1-pm and G1-ps as 'G1a' and 'G1b' respectively.

The progression through the correct sequence of cell-cycle events is governed by a set of cyclin-dependent kinases (Cdk's), whose activities rise and fall during the cell cycle as determined by a complex molecular regulatory network. For example, cyclin synthesis and degradation are controlled, respectively, by transcription factors and ubiquitin-ligating complexes whose activities are, in turn, regulated by cyclin/Cdk complexes.

Current models of the Cdk control system can be classified as either continuous or discrete. Continuous models track the changes of protein concentrations, $C_j(t)$ for $j = 1, 2, \ldots, N$, by solving a set of nonlinear ordinary differential equations (ODEs) of the form:

$$\frac{dC_j}{dt} = \sum_{r=1}^{R} v_{jr}\rho_r\left(C_1, C_2, \ldots, C_N\right) \tag{1}$$

where ρ_r is the rate of the rth reaction and v_{ir} is the stoichiometric coefficient of species i in reaction r. To each rate term is associated one or more kinetic constants that determine exactly how fast the reaction proceeds under specific conditions. These kinetic constants must be estimated from experimental data, and often there is insufficient kinetic data to determine their values. Nonetheless, continuous models, based on rate equations, have been used successfully to account for the properties of cell proliferation in a variety of cell typesyeast [3-5], fruit fly [6], frog egg [7-8], and cultured mammalian cells [9-11]. They have also proved successful in predicting novel cell-cycle characteristics [12-13].

Discrete models, on the contrary, represent the state of each regulatory protein as $B_j(\tau) = 0$ or 1 (inactive or active), and the state variables update from one discrete time step to the next ($\tau = 0, 1, 2, \ldots$ = ticks of a metronome) according to the rule:

$$B_j\left(\tau+1\right) = \Psi_j\left(B_1\left(\tau\right), B_2\left(\tau\right), \ldots, B_n\left(\tau\right)\right) \tag{2}$$

where $\Psi_j(\ldots)$ is a Boolean function (i.e., it equates to either 0 or 1) determined by the topology of the reaction network. For Boolean networks (BNs) there is no notion of reaction 'rate' and, hence, no need to estimate kinetic constants. The BN models of the Cdk regulatory network have been proposed for yeast cells [14,15] and for mammalian cells [16]. They have been used to study notions of 'robustness' of the cell cycle, but they have not been compared in detail to quantitative properties of cell cycle progression, and they have not been used as predictive tools.

In this chapter it is proposed to combine the strengths of both continuous and discrete modeling, while avoiding the weaknesses of each. The 'hybrid' model is inspired by the work of Li et al. [14], who proposed a BN for cell cycle controls. Their model employs 11 state variables that move around in a space of $2^{11} = 2048$ possible states.

Quite remarkably they found that 1764 of these states converge quickly onto a 'super highway' of 13 consecutive states that represent a typical cell cycle trajectory (G1b—S—G2—M—G1a). The results of Li et al. indicate that the cell cycle control network is 'robustly designed' in the sense that even quite large perturbations away from the usual sequence of cell cycle states are quickly restored to the super highway. In the model of Li et al., G1a is a stable steady state; they do not address the signals that drive cells past the restriction point (the G1a-to-G1b transition).

Despite their intuitive appeal, Boolean models have severe limitations. First of all, metronomic time in BN's is unrelated to clock time in the laboratory, so Boolean models cannot be compared to even the most basic observations of time spent by cells in the four phases of the division cycle [1]. Also, these models do not incorporate cell size, so they cannot address the evident importance of cell growth in driving events of the cell cycle [17-19]. Lastly, cyclins are treated as either absent or present (0 or 1), so Boolean models cannot simulate the continuous accumulation and removal of cyclin molecules at different stages of the cell cycle [20].

The goal is to retain the elegance of the Boolean representation of the switching network, while introducing continuous variables for cell size, cell age, and cyclin composition, in order to create a model that can be compared in quantitative detail to experimental measurements with a minimal number of kinetic parameters that must be estimated from the data. To this end, to keep the cyclin regulators as Boolean variables but model the cyclins themselves as continuous concentrations that increase and decrease due to synthesis and degradation. Next, replace the Boolean model's metronome with real clock time to account for realistic rates of cyclin synthesis and degradation, and for stochastic variability in the time spent in each Boolean state of the model. Finally, it introduced a cell size variable, M(t), which affects progression through late G1 phase. The M(t) increases exponentially with time as the cell grows and decreases by a factor of ~2 when the cell divides. (The assumption of exponential growth is not crucial; similar results are obtained assuming linear growth between cell birth and division.)

Since the pioneering work of Leon Glass [21,22], hybrid (discrete-continuous) models have been employed by systems biologists in a variety of forms and contexts [23-25]. Engineers have been modeling hybrid control systems for many years [26-28], and they have created powerful simulation packages for such systems [29]: SIMULINK [28], SHIFT [30-31] and CHARON [32], to name a few. It has not used these simulation packages because model can be solved analytically.

1.2 METHODS

1.2.1 Simulations

It simulate a flow cytometry experiment with hybrid model in two steps.

Step 1: Creating complete 'life histories' for thousands of cells. At the start of the simulation, we specify initial conditions at the beginning of the cycle (State 1) for a progenitor cell. It used the following initial values of the state variables: [CycA] = [CycB] = [CycE] = 1 and $M = 3$. The strategy is to follow this cell through its cycle until it divides into two daughters. Then choose one of the two daughters at random and repeat the process, continuing for 32,500 iterations. The first 500 cells discard, and keep a sample of 32,000 cells that have completed a replication-division cycle according to our model. In the second step, create a simulated sample of 32,000 cells chosen at random phases of the cell cycle, to represent the cells that were assayed by the flow cytometer.

Let us consider cell i ($1<i<32500$) at the time of its birth, t_{i0}. By definition, this cell is in State 1, and assume that know its birth mass, $M(t_{i0})$, and its starting concentrations of cyclins A, B and E. Denote the starting concentrations as [CycA(t_{i0})], [CycB(t_{i0})], [CycE(t_{i0})]. In the ensuing discussion, unless it is necessary for clarity, drop the i subscript, it being understood that are talking about a representative cell in the population. It will follow this cell until it divides to produce a daughter cell with known concentrations of cyclins.

According to Table 1, a cell in State 1 has no special conditions to satisfy before moving to State 2. Hence the residence time in State 1 is a random number T_1^T chosen from an exponential distribution with mean $\lambda_1 = 2$ h. The cell enters State 2 at $t_1 = t_0 + T_1^T$. Assuming exponential growth, its size at this time is $M(t_1) = M(t_0)\exp\{\gamma(t_1 - t_0)\} = M(t_0)\exp\{\gamma A_1\}$, where γ is the specific growth rate of the culture and $A_1 = t_1 - t_0$ is the age of the cell when it exits State 1. To illustrate how cyclin concentrations are computed at $t = t_1$, let us consider cyclin A as an example. During the interval $t_0<t<t_1$, [CycA] satisfies a linear ODE with effective rate constants $k_{sa1} = k'_{sa} = 5$ and $k_{da1} = k'_{da} + k'''_{da} = 1.4$, because $B_{TFE} = B_{TFB} = B_{Cdc20A} = 0$ and $B_{Cdh1} = 1$ for a cell in State 1. It can compute the concentration of cyclin A at any time during this interval from

$$[CycA(t)] = \frac{k_{sa1}}{k_{da1}} + \left([CycA(t_0)] - \frac{k_{sa1}}{k_{da1}}\right)e^{-k_{da1}(t-t_0)}, t_0 \le t \le t_1 \quad (3)$$

Setting $t = t_1$ in this equation gives the number we seek. In this fashion, to start tabulating the following information for each simulated cell:

Time	t_0	t_1	t_2	...
Enter State	1	2	3	...
Age	0	$A_1 = t_1 - t_0$	$A_2 = t_2 - t_0$	
Size	$M(t_0)$	$M(t_1)$	$M(t_2)$	...
Cyclin A	[CycA(t_0)]	[CycA(t_1)]	[CycA(t_2)]	...
Cyclin B	[CycB(t_0)]	[CycB(t_1)]	[CycB(t_2)]	...
Cyclin E	[CycE(t_0)]	[CycE(t_1)]	[CycE(t_2)]	...

Notice that, at $t = t_1$ when the cell enters State 2, the transcription factor (TFE) for cyclins E and A turns on, and these cyclins start to accumulate. The cell cannot leave State 2 until cyclin E accumulates to a sufficiently high level: $[CycE](t)\cdot M(t) = \theta_E$, according to Table 1. When this condition is satisfied, the cell leaves State 2 and enters State 3. The size dependence on this transition is a way to couple cell growth to the

DNA replication-division cycle. According to the parameter settings in Table 1, there is no stochastic component to the transition out of State 2.

To continue in this fashion until the cell leaves State 9 and returns to State 1, when cyclin B is degraded at the end of mitosis. This is the signal for cell division. The age of the cell at division is $A_9 = t_9 - t_0$, and the mass of the cell at division is $M(t_9) = M(t_0)\exp(\gamma \cdot A_9)$. The mass of the daughter cell at the beginning of her life history is $M_{daughter}(t_0) = \delta \cdot M_{mother}(t_9)$, where δ is a random number sampled from a normal distribution of mean 0.5 and standard deviation 0.0167 to allow for asymmetries of cell division.

Notice that simulating the life history of a single cell only requires generating about a dozen random numbers and performing a handful of algebraic calculations. At no point do we need to solve differential equations numerically. Hence, one can quickly calculate the life histories of tens of thousands of cells.

Step 2: Finding the DNA and cyclin levels of each cell in an asynchronous sample. In the flow cytometry experiments of Yan et al. [42], a random sample of cells is taken from an asynchronous population, the cells are fixed and stained, and then run one-by-one through laser beams where fluorescence measurements are made. So each data point consists of measurements of light scatter (related to cell size) and fluorescence proportional to DNA and cyclin content for a single cell taken at some random point in the cell cycle. To simulate this experiment we must assign to each of 32,000 simulated cells a number φ_i selected randomly from the interval [0,1], where φ_i refers to the fraction of the cell cycle completed by cell i when it was fixed and stained for measurement. Because, each mother cell divides into two daughter cells, the density of cells at birth, $\varphi = 0$, is twice the density of cells at division, $\varphi = 1$. The 'ideal' probability density for an asynchronous population of cells expanding exponentially in number is

$$f(\varphi) = (\ln 2) \times 2^{1-\varphi} \quad (4)$$

According to the 'transformation method' [47, Chapter 7.2], to compute φ as

$$\varphi = \log_2\left(\frac{2}{2-r}\right) \quad (5)$$

where r is a random number chosen from a uniform distribution on [0,1]. In this way, to generate 32,000 fractions, φ_i.

If φ_i is the cell-cycle location of the ith cell when it is selected for the flow cytometry measurements, then its age at the time of selection is $a_i = \varphi_i \cdot A_{i9}$, where A_{i9} is the age of the ith cell at division. Given a value for a_i, we then find the state n (= 1, 2, … or 9) of the ith cell at the time of its selection:

$$t_{i,n-1} \le t_{i0} + a_i < t_{i,n} \quad (6)$$

where $t_{i,n}$ (as defined above) is the time at which the ith cell left state n to enter state $n+1$.

Once to know the state n of the cell, one can compute the concentration of each cyclin in the cell at its exact age a_i by analogy to Eq. (3):

$$[CycA(a_i)] = \frac{k_{sa,n}}{k_{da,n}} + \left([CycA(t_{i,n-1})] - \frac{k_{sa,n}}{k_{da,n}}\right)e^{-k_{da,n}(t_{i0}+a_i-t_{i,n-1})} \tag{7}$$

where $k_{sa,n}$ and $k_{da,n}$ are the synthesis and degradation rate constants for cyclin A in state n. This is a straightforward calculation because in Step 1 stored the values of t_n and $[CycA(t_n)]$ for every state of each cell. It can also calculate the mass of cell i at the time of its selection:

$$M(a_i) = M(t_{i0}) \cdot \exp(\gamma \cdot a_i) \tag{8}$$

where $M(t_{i0})$ is the mass at birth of cell i and γ is the specific growth rate of the culture. Because the flow cytometer measures the total amount of fluorescence proportional to all cyclin A molecules in the ith cell, we take as our measurable the product of $[CycA(a_i)]$ times $M(a_i)$.

Lastly, to determine the DNA content of cell i at age a_i according to:

1. DNA = 1 for $t_{i0} \leq t_{i0} + a_i < t_{i3}$ = entry of ith cell into S phase
2. DNA = $1 + (t_{i0} + a_i - t_{i3})/(t_{i4} - t_{i3})$ for $t_{i3} \leq t_{i0} + a_i < t_{i4}$ = exit of ith cell from S phase
3. DNA = 2 for $t_{i4} \leq t_{i0} + a_i < t_{i9}$

Now, one simulated values for the measurable quantities of each cell at the time point in the cell cycle when it was selected for analysis. Before, plotting these numbers, one should take into account experimental errors, such as probe quality, fixation, staining and measurement. To do so by multiplying each measurable quantity (DNA content and cyclin levels) by a random number chosen from a Gaussian distribution with mean 1 and standard deviation = 0.03 for DNA measurements and 0.15 for cyclin measurements. These choices give scatter to the simulated data that is comparable to the scatter in the experimental data.

1.2.2 Cells, Culture, and Fixation

Culture and fixation of RKO cells were described [42]. The immortalized HUVEC cells [48] at passage 93 were seeded at 2.5×10^3 cells/cm^2 in 10 ml EGM-2 media with 2% fetal bovine serum (Lonza, Basel). Duplicate plates were prepared for each time point at days 1, 2, 3, 4, 5, 6, 7, 10, and 15. The cells were fed every other day by replacing half the volume of used media. At the indicated times, cells were trypsinized, washed, and cell counts performed with a Guava Personal Cytometer (Millipore, Billerica, MA). Fixation was as previously described [49]; briefly, cells were treated with 0.125% formaldehyde (Polysciences, Warrington, PA) for 10 min at 37°C, washed, then dehydrated with 90% Methanol. Cells were fixed in aliquots of 1×10^6 cells (days 1–3) or 2×10^6 (days 4–15). Fixed cell samples were stored at −20°C until staining for cytometry.

1.2.3 Immunofluorescence Staining, Antibodies, Flow Cytometry

Staining and cytometry for RKO cells were described [42]. Briefly, cells were trypsinized, fixed with 90% MeOH, washed with phosphate buffered saline, then stained with monoclonal antibodies reactive with cyclin B1, cyclin A, phospho-S10-histone H3, and with 4′,6-diamidino-2-phenylindole (DAPI). For a detailed, updated version of antibodies, staining, and cytometry for cyclins A2 and B1, phospho-S10-histone H3, and DNA content, see Jacobberger et al. (38).

1.2.4 Data Pre-Processing

Data pre-processing was performed with WinList (Verity Software House, Topsham, ME). Doublet discrimination (peak versus area DAPI plot) was used to limit the analysis to singlet cells; non-specific binding was used to remove background fluorescence from the total fluorescence related to cyclin A2 and B1 staining. The phycoerythrin channel (cyclin A2) was compensated for spectral overlap from FITC or Alexa Fluor 488. For simplification, very large 2C G1 HUVEC cells and any cells cycling at 4C→8C were removed from the analysis. These were present at low frequency.

1.3 DISCUSSION

It had constructed a simple, effective model of the cyclin-dependent kinase control system in mammalian cells and used the model to simulate faithfully the accumulation and degradation of cyclin proteins during asynchronous proliferation of RKO (colon carcinoma) cells. The model is inspired by the work of Li et al. [14], who proposed a robust Boolean model of cell cycle regulation in budding yeast. The goal was to retain the elegance of the Boolean representation of the switching network, while introducing continuous variables for cell size, cell age, and cyclin composition, in order to create a model that could be compared in quantitative detail to experimental measurements.

It was shown that this model can accurately simulate flow-cytometric measurements of cyclin abundances in asynchronous populations of growing-dividing mammalian cells. The parameters in the model that allow for a quantitative description of the experimental measurements are easily estimated from the data itself. Now that the model is parameterized and validated for wild-type cells, are currently extending it to handle the behavior of cell populations perturbed by drugs and by genetic interference. In some cases, only modest extensions of the model are required; in other cases, a more thorough overhaul of the way the discrete and continuous variables interact with each other is necessary.

To choose parameter values in model to capture the major features of cyclin fluctuations as measured by flow cytometry during the somatic division cycle of mammalian cells. A human tumor cell line used to calibrate model. Between cell lines and normal human cultured cells, there are differences in the expressions of A and B cyclins [43]; however, when the levels of cyclin B1 were rigorously compared for HeLa, K562, and RKO cells, both the patterns and magnitudes of expression are remarkably similar, apparently dependent to some degree on the rate of population growth [44]. In addition, the patterns of expression of cyclins A2 and B1 are similar for these human tumor cell lines and stimulated normal human circulating lymphocytes (Supporting Fig. S2). Overall, the simulation outputs have satisfying similarity both in pattern and magnitude to the real data for RKO cells, and simulated expression patterns of cyclins A, B and E for the tumor cell line are quite similar to the simulated expression patterns in HUVEC cells (see Supporting Fig. S1).

However, there remain some inconsistencies between mathematical simulations and our experimental observations that point out where future modifications to the model are needed. For example, in the model DNA synthesis starts when cyclin A has accumulated to ~8% of its maximum level (see arrow in Fig. 1D; 50/600≈8%), whereas in our measurements DNA synthesis starts when cyclin A is ~5% of its maximum level (arrow in Fig. 1C). This discrepancy is tempered by the fact that we are not confident of the quantitative accuracy of cyclin A expression levels below ~4% of its maximum level in Fig. 1C. Where we place the minimum expression level of cyclin A in Fig. 1D affects our estimate of the cyclin A level at onset of DNA synthesis (50 AU at present). By lowering the minimum expression level of cyclin A below 10 AU in Fig. 1D (e.g., by lowering k'_{sa}), we could line up the two arrows in Figs. 1C and D. Nonetheless, it observe (supporting Fig. S3) that cyclin A expression correlates highly with BrdU incorporation, suggesting that significant accumulation of cyclin A begins simultaneously with the onset of DNA synthesis, whereas in our model cyclin A production begins in mid-G1 phase. This discrepancy could be minimized by lowering the cyclin A threshold (θ_A) in the model.

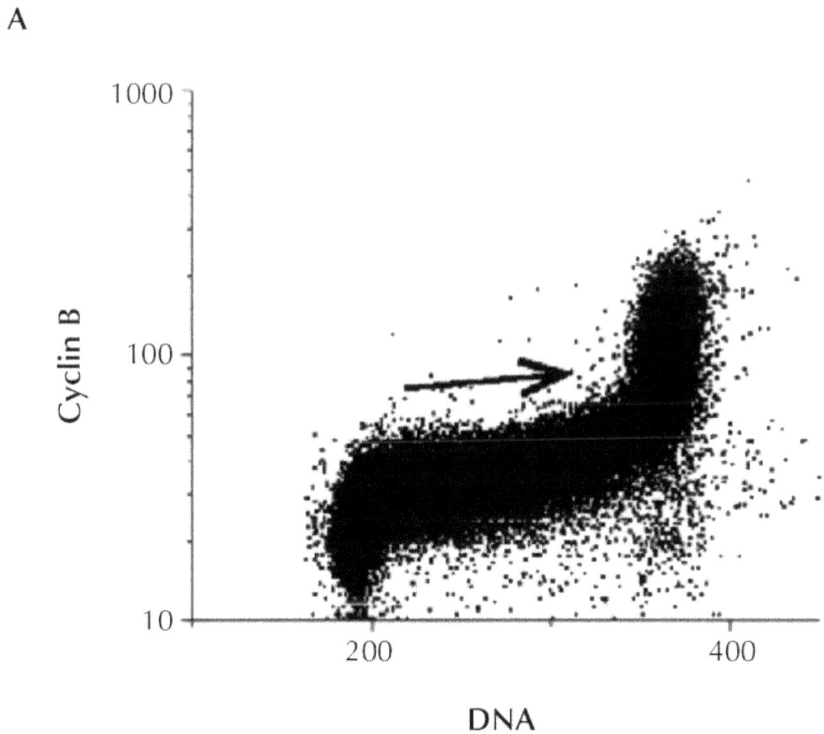

FIGURE 1 *(Continued)*

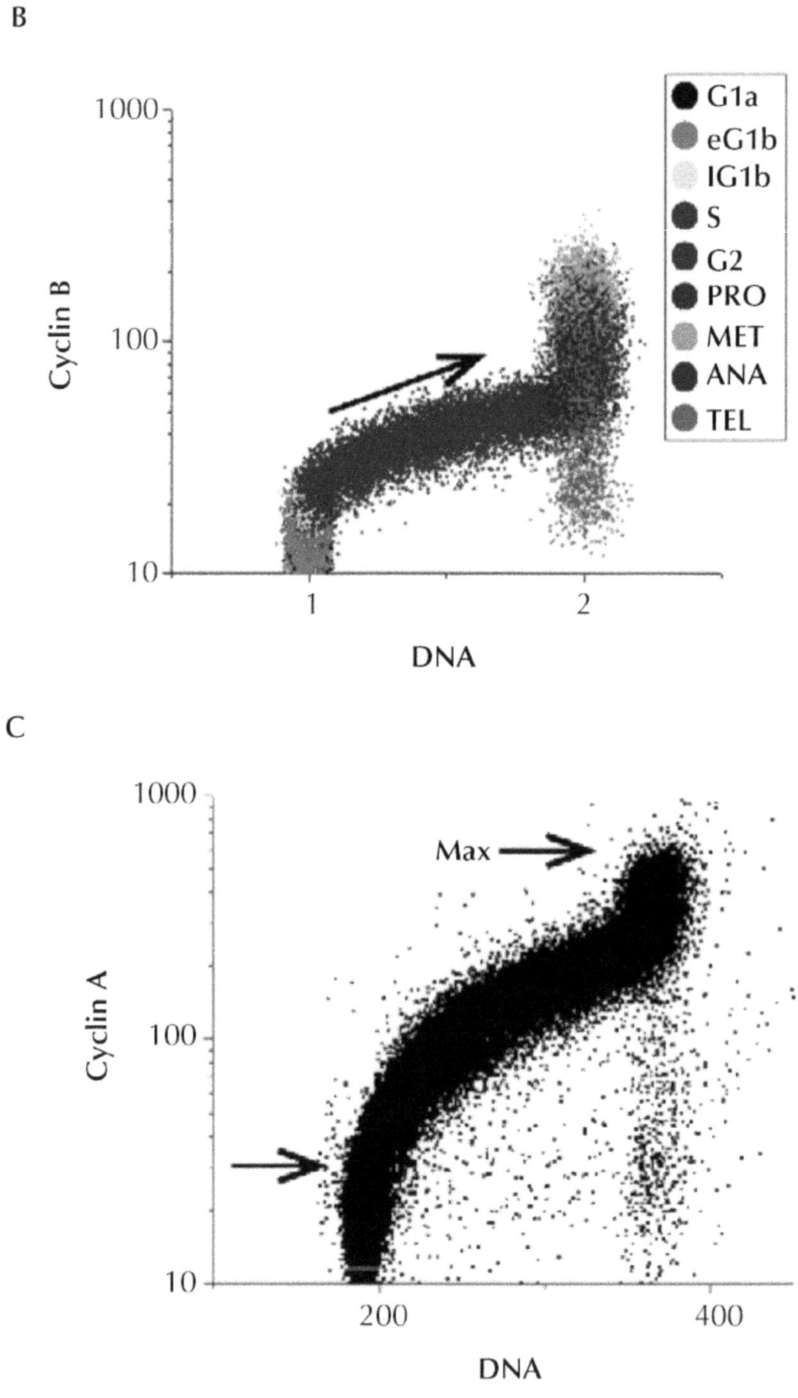

FIGURE 1 *(Continued)*

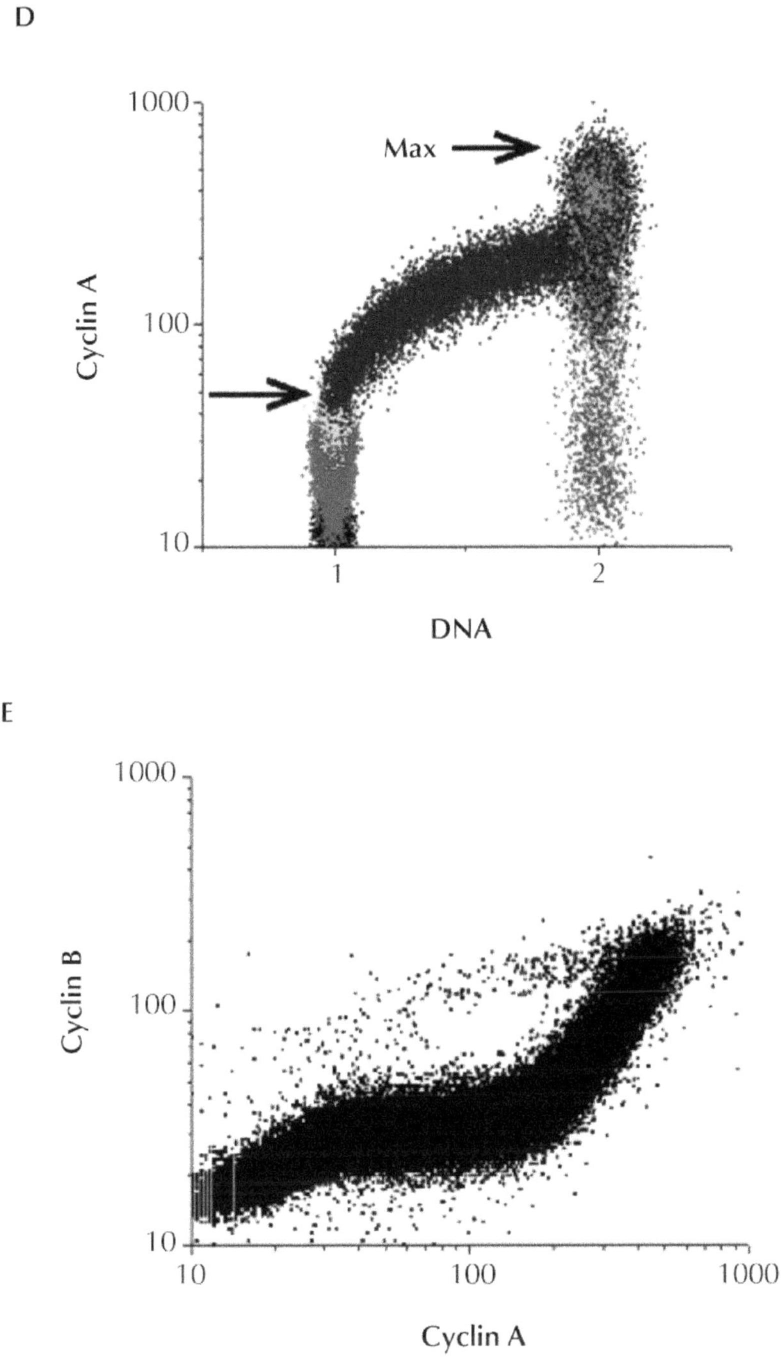

FIGURE 1 *(Continued)*

F

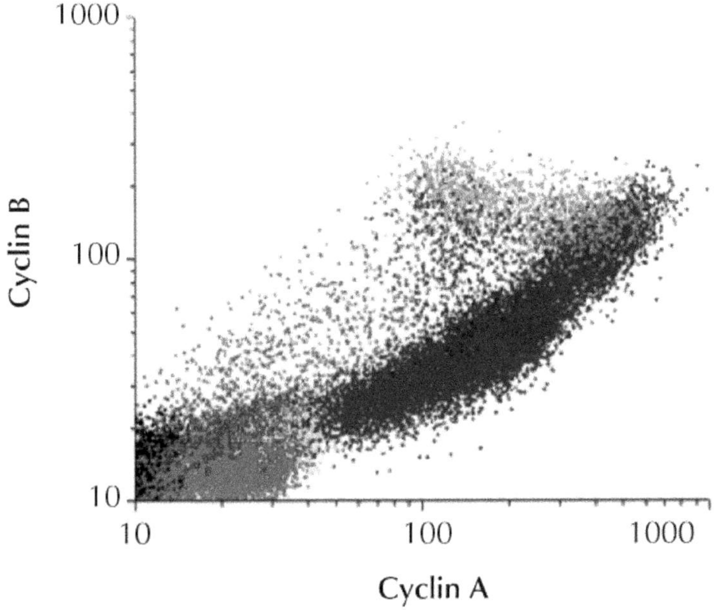

FIGURE 1 Scatter plots. (A,C,E) Flow cytometry data from Yan et al [42]. DNA = 190 corresponds to G1 and DNA = 380 corresponds to G2/M. (B,D,F) simulations. We are plotting the total amount of cyclin A and cyclin B per cell, i.e., [CycA]·M(t) and [CycB]·M(t). DNA = 1 in G0/G1 phase; = 2 in G2/M phase. Some 'instrumental noise' has been added to the calculated levels of cyclins and DNA, as described in the Methods. The arrows in (A, B) indicate the rate of cyclin B accumulation in S phase in the measurements and in the model. The arrows in (C, D) indicate the cyclin A level at the onset of DNA synthesis, compared to the maximum expression level of ~600 AU.

The simulation (Fig. 1B) captures the observed accummulation of cyclin B in late G1 (when Cdh1 turns off), but the simulated rise in cyclin B during S phase appears to be faster than the observed rise [45](compare the arrows in Figs. 1A and B). The simulation does capture the rapid accumulation of cyclin B observed in G2. Finally, while it did not calibrate the cyclin E expression parameters to any specific dataset, the pattern of expression in Fig. 2A is quite similar to expected expression patterns for normal human somatic cells and some human tumor cell lines [46].

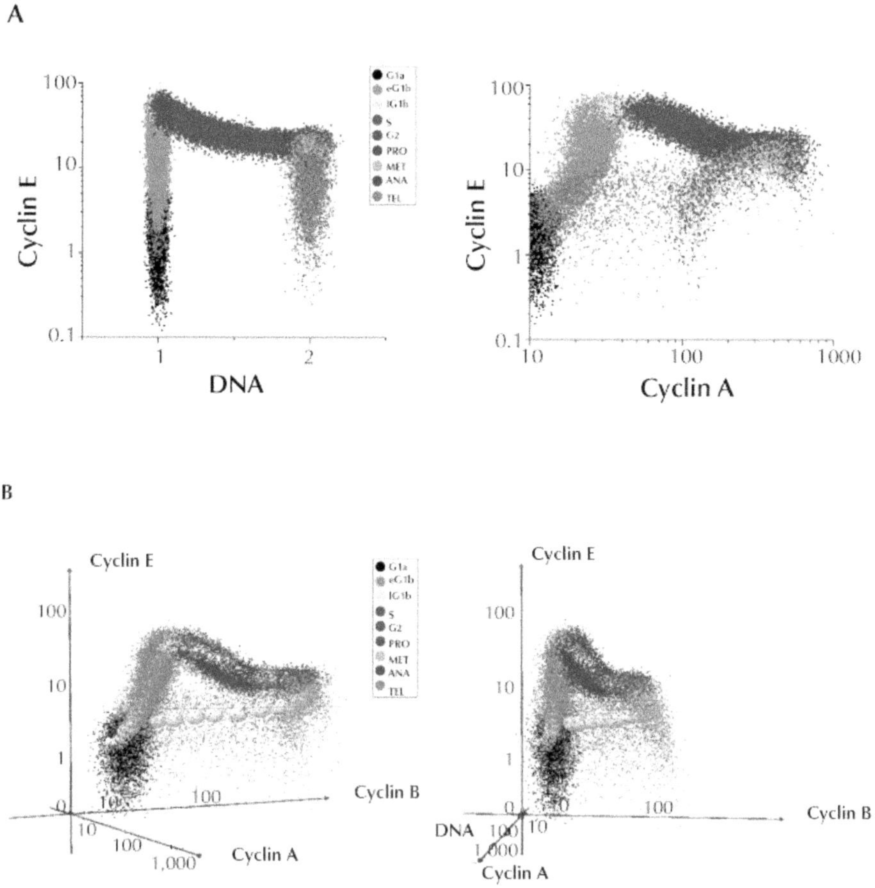

FIGURE 2 Model predictions of cyclin E dynamics. (A) Scatter plots. (B) Stochastic limit cycle in the state space of cyclins A, B and E. We provide two different perspectives of this three dimensional figure to help visualize how the cyclin levels go up and down. In addition, we have added golden-colored balls to help guide the eye along the cell cycle trajectory. Each ball represents the average of the cyclin levels of all the cells binned over a hundredth of the φi interval [0,1], where φi refers to the fraction of the cell cycle completed by cell i (as described in the Methods section). Finally, it may help to recognize that Fig. 1E is a projection of the data on the CycA-CycB plane, and Fig. 2B is a projection on the CycA-CycE plane.

It is believe that hybrid approach will be generally useful for modeling macromolecular regulatory networks in cells, because it combines the qualitative appeal of Boolean models with the quantitative realism of reaction kinetic models.

1.4 RESULTS

1.4.1 Hybrid Modeling Approach

The modeling approach is hybrid in two senses. First, employ both continuous and discrete variables, and second to allow for both deterministic and stochastic processes. The concerning the components of the control system, to track cyclin levels as continuous concentration variables, but to use discrete Boolean variables to represent the activities ('on' or 'off') of the regulatory proteins (transcription factors and ubiquitinating enzymes) that control cyclin synthesis and degradation. This distinction is equivalent to a presumed 'separation of time scales' the activities of the regulatory proteins change rapidly between 0 and 1, while the concentrations of cyclins change more slowly due to synthesis and degradation. The Boolean variables, to assume, proceed from one state to the next according to a fixed sequence corresponding roughly to the super highway of Li et al. [14]. The time spent in each state, however, is not a 'tick' of the metronome but rather the sum of a deterministic execution time (which may be 0) plus a random, exponentially distributed waiting time. In this sense, the model combines deterministic and stochastic processes.

In its present version, model is not fully autonomous. The discrete variables do not update according to Boolean functions of the current state of the network. Rather, they go through a fixed sequence of states predetermined by the Boolean network model of Li et al. [14]. The discrete variables determine the rates of synthesis and degradation of the continuous variables (the cyclins), and the cyclins feedback on the discrete variables by determining how much time is spent in some of the Boolean states. This strategy keeps the model simple and is appropriate for the cases, considered in this chapter, of unperturbed cycling of 'wild type' cells, which travel serenely along the super highway of Li et al. To consider more complicated cases, of mutant cells that travel a different route through discrete state space or of cells that are perturbed by drugs or radiation, it will have to elaborate on this basic model with additional rules governing the interactions of the discrete and continuous variables. It is currently working on alternative strategies to adapt this basic modeling paradigm to more complex situations.

The model (Fig. 3) tracks three cyclin species (A, B and E), two transcription factors ('TFE' and 'TFB') and two different E3 ubiquitin-ligase complexes (APC-C and SCF). The TFE drives the synthesis of cyclins E and A early in the cell cycle (comparable to the E2F family of transcription factors) [33], and TFB drives the synthesis of cyclins B and A late in the cell cycle (comparable to FoxM1 and Myc) [34-35]. The Anaphase Promoting Complex—Cyclosome (APC-C) is active during M phase and early G1, when it combines with Cdc20 and Cdh1 to label cyclins A and B for degradation by proteasomes. It make a further distinction between Cdc20 activity on cyclin A (Cdc20A, active throughout mitosis) from Cdc20 activity on cyclin B (Cdc20B, activated at anaphase). The SCF labels cyclin E for degradation via ubiquitination, but only when cyclin E is phosphorylated [36], which assume is correlated primarily with cyclin A/Cdk2 activity [37].

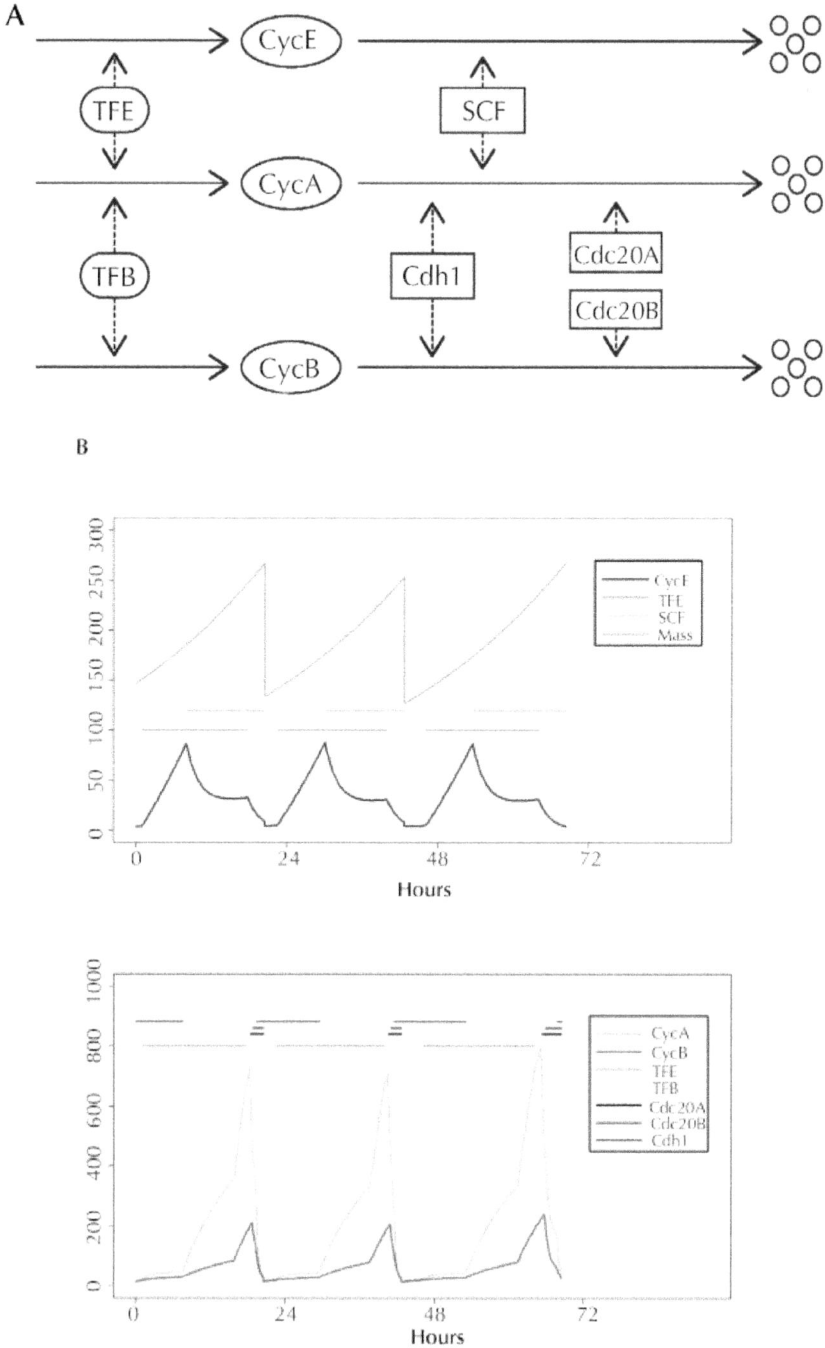

FIGURE 3 The model. (A) The synthesis and degradation of cyclin proteins is regulated by transcription factors (TFE and TFB) and by ubiquitination machinery (SCF, Cdc20 and Cdh1). (B) Three successive cell cycles are simulated as explained in the Methods.

In model, the two transcription factors and the four ubiquitination factors are each represented by a Boolean variable, B_{TFE}, etc. For each cyclin component to write an ordinary differential equation, $d[CycX]/dt = k_{sx} - k_{dx}[CycX]$, where the rate 'constants' for synthesis and degradation, k_{sx} and k_{dx}, depend on the Boolean variables (see Table 1). Hence, each cyclin concentration is governed by a piecewise linear ODE. The parameters in the model ($(k'_{sx}, k''_{sx}$, etc.) are assigned numerical values (Table 1), chosen to fit observations of how fast cyclins accumulate and disappear during different phases of the cell cycle.

Next, one must assign rules for updating the Boolean variables in the model. It assume that the Boolean variables follow a strict sequence of states (see Table 1) that corresponds roughly to the super highway discovered by Li et al. [14]. This sequence of states conforms to current ideas of how the mammalian cell cycle is regulated. Newborn cells are said to be in 'G1a' state, because they are not yet committed to a new round of DNA synthesis and mitosis. The transcription factors, TFE and TFB, are silent, and Cdh1/APC-C is active, so the levels of cyclins A, B and E are low in newborn cells. For a mammalian cell to leave the G1a state and commit to a new round of DNA replication and division, it must receive a specific set of extracellular signals (growth factors, matrix binding factors, etc.), which up-regulate the activity of TFE. It assume that these 'proliferation signals' are present and that our (simulated) cell spends only a few hours in G1a before transiting into G1b. In model, the time spent in G1a is an exponentially distributed random variable with mean = 2 h. When the cell passes the 'restriction point' and enters G1b, TFE is activated and CycE begins to accumulate. Among other chores, Cdk2/CycE inactivates Cdh1/APC-C, allowing Cdk2/CycA dimers to accumulate. In model, the transition from early G1b to late G1b is weakly size dependent, because the condition for this transition is that [CycE]*Mass exceeds a certain threshold (θ_E). Because this transition depends on cell mass, those cells that are larger than average tend to make the transition sooner, and cells that are smaller than average tend to make the transition later. This effect allows the cell population to achieve a stable size distribution. In the late G1b state, CycA/Cdk2 level rises to a certain threshold (θ_A), when it triggers entry into S phase. Cdk2/CycA also promotes the degradation of cyclin E by SCF during S phase. It assume that DNA synthesis requires at least 7 h.

Cyclin B begins to accumulate in late G1 and S, after Cdh1 is inactivated, but the major accumulation of cyclin B protein occurs in G2 phase, after DNA synthesis is completed and TFB is activated. The G2—M transition is delayed until enough Cdk1/CycB dimer accumulates ($[CycB]>\theta_B'$) to promote entry into prophase and the appearance Cdc20A/APC-C, which begins the process of cyclin A degradation [38-40]. Cdc20B/APC-C is activated at the metaphase—anaphase transition, where it promotes three crucial tasks (1) separation of sister chromatids by the mitotic spindle, (2) partial degradation of cyclin B, and (3) re-activation of Cdh1. Cdh1/APC-C degrades Cdc20 [41], and then finishes the job of cyclin B degradation (telophase). When [CycB] drops below the threshold θ_B'', the cell finishes telophase and divides into two newborn daughter cells in G1 phase (unreplicated chromosomes) with low levels of cyclins A, B and E.

TABLE 1 Hybrid model of mammalian cell cycle control.

$\frac{d[CycA]}{dt} = k_{sa} - k_{da}[CycA]$	$k_{sa} = k'_{sa} + k''_{sa} B_{TFE} + k'''_{sa} B_{TFB}$	$k'_{sa} = 5$	$k''_{sa} = 6$	$k'''_{sa} = 20$
	$k_{da} = k'_{da} + k''_{da} B_{Cdc20A} + k'''_{da} B_{Cdh1}$	$k'_{da} = 0.2$	$k''_{da} = 1.2$	$k'''_{da} = 1.2$
$\frac{d[CycB]}{dt} = k_{sb} - k_{db}[CycB]$	$k_{sb} = k'_{sb} + k''_{sb} B_{TFB}$	$k'_{sb} = 2.5$	$k''_{sb} = 6$	
	$k_{db} = k'_{db} + k''_{db} B_{Cdc20B} + k'''_{db} B_{Cdh1}$	$k'_{db} = 0.2$	$k''_{db} = 1.2$	$k'''_{db} = 0.3$
$\frac{d[CycE]}{dt} = k_{se} - k_{de}[CycE]$				
	$k_{se} = k'_{se} + k''_{se} B_{TFE} k_{se} = k'_{se} + k''_{se} B_{TFE}$	$k'_{se} = 0.02$	$k''_{se} = 2$	
	$k_{de} = k'_{de} + k''_{de} B_{SCF}$	$k'_{se} = 0.02$	$k''_{de} = 0.5$	
$\frac{dM}{dt} = \gamma \cdot M$	$M = \delta \cdot M$ at division	$\gamma = 0.029\ hr^{-1}$	$\delta = 0.5 \cdot G^{a}$	
			$G: \mu = 1, \sigma = 3.33$	

State	Phase	B_{TFE}	B_{SCF}	B_{TFB}	B_{Cdc20A}	B_{Cdc20B}	B_{Cdh1}	Condition for exit	λ (h)
1	G1a	0	0	0	0	0	1	none	2
2	Early G1b	1	0	0	0	0	1	$[CycE]^{*}M = \theta_E$	0
3	Late G1b	1	0		0	0	0	$[CycA] > \theta_A$	0.01
4	S	1	1	0	0	0	0	$T_{min} = 7$ h	1
5	G2	1	1	1	0	0	0	$[CycB] > \theta'_B$	0.5
6	Prophae	0	1	1	0	0	0	none	0.75
7	Metaphase	0	1	1	1	0	0	none	1.5
8	Anaphase	0	1	1	1	1	0	none	0.5
9	Telophase	0	1	0	1	1	1	$[CycB] < \theta''_B$	0.025

[a] G is a Gaussian random variable with mean = 1, $\sigma = 3.3\%$.
$\theta_A = 12.5$, $\theta'_B = 21.25$, $\theta''_B = 3$, $\theta_E = 80$.

doi:10.1371/journal.pcbi.1001077.t001

It assume that cell division is symmetric, with some variability; i.e., the mass of the two daughter cells at birth are δM_{div} and $(1-\delta)M_{div}$, where M_{div} = mass of mother cell at division, and δ is a Gaussian-distributed random variable with mean = 0.5 and standard deviation = 0.0167. In all simulations reported here we assume that cells grow exponentially between birth and division. However, we have also simulated linear growth, and the results are not significantly different.

It introduce stochastic effects into the model by assuming that the time spent in each state of the Boolean subsystem, as it moves along the super highway, has a random component (T_i^r) as well as a deterministic component (T_i^d): $T_i = T_i^d + T_i^r$. From Table 1, we see that $T_i^d = 0$ for i = 1, 6, 7, 8, and $T_4^d = 7$ h. For the remaining cases (i = 2, 3, 5, 9), T_i^d is however long it takes for the cyclin variable to reach its threshold. The stochastic component for each transition is a random number chosen from an exponential distribution with mean = λ_i. The random time delay is calculated from a uniform random deviate, r, by the formula $T_i^r = -\lambda_i \ln(r)$. The values chosen for the λ_i's are given in Table 1.

In the Methods section, it describes how to simulate the progression of a single cell through its DNA replication/division cycle. Because the model's differential equations are piecewise linear, they can be solved analytically, and an entire 'cell cycle trajectory' can be determined by computing a few random numbers and solving some algebraic equations. A typical result of such simulations, over three cell cycles, is illustrated in Fig. 3B. Not surprisingly, the accumulation and loss of the cyclins correlate with the activities of the cyclin regulators. At the beginning of each cycle, the cell starts in State 1 (G1a phase in Table 1), with low levels of all cyclin because TFE and TFB are off and Cdh1 is on. When the cell leaves G1a, TFE turns on and cyclin E rises rapidly, but cyclin A increases only modestly, because Cdh1 is still active in early G1b. Cdh1 turns off when cyclin E level crosses θ_E, allowing cyclin A to increase dramatically in late G1b and drive the cell into S phase (State 4). Cyclin B increases modestly in late G1 and S phase, because Cdh1 is off but TFB has not yet turned on. Cyclin E is degraded in S phase, because SCF is now active. When the cell finishes DNA synthesis, TFB turns on, causing further increase of cyclins A and B. When cyclin B level rises above its first threshold, θ_B', the cell enters prophase (State 6) and then prometaphase-metaphase (State 7). During State 7, cyclin A level drops precipitously because Cdc20A is turned on. After the replicated chromosomes are fully aligned on the mitotic spindle, Cdc20B turns on (State 8) and cyclin B is partially degraded. Cdc20B activates Cdh1 (State 9) and cyclin B is degraded even faster. When cyclin B level drops below its second threshold, θ_B'', the cell divides and returns to G1a (State 1).

1.4.2 Cyclin Distributions in an Asynchronous Culture

The first test for the hybrid model is to simulate flow cytometry measurements of the DNA content and cyclin levels in an asynchronous population of RKO (colon carcinoma) cells [42]. In the data set, a typical scatter plot has about 65,000 data points, each point displaying the measurements of two observables in a single cell chosen at random from the cell cycle (Fig. 1). When the data are plotted in this way, they form a cloudy tube of points through a projection of the state space (say, cyclin B versus cyclin A). Because there will be some cells from every phase of the cell cycle, the tube closes on itself. If the systems were completely deterministic and the measurements were absolutely precise, the data points would be a simple closed curve (a 'limit cycle') in the state space. The data actually present a fuzzy trajectory that snakes through state space before closing on itself. The indeterminacy of the points comes (presumably) from two sources: intrinsic noise in the molecular regulatory system (modeled by the random waiting times, T_i^r) and extrinsic measurement errors, which shall introduce momentarily. The strategy for simulating flow-cytometry data is explained in more detail in the Methods section.

In Fig. 1 compare the simulated flow-cytometry scatter plots with experimental results of Yan et al.[42]. The color-code used, each cell in the simulated plot according to which Boolean State (Table 1) the cell is in at the time of fixation. In Fig. 2 plot cyclin E fluctuations, as predicted by model, along with a projection of the cell cycle trajectory in a subspace spanned by the three cyclin variables (A, B and E).

1.4.3 Contact Inhibition of Cultured Cells

As a further test of the utility of this modeling approach used the hybrid model to simulate an exponentially growing population of an immortalized Human Umbilical Vein Endothelial cell line (HUVEC). In the experiment (Fig. 4A; see Methods), a culture is seeded with 5×10^4 cells on 'Day 0' and allowed to grow. At Day 6, it reaches confluence and cell number plateaued at a constant level.

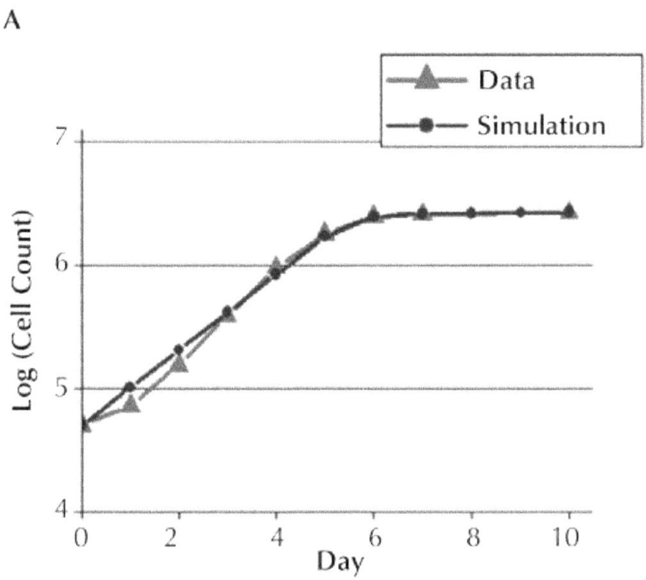

FIGURE 4 *(Continued)*

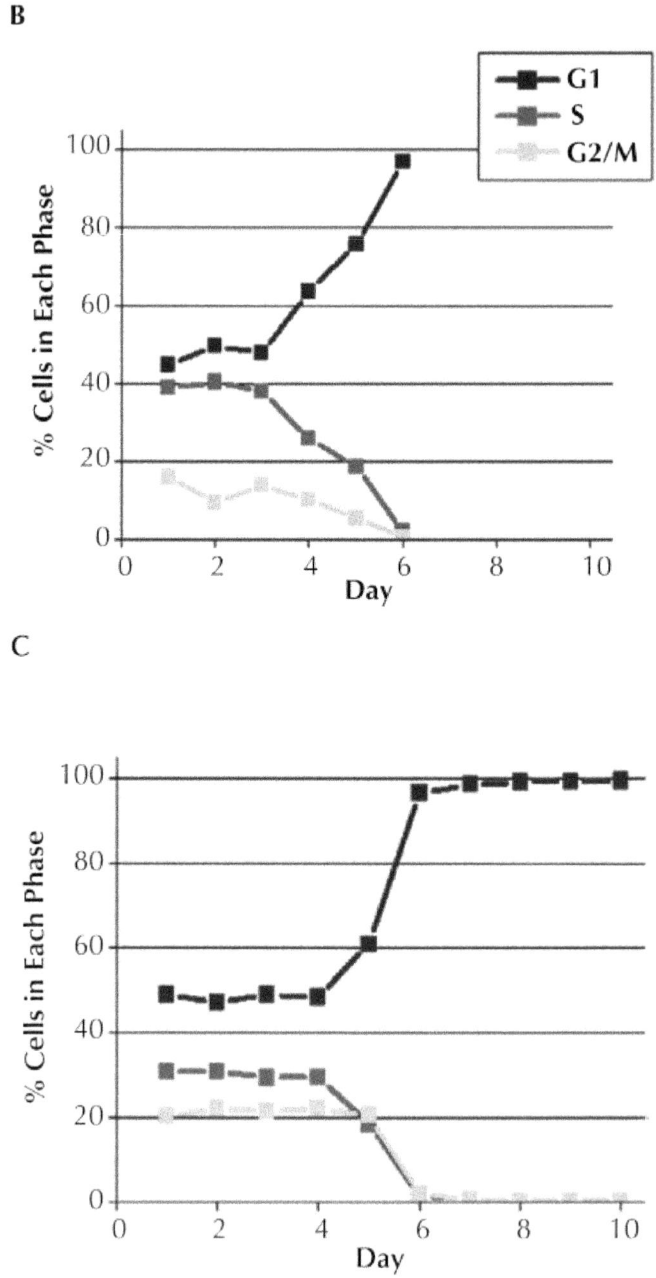

FIGURE 4 Contact inhibition of a culture of human umbilical vein endothelial cells. (A) Growth curve for the HUVEC population over 10 days, showing the base-10 logarithm of the cell count for both experimental data and our simulation (with N0 = 11000 and N1 = 500). (B) Daily distribution of cells across the phases of the cell cycle, from experimental data. (C) Model simulation of the phase distributions.

To apply the hybrid model to this data, it had to devise a way to model contact inhibition, which arrests cells in a stable quiescent state. To this end, it assume that the transition probability, p, for exiting State 1 is a function of the number of cells alive at that time, N:

$$p = \frac{p_0}{1 + \exp\left(\frac{N - N_0}{N_1}\right)} \tag{9}$$

For $0<N_1 \ll N_0$, p is a sigmoidal function of N that drops abruptly from p_0 to 0 for $N>N_0$. For each cell in this simulation, to set λ_1 (the mean for the random time spent in G1a) to $1/p$, and we choose $p_0 = 0.5\ h^{-1}$ to conform to the value of λ_1 in Table 1. As the population size N increases, the time spent in G1a phase increases until cells eventually arrest in State 1, and the growth curve, $N(t)$, levels off. In this case, State 1 in model corresponds to a quiescent state (G0) in which cells are alive but not proliferating.

To make the simulation more tractable, to start off with 500 cells (instead of 50,000 cells) and follow the lineage of each initial cell until Day 10. Every 24 hours, to compute the number of cells alive at that point of time and plot the results in Fig. 4A, along with the experimental data (scaled down by a factor of 100). The parameter values, $N_0 = 11{,}000$ and $N_1 = 500$, are chosen to fit the simulation to the observed growth curve. From the model we can also compute the percentage of cells in G0/G1, S and G2/M phases on each day (Fig. 4C), and the results compare favorably with the experimental observations (Fig. 4B). Lastly, it also simulates the patterns of cyclin A2 and cyclin B1 expression on each day for the growing population of HUVEC cells (see Supporting Fig. S1).

KEYWORDS

- **Asynchronous Culture**
- **Budding Yeast**
- **Cell Division**
- **Cyclin**
- **Data pre-processing**

Supporting Information

Figure S1 Patterns of Cyclin A and Cyclin B expression in simulated populations of HUVECs growing toward confluence over days 0–10.

Figure S2 Pattern of Cyclin A expression in stimulated human circulating lymphocytes.

Figure S3 Correlation of Cyclin A expression with BrdU labeling.

ACKNOWLEDGMENTS

We thank Keith E. Schultz for collecting the data in Fig. 4, as well as Kathy Chen and Tongli Zhang for helpful suggestions. The cyclin A2 antibody was a gift from Vincent Shankey, Beckman-Coulter (Miami).

AUTHOR CONTRIBUTIONS

Conceived and designed the experiments: RMS JWJ. Performed the experiments: RMS. Analyzed the data: RS JWJ JJT. Wrote the paper: RS RMS JWJ JJT. Conceived modeling approach: JJT. Implemented the model and ran all simulations: RS. Designed the model: JWJ.

REFERENCES

1. Mitchison, J. M. *The Biology of the Cell Cycle*. Cambridge UK: Cambridge University Press. 313 p. (1971).
2. Zetterberg, A., Larsson, O., and Wiman, K. G. What is the restriction point? *Curr Opin Cell Biol* **7**: 835–842. (1995).
3. Chen, K. C., Csikasz-Nagy, A., Gyorffy, B., Val, J., and Novak, B., Kinetic analysis of a molecular model of the budding yeast cell cycle. *Mol Biol Cell* **11**: 369–391. (2000).
4. Novak, B., Pataki, Z., Ciliberto, A., and Tyson, J. J. Mathematical model of the cell division cycle of fission yeast. *Chaos* **11**: 277–286. (2001).
5. Chen, K. C., Calzone, L., Csikasz-Nagy, A., Cross, F. R., and Novak, B., Integrative analysis of cell cycle control in budding yeast. *Mol Biol Cell* **15**: 3841–3862. (2004).
6. Calzone, L., Thieffry, D., Tyson, J. J., and Novak, B. Dynamical modeling of syncytial mitotic cycles in Drosophila embryos. *Mol Syst Biol* **3**: 131–141. (2007).
7. Novak, B. and Tyson, J. J. Numerical analysis of a comprehensive model of M-phase control in *Xenopus* oocyte extracts and intact embryos. *J Cell Sci* **106**: 1153–1168. (1993).
8. Pomerening, J. R., Kim, S. Y., and Ferrell, J. E. Jr. Systems-level dissection of the cell-cycle oscillator: bypassing positive feedback produces damped oscillations. *Cell* **122**: 565–578. (2005).
9. Aguda, B. D. and Tang, Y. The kinetic origins of the restriction point in the mammalian cell cycle. *Cell Prolif* **32**: 321–335. (1999).
10. Qu, Z., Weiss, J. N., and MacLellan, W. R. Regulation of the mammalian cell cycle: a model of the G1-to-S transition. *Am J Physiol Cell Physiol* **284**: C349–C364. (2003).
11. Novak, B., and Tyson, J. J. A model for restriction point control of the mammalian cell cycle. *J Theor Biol* **230**: 563–579. (2004).
12. Sha, W., Moore, J., Chen, K., Lassaletta, A. D., and Yi, C-S. Hysteresis drives cell-cycle transitions in *Xenopus laevis* egg extracts. *Proc Natl Acad Sci USA* **100**: 975–980. (2003).
13. Pomerening, J. R., Sontag, E. D., and Ferrell, J. E. Jr. Building a cell cycle oscillator: hysteresis and bistability in the activation of Cdc2. *Nature Cell Biol* **5**: 346–351. (2003).
14. Li, F., Long, T., Lu, Y., Ouyang, Q., and Tang, C. The yeast cell-cycle network is robustly designed. *Proc Natl Acad Sci USA* **101**: 4781–4786. (2004).
15. Davidich, M. I. and Bornholdt, S. Boolean network model predicts cell cycle sequence of fission yeast. *PLoS One 3*: e1672. (2008).
16. Faure, A., Naldi, A., Chaouiya, C., and Thieffry, D. Dynamical analysis of a generic Boolean model for the control of mammalian cell cycle. *Bioinformatics* **22**: e124–131. (2006).
17. Fantes, P. A. and Nurse, P. Division timing: controls, models and mechanisms. In: Ed. John PCL. *The Cell Cycle*.: Cambridge University Press. Cambridge UK pp. 11–33. (1981).
18. Tyson, J. J. The coordination of cell growth and division–intentional or incidental? *Bioessays* **2**: 72–77. (1985).
19. Tyson, J. J. Size control of cell division. *J Theor Biol* **126**: 381–391. (1987).

20. Darzynkiewicz, Z., Crissman, H., and Jacobberger, J. W. Cytometry of the cell cycle: cycling through history. *Cytometry* A **58**: 21–32. (2004).
21. Glass, L. and Kauffman, S. A. The logical analysis of continuous, non-linear biochemical control networks. *J Theor Biol* **39**: 103–129. (1973).
22. Glass, L, and Pasternack, J. Stable oscillations in mathematical models of biological control systems. *J Math Biol* **6**: 207–223. (1978).
23. Matsuno, H., Inouye, S. T., Okitsu, Y., Fujii, Y., and Miyano, S. A new regulatory interaction suggested by simulations for circadian genetic control mechanism in mammals. *J Bioinform Comput Biol* **4**: 139–153. (2006).
24. Bosl, W. J. Systems biology by the rules: hybrid intelligent systems for pathway modeling and discovery. *BMC Syst Biol* **1**: 13. (2007).
25. Li, C., Nagasaki, M., Ueno, K., and Miyano, S. Simulation-based model checking approach to cell fate specification during Caenorhabditis elegans vulval development by hybrid functional Petri net with extension. *BMC Syst Biol* **3**: 42. (2009).
26. Alur, R., Dang, T., Esposito, J. M., Fierro, R. B., and Hur, Y. Hierarchical Hybrid Modeling of Embedded Systems. In: Ed. Henzinger, T. A., Kirsch, C. M. *Embedded Software*: Proceedings of the First International Workshop: Springer Berlin. pp. 14–31. (2001).
27. Fishwick, P. A. *Handbook of dynamic system modeling*. Boca Raton: Chapman & Hall/CRC. (2007).
28. Klee, H. and Allen, R. Simulation of dynamic systems with MATLAB and Simulink. Boca Raton, FL: CRC Press. 840 p. (2011).
29. Mosterman, P. An Overview of Hybrid Simulation Phenomena and Their Support by Simulation Packages. In: Ed. Vaandrager, F., van Schuppen, J. Hybrid Systems: Computation and Control. Springer Berlin. pp. 165–177. (1999).
30. Deshpande, A., Gollu, A., and Varaiya, P. SHIFT: A Formalism and a Programming Language for Dynamic Networks of Hybrid Automata. In: Ed. Antsaklis, P., Kohn, W., Nerode, A., Sastry, S. Hybrid Systems IV. Springer Berlin. pp. 113–133. (1997).
31. Deshpande, A., Gollu, A., and Semenzato, L. The SHIFT programming language and run-time system for dynamic networks of hybrid systems. *IEEE Trans Automat Contr* **43**: 584–587. (1998).
32. Alur, R., Grosu, R., Hur, Y., Kumar, V., and Lee, I. Modular Specification of Hybrid Systems in CHARON. In: Ed. Lynch, N., Krogh, B. H. Hybrid Systems: Computation and Control. Springer Berlin. pp. 6–19. (2000).
33. Trimarchi, J. M. and Lees, J. A. Sibling viralry in the E2F family. *Nat Rev Mol Cell Biol* **3**: 11–20. (2002).
34. Laoukili, J., Kooistra, M. R., Bras, A., Kauw, J., and Kerkhoven, R. M. FoxM1 is required for execution of the mitotic programme and chromosome stability. *Nat Cell Biol* **7**: 126–136. (2005)
35. Wierstra, I. and Alves, J. FoxM1, a typical proliferation-associated transcription factor. *Biol Chem* **388**: 1257–1274. (2007).
36. Cardozo, T. and Pagano, M. The SCF ubiquitin ligase: insights into a molecular machine. *Nat Rev Mol Cell Biol* **5**: 739–751. (2004).
37. Welcker, M., Singer, J., Loeb, K. R., Grim, J., and Bloecher, A. Multisite phosphorylation by Cdk2 and GSK3 controls cyclin E degradation. *Mol Cell* **12**: 381–392.
38. Harper, J. W., Burton, J. L., and Solomon, M. J. The anaphase-promoting complex: it is not just for mitosis any more. *Genes Dev* **16**: 2179–2206. (2002).
39. Peters, J. M. The anaphase-promoting complex proteolysis in mitosis and beyond. *Mol Cell* **9**: 931–943. (2002).
40. Geley, S., Kramer, E., Gieffers, C., Gannon, J., and Peters, J. M. Anaphase-promoting complex/cyclosome-dependent proteolysis of human cyclin A starts at the beginning of mitosis and is not subject to the spindle assembly checkpoint. *J Cell Biol* **153**: 137–148. (2001).
41. Pfleger, C. M. and Kirschner, M. W. The KEN box: An APC recognition signal distinct from the D box targeted by Cdh1. *Genes Dev* **14**: 655–665. (2000).

42. Yan, T., Desai, A. B., Jacobberger, J. W., Sramkoski, R. M., and Loh, T. CHK1 and CHK2 are differentially involved in mismatch repair-mediated 6-thioguanine-induced cell cycle checkpoint responses. *Mol Cancer Ther* **3**: 1147–1157. (2004).
43. Gong, J., Ardelt, B., Traganos, F., and Darzynkiewicz, Z. Unscheduled expression of cyclin B1 and cyclin E in several leukemic and solid tumor cell lines. *Cancer Res* **54**: 4285–4288. (1994)
44. Frisa, P. S. and Jacobberger, J. W. Cell cycle-related cyclin B1 quantification. *PLoS One* **4**: e7064. (2009).
45. Jacobberger, J. W., Sramkoski, R. M., Wormsley, S. B., and Bolton, W. E. Estimation of kinetic cell-cycle-related gene expression in G1 and G2 phases from immunofluorescence flow cytometry data. *Cytometry* **35**: 284–289. (1999).
46. Darzynkiewicz, Z., Gong, J., Juan, G., Ardelt, B., and Traganos, F. Cytometry of cyclin proteins. *Cytometry* **25**: 1–13. (1996).
47. Press, W. H., Teukolsky, S. A., Vetterling, W. T., and Flannery, B. P. Numerical recipes in C. *The art of scientific computing*. Cambridge University Press. Cambridge. pp. 707–752. (1992).
48. Freedman, D. A. and Folkman, J. CDK2 translational down-regulation during endothelial senescence. *Exp Cell Res* **307**: 118–130. (2005).
49. Schimenti, K. J. and Jacobberger, J. W. Fixation of mammalian cells for flow cytometric evaluation of DNA content and nuclear immunofluorescence. *Cytometry* **13**: 48–59. (1992).

2 Spindles and Active Vortices

David A. Head, W. J. Briels, and Gerhard Gompper

CONTENTS

2.1 Introduction 25
2.2 Methods 27
2.3 Discussion 29
2.4 Results 31
 2.4.1 Stationary States 31
 2.4.2 Dynamics and Vortices 36
 2.4.3 Enhanced Detachment from Ends 42
 2.4.4 Controlled Volume 43
2.5 Conclusion 45
Competing Interests 45
Authors' Contributions 45
Keywords 45
Acknowledgements 46
References 46

2.1 INTRODUCTION

Robust self-organization of subcellular structures is a key principle governing the dynamics and evolution of cellular life. In *fission* yeast cells undergoing division, the mitotic spindle spontaneously emerges from the interaction of microtubules, motor proteins and the confining cell walls, and asters and vortices have been observed to self-assemble in quasi-two dimensional microtubule-kinesin assays. There is no clear microscopic picture of the role of the active motors driving this pattern formation, and the relevance of continuum modeling to filament-scale structures remains uncertain. Here results of numerical simulations of a discrete filament-motor protein model confined to a pressurised cylindrical box. Stable spindles, nematic configurations, asters and high-density semi-asters spontaneously emerge, the latter pair having also been observed in cytosol confined within emulsion droplets. State diagrams are presented delineating each stationary state as the pressure, motor speed, and motor density are varied. The parameter regime where vortices form exhibiting collective rotation of all filaments have a finite life-time before contracting to a semi-aster. The quantifying the distribution of life-times suggests this contraction is a Poisson process. Equivalent systems with fixed volume exhibit persistent vortices with stochastic switching in the direction of rotation, with switching times obeying similar statistics to contraction times in pressurised systems. Furthermore, we show that increasing the detachment rate of motors from filament plus-ends can both destroy vortices and turn some asters into vortices.

The discrete filament-motor protein models provide new insights into the stationary and dynamical behavior of active gels and subcellular structures, because many phenomena occur on the length-scale of single filaments. Based on findings, it is argued that the need for a deeper understanding of the microscopic activities underpinning macroscopic self-organization in active gels and urge further experiments to help bridge these lengths.

Filamentous proteins are prevalent within eukaryotic cells and perform a variety of crucial tasks relating to cellular integrity, locomotion, transport, and division [1,2]. Such tasks are often *active* in that they can only proceed in concert with energy-consuming mechanisms, including directed filament growth and motor protein-generated tension, placing such processes outside the realm of equilibrium thermodynamics [3]. Self-organisation of motor protein-filament mixtures will be selected for when it robustly reproduces static or dynamic structures beneficial to the cell's viability. An example is the mitotic spindle that forms during division of *fission yeast* cells. It has been shown that this bipolar structure, consisting of microtubules emanating from spindle pole bodies towards an overlapping midplane region, exists and functions essentially as normal even in cells with no nucleus-associated microtubule organizing center [4,5]. The plausible conclusion is that the interaction between filaments and motor proteins in the confined cell geometry controls the location of the pole bodies. For *budding yeast* this self-organisation scenario has been reinforced by the evolution of more sophisticated regulatory mechanisms [6]. Also, egg cell extracts from the amphibious genus *Xenopus* can generate a well-formed spindle apparatus despite entirely lacking cell walls [7-9]. Nonetheless an understanding of the principles underlying self-organization of bioflaments driven by motor proteins in confined spaces is of direct relevance to many organisms [10].

Given the complexity of real cells it is often advantageous to consider simplified model systems, and this approach has been adopted to investigate the role of confinement in filament-motor mixtures. Experiments on growing microtubules confined to spherical emulsion droplets revealed a droplet-size dependency on the observed structure [11]: Droplets larger than $\approx 29\ \mu m$ in diameter contained *asters* with the polar microtubules pointing towards the centre, controlled by the motor protein dynein, whereas smaller droplets were found to contain *semi-asters* with the aster's focus near the interface. These findings demonstrate that the degree of confinement can partly determine structure formation, but as motor density and speed were not control variables in these experiments their influence could not be assayed.

A strikingly non-equilibrium property of filament-motor mixtures is their ability to spon-taneously generate flows due to their active components, even in the absence of boundary driving forces[12-14]. Assays of microtubule-oligomeric kinesin mixtures in a quasi-two dimensional geometry with flat, parallel confining walls found a dynamic rotating structure denoted a *vortex* [15,16]. Accompanying simulations of semiflexible filaments [16] and subsequent hydrodynamic theories [17,18] appeared to reproduce the observed structures. However [19], it is unlikely that the simulations of Surrey *et al.* [16] and the theories and simulations [17,18] describe the same type of vortex, because the hydrodynamic theories are based on a nematic order-parameter description, while simulations of semi-flexible filaments [16] neglect self-avoidance (and thus nematic order). Mesoscopic models based on the Smoluchowski equations have not resolved this issue [20,21]. The simulations of self-avoiding filaments strictly in two dimensions showed no evidence of a vortex state [19]. The microscopic picture underlying vortex formation thus remains unknown. Gliding assays of filaments along motor beds permit quantitative comparison to models [9,22] and at high concentration exhibit vortex-like 'swirls' [23,24], although in this situation the active forces are unbalanced monopoles, unlike dipoles generated by motors connecting two filaments in the bulk [25]. Vortex-like motion is often observed in self-propelled systems such as bacterial swimmers [26-28], but with differing microscopic mechanisms.

It is apparent that the combined influence of confinement and activity on structure formation and spontaneous flows in filament-motor mixtures is presently not well understood. The aim here is to acquire a deeper understanding of this problem in a broad sense, not restricted to any one biological realisation, *i.e.* microtubule-dynein or actin-myosin. It is therefore desirable to study model systems in which all parameters can be freely varied. The application of continuum equations, which are coarsegrained over lengths much larger than the filament length L, to structures of only a few L in spatial extent is not guaranteed to be successful. Therefore adopt a discrete numerical model in which motors and filament segments are explicitly represented, and all physical mechanisms that are potentially relevant (steric hinderance, thermal fluctuations *etc.*) are incorporated. This model is an extension of one previously employed in two dimensions [19], where it was found to produce some signatures of active gels such as super-diffusion and anomalous small wavelength density fluctuations, but not vortices.

Consider arrays of filaments confined to a quasi- two dimensional cylinder, with a height of a few filament diameters which permits filament overlap, and an external pressure at the curved walls. Then systematically vary the motor density, speed, and applied pressure. Four steady-state configurations arise within the covered parameter space, including an aster and semi-aster as observed in confined emulsion droplets [11], and also a spindle-like state that spontaneously emerges from the motor-filament interaction in the confined geometry, possibly reproducing the fission yeast observations [4,5]. These states are described along with a fourth nematic state that links to known equilibrium phases. It is also find a fifth, vortex state associated with a definite rotation of filaments about a fixed center that appears to be always transient. The existence and properties of these vortices are characterised. To highlight the important role played by motors at filament plus-ends, independently vary the detachment rate of motors from plus-ends and show that vorticity is associated with a critical fraction of plus-ended motors. The observation of vortices in fixed volume systems described, confirm that they are driven at least partly by motor motion and not boundary fluctuations.

2.2 METHODS

Consider a system of N semiflexible polar filaments, which can be connected by motor proteins. Each filament consists of $M = 30$ monomers separated by a bond length b with Hookean bond potentials with a spring constant 100 k_{BT}/b^2. Self-avoidance of filaments is introduced by repulsive Lennard Jones potentials with diameter σ and energy parameter $\varepsilon = 5\ k_{BT}$. A natural choice is $\sigma = b$. Semi-flexibility is described by curvature elasticity with bending rigidity $\kappa = 200bk_{BT}$ such that the persistence length $\ell p = \kappa/kB$T is $\ell p = 20L/3$ with $L = Mb$ the filament length.

Only motors simultaneously connected to two different filaments are explicitly represented. The concentration of free motors in solution is assumed to be spatiotemporally uniform, which is a valid assumption for rapidly-diffusing free motors when the ratio of attached to free motors is small. This concentration is renormalized into a constant rate of attachment as discussed below. Motors are modeled as two-headed Hookean springs with a spring constant k_{BT}/b^2 and dynamics defined by four rates as shown in Figure 1(a): (i) The attachment rate k_A for a motor to attach to two monomers within a predefined range, here taken to be the excluded volume radius $2^{1/6\sigma}$ (sok_A is the product of a molecular attachment rate and the free motor concentration); (ii) the detachment rate k_D of each head independently from its filament (detachment of either head results in removal of the whole motor from the system); (iii) the movement rate k_M of each head independently towards the filament's [+]-end, and (iv) the detachment rate k_E for motor heads already at a [+] end. The movement rate is attenuated by an exponential factor $e\_\Delta E/kBT$ with ΔE the change in motor spring energy for the trial move. Except where otherwise stated, $k_E = k_D$.

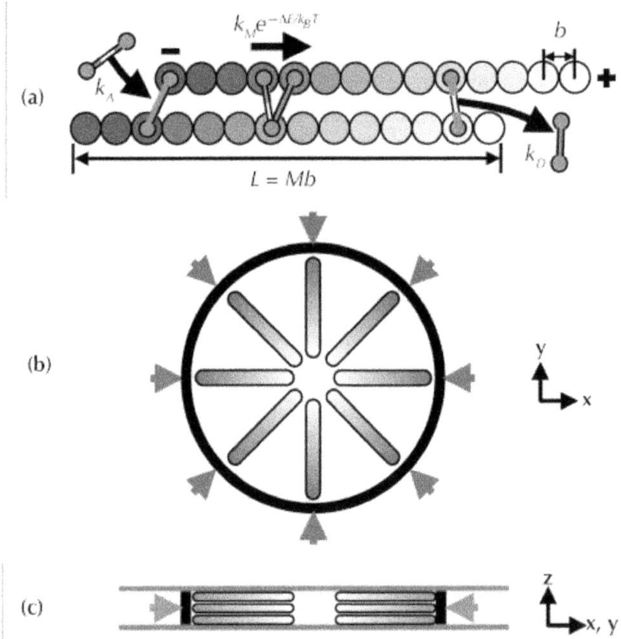

FIGURE 1 Model definition. (a) Summary of key model parameters including the rates of motor attachment k_A and detachment k_D, and the bare stepping rate k_M. See text for details. (b) Plan view showing the filaments oriented with their light-shaded [+]-ends towards the center. The arrows denote the external pressure acting on the circular elastic wall. (c) Side view of the same, showing the confining walls perpendicular to the z-axis.

Simulating 3D filament gels in a spherical or cubic box at physiologically-relevant densities is computationally prohibitive when excluded volume interactions are included. To reduce computational demands while still permitting filament overlap, we therefore adopt a quasi-2D simulation cell with parallel confining walls normal to the z-axis spaced $5b \ll L$ apart; see Figures 1(b) and 1(c). This is a similar geometry to the microtubule experiments [15, 16]. Furthermore we only consider a single ring of filaments driven by an inward-acting pressure, intended to describe confinement in a cell, or other filaments nearby. This external pressure acts through a flexible, elastic wall as evident from Figures 1(b) and 1(c) .

All walls repel the monomers with the same Lennard Jones non-bonding potential as for filaments. $N = 175$ filaments are placed in a radial aster configuration with all [+]-ends pointing towards the center, in three parallel layers with roughly 6668 filaments per layer (note there is some stochasticity in the initial conditions). This initial condition was chosen to promote the formation of asters and vortices, but does not inhibit other structures. The system is surrounded by an elastic wall, which initially is circular with radius 40b. The wall is discretized into 80 nodes that are initially regularly spaced. Changes in node separation from the initial value ℓ_0 to $\ell_0 + \delta\ell$ incur an energy cost *12kBTZδl2l_30* per adjacent node pair, with $Z = 5$b the wall height. Similarly, changes in the local curvature between adjacent node triplets from the initial value κ_0 to$\kappa_0 + \delta\kappa$ incur an energy $10^{3k}{}_{\mathrm{B}T\ell 0Z\delta\kappa}{}^{2}$ per triplet. The chosen coefficients ensure an approximately circular wall shape throughout the deformation without significantly countering the external pressure for typical filament densities, as confirmed by the far smaller final wall radii measured when the filaments were absent.

The filaments, motors, and elastic wall are all updated stochastically. The filaments obey Brownian dynamics [29] governed by an effective monomer friction coefficient γ for hydrodynamically anisotropic slender elements, *i.e.* with a 2:1 ratio between implicit solvent drag perpendicular to the filament axis ($= 2\gamma$) to the parallel direction ($= \gamma$) [30]. Wall nodes move by Monte Carlo Metropolis moves applied to the (x, y) coordinates of 80 nodes initially equispaced along its contour. The energy for these moves includes the wall elastic and wall-filament interaction energies, and a pressure-volume term PV where $P > 0$ is a fixed parameter for each run. To check for convergence with time, various scalar quantities, such as the number of motors per filament, were checked to be constant within noise when plotted against log(t). In addition, the mean squared rotation $(\Delta\theta)_2(t, t+\Delta t) = n-1\sum_{ni=1}\|\theta_i(t+\Delta t)-\theta_i(t)\|_2$, with θi the angle between filament i's centre-of-mass with respect to the nominal centre of the box (more precisely, the mean of all wall nodes) and some fixed axis, was checked to no longer to vary with t to within noise. Stationarity was not achieved for the vortex states, for which alternative measures were employed.

2.3 DISCUSSION

It has been demonstrated that the vortices described here involve the collective rotation of the filaments about a fixed centre. This was predicted by the nematodynamics theory of Kruse *et al.*[17], but contrasts with the simulations of intersecting filaments [16] for which movies appear to show no actual filament rotation, rather the motors run along a *static* vortex configuration of the growing filaments. It is possible that our inclusion of excluded volume interactions, which are necessary to generate the nematic elasticity required by the theory but were absent in these earlier simulations, may explain this discrepancy. Those simulations also employed growing filaments, whereas here as in the continuum modeling these lengths were fixed, suggesting a further potential source of discrepancy. Unlike the continuum theories, however, our vortices are only one filament in radius so there is no radial gradient in the polarity field, making direct comparison problematic. It is concluded that the challenging task remains to demonstrate a definitive link between macroscopic vortices and microscopic filament-motor interactions. An important aspect could be the system size of the microscopic models, because a minimum size much larger than currently accessible by simulations may be needed to see vortices in a bulk system. Further experiments might elucidate the underlying mechanism. For instance, fluorescently tagging a small fraction of microtubules would allow individual filament rotation (if any) to be visualized.

The variation of steady-state structure with motor speed and density shown in Figure 3 can in large part be understood as due to a non-uniform density of motors along the filament, in particular the fraction of motors at a [+]-end which dwell there before detaching. Varying the end-detachment rate k_E confirms that a strong binding at [+]-ends can stabilise an aster relative to a vortex or semi-aster. The motor speed k_M plays a role in selecting the distribution of motors along the filament, but also contributes to the rotation of vortices as inferred from the fixed volume system. Therefore, it is claimed that the observed vortex is a genuine non-equilibrium state powered at least partially by the unidirectional motion of energy-consuming motor heads along the filaments, although at constant pressure they appear to be transient. It is not clear if varying some other parameters may produce stable vortices.

Systematically quantifying the role of all of the model parameters is clearly challenging for such a high-dimensional parameter space, and here adopted the pragmatic approach of holding most parameters fixed while varying those deemed most likely to be critical. Eventually, the impact of all parameters on structure and dynamics will need to be quantified if a broad description of active gels is to be attained. Here, highlight two parameters likely to reveal novel or interesting behaviour. First, the filament length $L = Mb$ was fixed at $M = 30$ monomers throughout, whereas extensive simulations with $M = 25$ revealed similar steady-state diagrams as Figure 3 but *no vortices*. Increasing the aspect ratio therefore seems to enhance vorticity, and it would be interesting to quantify this effect. Secondly, the elastic parameters for the wall were set to maintain a roughly circular shape, as in the emulsion experiments of Pinot *et al.* [11]. However, in those experiments flexible vesicles were also considered that produced a richer array of observed structures, and this effect could be easily investigated within our model by lowering the bending stiffness of the wall.

While it was always intention to model filament-motor systems as generally as possible, it is nonetheless insightful to consider the corresponding parameters for an actual system. Taking the filament diameter to be b = 10 nm, intermediate between microtubules (about 25 nm) and actin (about 7 nm), the filament length in our simulations becomes 0.3 μm, and the forces generated by our motors correspond to $k_{BT}/b \approx 0.4$pN. These values are smaller than but comparable to real systems, *e.g.* for actin-myosin complexes, filament lengths are typically around 1 μm and myosin proteins generate approximately 1.5pN [33]. Mapping our inverse movement rate k–1M to the typical motor cycle time of 2040 ms [33] suggests that our average simulation run extended to about 1 m, again comparable to but shorter than typical experimental times. The vortex rotation times will also be of the order of a minute. The findings should thus be experimentally accessible, and we predict that the steady states observed here will be reproduced if a confining geometry of size comparable to the mean filament length can be engineered in *in vitro* actin-myosin or microtubule-kinesin experiments. The chances of success will be increased if motor concentrations capable of generating around 10100 or more motors per filament are chosen, and perhaps reducing the mean filament length to give aspect ratios around 3050.

In very recent experiments, on actin (without myosin) in confined geometries [34], pattern formation strikingly similar to our nematic state in Figure 2(c) was observed. Since we find such states for a low density of slow motors, this is entirely consistent with the approach to the passive systems investigated in these experiments.

2.4 RESULTS

Results are presented here in terms of the normalised attachment rate k_A/k_D, the normalised motor rate $k_{M\tau b}$ where $\tau b = Lb\gamma/4k_{BT}$ is the approximate time for a filament to freely diffuse one monomer distance; the normalised pressure P/P_0 with $P_0 = \varepsilon/b^3$ with ε = 5 k_BT the LennardJones repulsion energy, and, where relevant, the scaled end-detach rate k_E/k_D.

2.4.1 Stationary States

For the parameter space considered, it is observed that four classes of steady-state configuration as shown in Figure 2. For a low density of fast motors, *spindles* are observed as in Figure 2(a), which crossover to a radially-symmetric *aster* as the motor density is increased as show in Figure 2(b). For slower motors, we observe a *nematic* at low motor densities and *semi-asters* at high motor densities, as shown in Figures 2(c) and 2(d) resp. Semi-asters typically arise for higher pressures than asters and are more compressed, consistent with the emulsion experiments of Pinot *et al.*[11] and justifying use of the term. Movies demonstrating the spontaneous emergence of all of these states from the initial conditions for exactly the same parameters are available as Additional File 1 (spindle, corresponding to Figure 2(a)), Additional File 2 (aster, corresponding to Figure2(b)), Additional File 3 (nematic, corresponding to Figure 2(c)) and Additional File 4 (semi-aster, corresponding to Figure 2(d)).

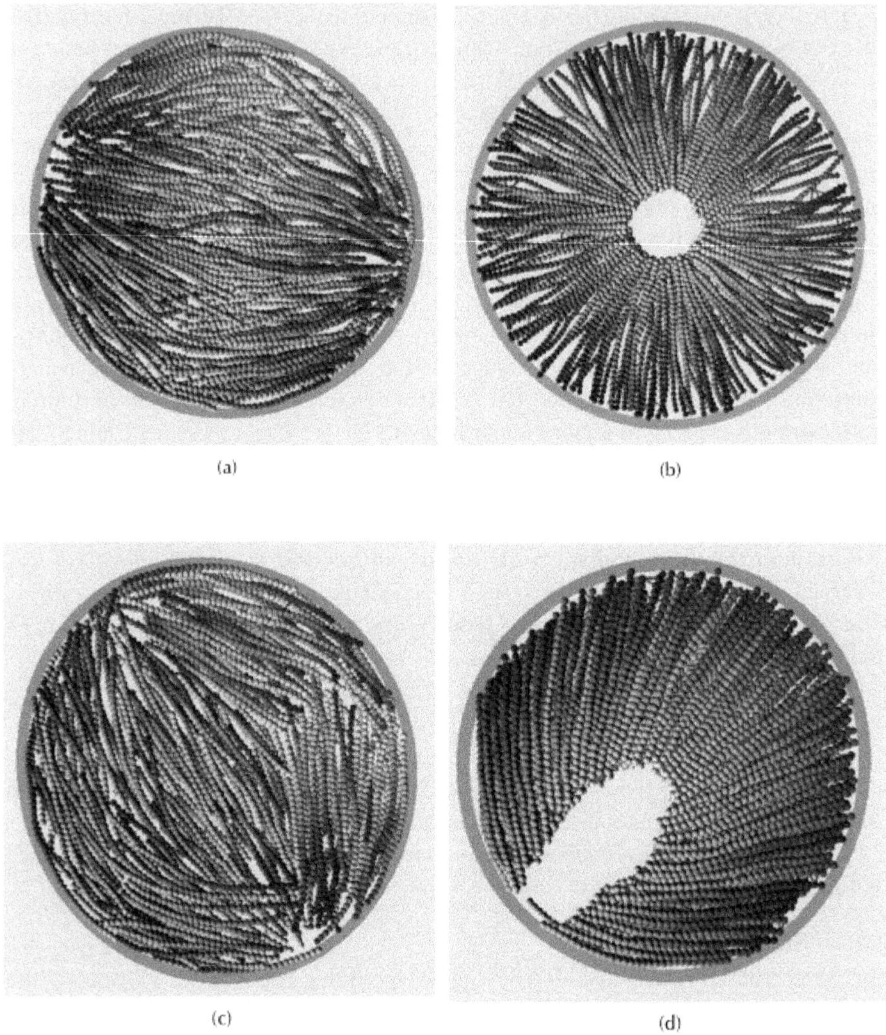

FIGURE 2 **Snapshots of steady-states.** Snapshots of steady-states for a low motor attachment rate $k_A/k_D = 1$ (left) and a higher rate $k_A/k_D = 30$ (right). Conversely, the top line is for fast motors $k_{M\tau b} = 3.75 \times 10^{-2}$and the bottom line for motors 10 times slower. Filaments are shaded light (dark) towards their plus (minus) ends, respectively. These states are referred to as (a) spindle, (b) aster, (c) nematic and (d) semi-aster. The other parameters are $P/P_0 = 0.03$ and $k_E = k_D$. Movies of the same parameter values are available from the supplementary information.

To quantify to which state a system belongs, each filament's polarity vector is projected onto the$x$-y plane to give a two-dimensional unit vector aligned towards the

[+]-end. This is averaged over all filaments whose centers of mass have azimuthal angle θ with respect to the center of the system, giving rise to the mean orientation $\hat{\mathbf{p}}(\theta)$. This is then decomposed into angular mode vectors **a***m* and **b***m*,

$$\hat{\mathbf{p}}(\theta) = \frac{1}{2\pi}\mathbf{a}_0 + \frac{1}{\pi}\sum_{m=1}^{\infty}\{\mathbf{a}_m \cos m\theta + \mathbf{b}_m \sin m\theta\} \quad (1)$$

from which can be defined the mode amplitudes *Qm*,

$$Q_m = \frac{1}{2\pi^2}(a_m^2 + b_m^2) \quad (2)$$

The *Qm* are invariant under global rotations of the whole box.

To determine the corresponding state, the measured Q_m up to $m = 3$ are compared to known values for ideal states, and that with the closest Euclidean distance is taken to be the state. The values for pure asters and nematic phases are easy to derive; for an aster $\hat{\mathbf{p}}(\theta) = (-\cos\theta, -\sin\theta)$, $(Q_0, Q_1, Q_2, Q_3) = (0,1,0,0)$, whereas for the nematic state where $\hat{\mathbf{p}}(\theta) = (0, 0)$, $(Q_0, Q_1, Q_2, Q_3) = (0,0,0,0)$. Note that while an isotropic state would give the same Q_m as for the nematic, such states only arise for k_A and P well below the considered ranges. For spindles and semi-aster states there is a degree of choice in how the target Q_m are calculated, so we choose simple forms that permit exact evaluation of the *Qm*. For the spindle, $\hat{\mathbf{p}}(\theta) = (-\cos\theta, \sin\theta)$ for $\theta \in (-\pi/4, \pi/4)$ or $\theta \in (3\pi/4, 5\pi/4)$ and zero otherwise, for which $(Q_0, Q_1, Q_2, Q_3) = (0, 2/\pi^2 + 1/2, 0, 2/\pi^2)$. For the semi-aster, $\hat{\mathbf{p}}(\theta) = (-\cos[\theta/3], -\sin[\theta/3])$ for $\theta \in (3\pi/4, 3\pi/4)$ and zero otherwise, for which $(Q_0, Q_1, Q_2, Q_3) = (9/\pi^2, 9/4\pi^2, 333/1715\pi^2, 9/100\pi^2)$. Variations in these forms have been tested and although the boundaries between states shift slightly, the underlying trends remain the same.

The occurrence of the four steady-states, plus a fifth 'vortex' state to be discussed below, with motor density and speed are presented in Figure 3 for two different pressures. The observed configurations correlate with the density and distribution of motors along the filaments. The mean number of motors per filament n_{mot}/N is plotted in Figure 4 and shows an expected increase with the attachment rate k_A as well as the pressure. The increase with pressure can be understood as due to the closer packing of the filaments, increasing the number of potential attachment points for motors and hence n_{mot}. The approximate scaling $n_{mot} \sim k_A^{3/2}$ is faster than the linear relationship measured for constant volume, two-dimensional simulations [19], presumably due to similar reasons: as k_A increases so does the motor density which, in this constant-pressure ensemble, allows the system to contract, presenting more potential attachment points between monomers and hence further increasing the motor density. A derivation of the value 3/2 of the exponent is not available so far.

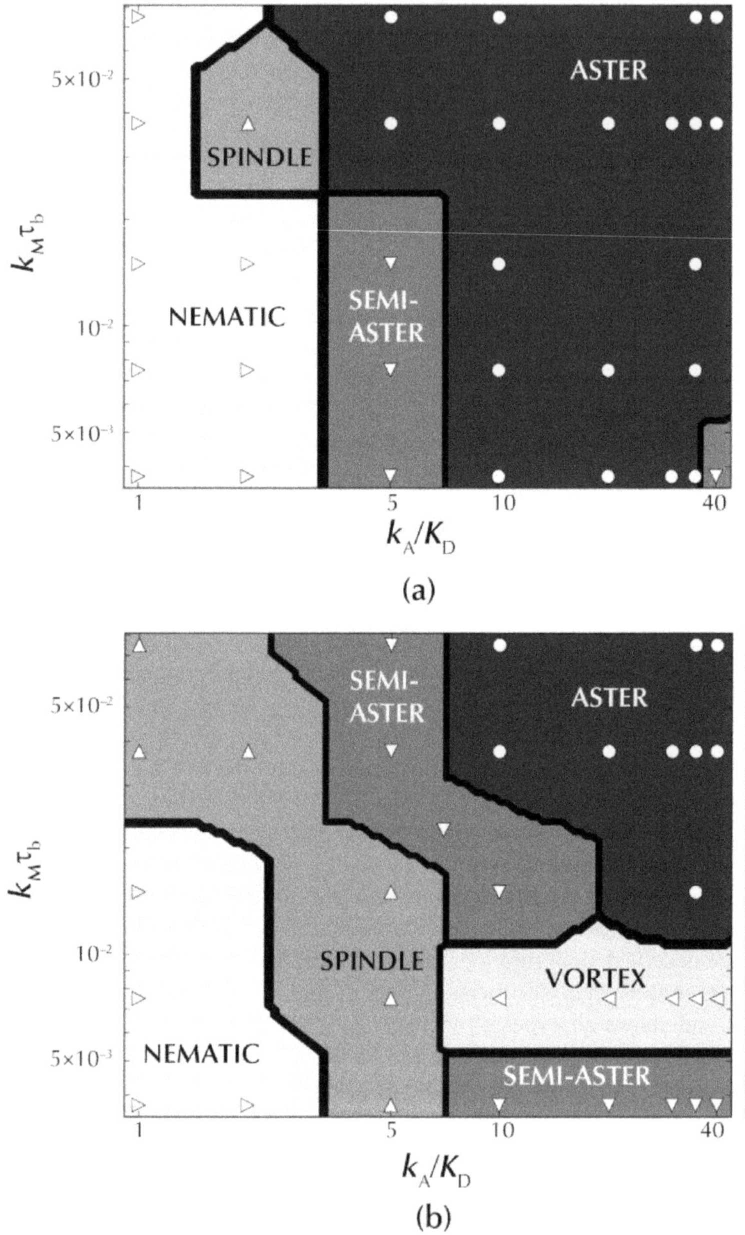

FIGURE 3 State diagrams. Variation of steady-state with motor density and speed for (a) $P/P_0 = 0.01$ and (b) $P/P_0 = 0.024$. Markers denote actual states determined as described in the text and the boundaries are equidistant between pairs of data points. The vortex region in (b) is a transient configuration that is explained in section 3.2 and is delineated as those states with a vorticity υ exceeding 0.7. Since it eventually contracts to a semi-aster, the two distinct semi-aster regions in (b) become contiguous in steady-state.

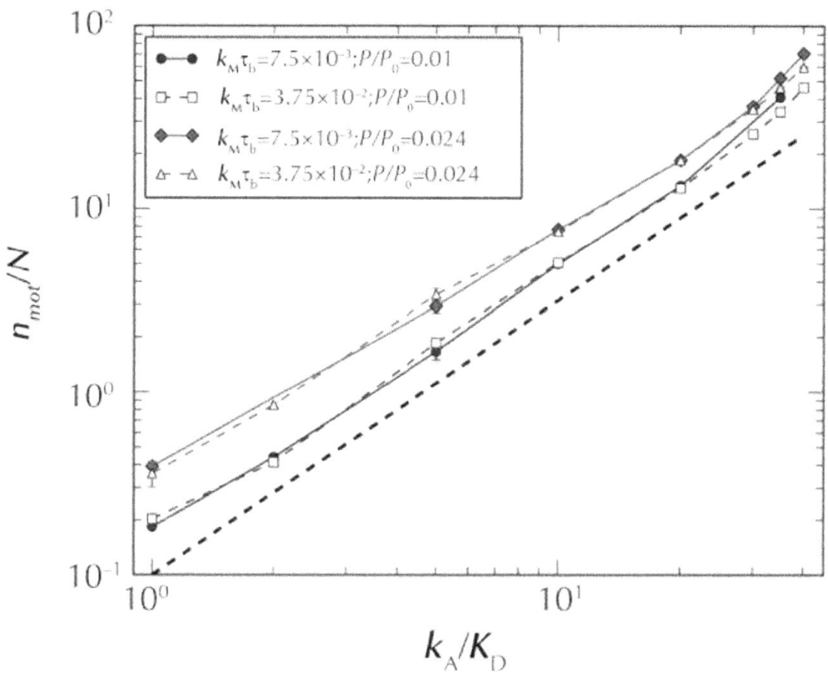

FIGURE 4 Motor density. No. of motors per filament *versus* attachment rate k_A/k_D for the motor speeds and external pressure denoted in the legend. The thick black dashed line has a slope of 3/2. Where data for the required P was not available, n_{mot} was interpolated from runs with Pslightly higher and lower than the target value.

Motors move to the [+]-end and dwell there until detaching, thus a greater fraction are expected to occupy filament [+]-ends, potentially resulting in tight binding mediated *via* many motors. Plotted in Figure 5 is the fraction of motors with at least one head at a filament's [+]-end, $n_{mot}^{[+]} / n_{mot}$ for the same runs as in Figure 4. By comparing to the configuration plots in Figure 3it is possible to infer signatures of crossovers between states in the inflection points in these curves. For the slower motors, there is an increase in $n_{mot}^{[+]} / n_{mot}$ as the nematic state changes to a state with a greater degree of polar ordering (spindle or semi-aster depending on the pressure). For the faster motors there is a marked increase in $n_{mot}^{[+]} / n_{mot}$ with k_A, which corresponds to the crossover to the aster state with a high degree of [+]-end binding. Further indication of the importance of [+]-end binding is presented in section 3.3 where enhanced end-unbinding rates $k_E > kD$ are considered.

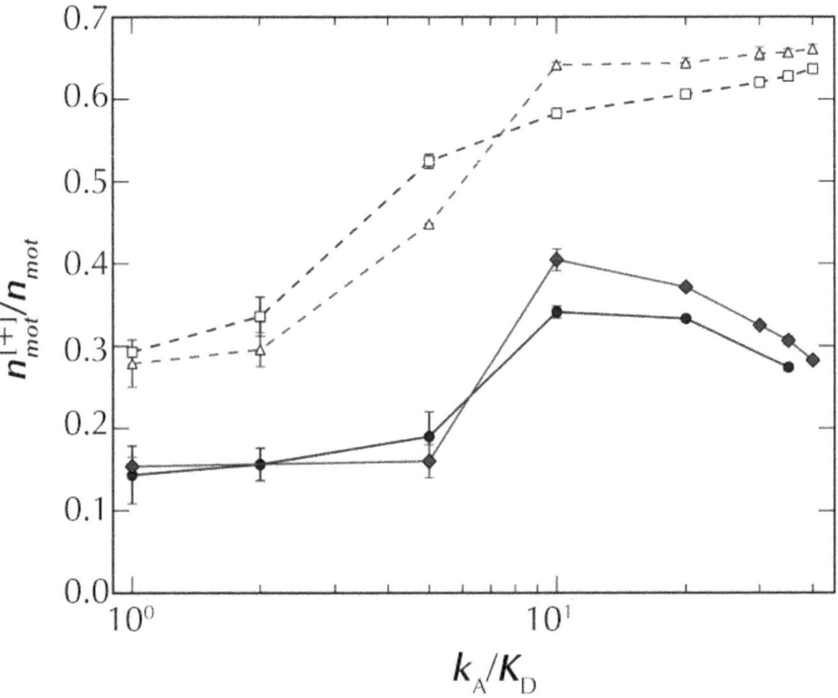

FIGURE 5 Plus-end motors. Fraction of motors with at least one head at a filament's [+]-end versus k_A/k_D for the same data as in Figure 4, so the top two lines correspond to fast motors and the bottom to slow motors.

2.4.2 Dynamics and Vortices

The stationary states described above admit no spontaneous non-equilibrium flows, despite the motor motion generating a positive energy flux: the increase in the stored motor elastic energy due to motor motion and thermal drift of the connected filaments is balanced by the loss due to detachment, with no observed net translocation or rotation of the filaments in steady-state. Collective rotation of all filaments about a fixed center arises for one region of the considered parameter ranges, but appears to be a transient flow that irreversibly contracts to a non-rotating semi-aster configuration. These states are referred to here as *vortices* due to their superficial similarity with the rotational modes observed in microtubule-kinesin assays [15, 16] and are described in detail in this section. A snapshot is given in Figure 6 and movies are provided as Additional File 5 (all filaments shown) and Additional File 6 (one filament highlighted for the same run as Additional File 5).

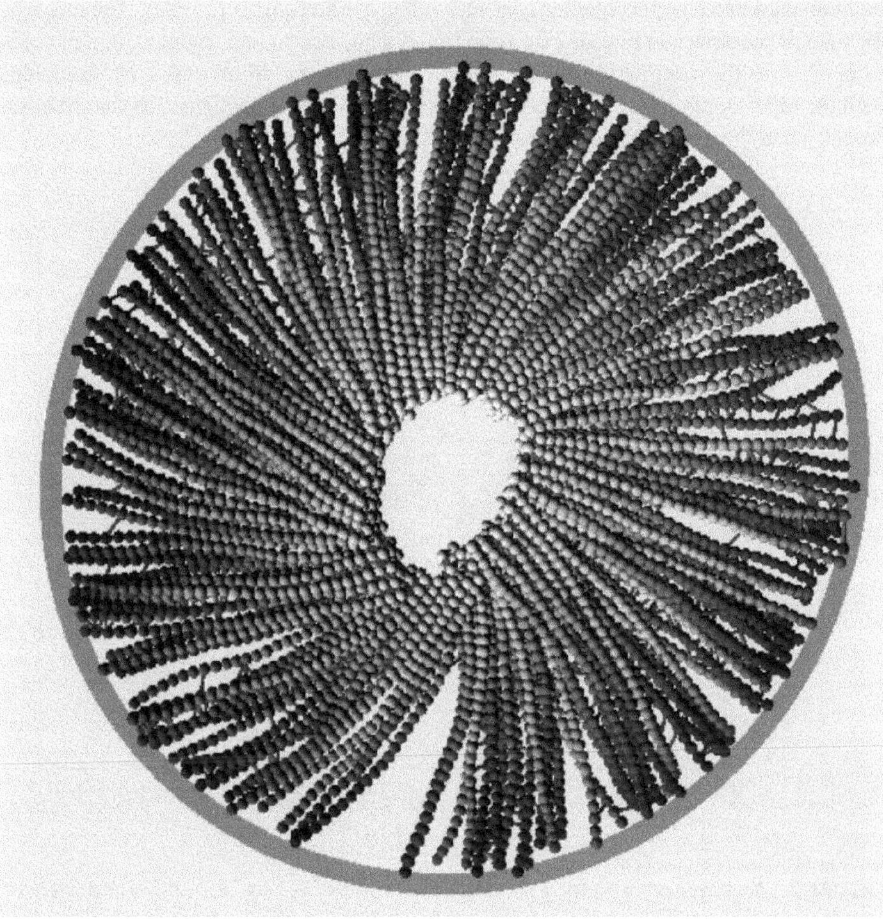

FIGURE 6 Snapshot of a vortex. Snapshot of a vortex rotating in the anti-clockwise direction as presented, for parameters $P/P_0 = 0.024, k_A/k_D = 35$ and $k_{M\tau b} = 7.5 \times 10^{-3}$ taken at a time $t/\tau_b \approx 5.3 \times 10^3$. The colour code is the same as in Figure 2.

Collective rotation of the whole system can be quantified by the mean angular velocity of filament centre-of-mass vectors r relative to the system center, or alternatively by the net transverse velocity of each filament's centre-of-mass relative to its polarity, $\langle(\mathbf{v} \times \hat{\mathbf{p}})z\rangle$.

Here to employ the latter as it is available for all of runs, but confirmed that it closely tracks the angular velocity in those runs for which both were measured. Examples of $\mathbf{v} \times \hat{\mathbf{p}}$ for three independent runs are given in Figure 7, and show finite rotation of either sign until the system irreversibly contracts to a semi-aster state and rotation ceases. This contraction time can be confirmed by visual observation of system states, and can be precisely located by fitting the system radius as a function of time, R(t), to the four-parameter hyperbolic tangent $R(t) = R_{min} + \Delta R \tanh[(t - t^{cont})/\Delta t]$. The mean of $\langle(\mathrm{v} \times \hat{\mathrm{p}})_z\rangle$ is presented in Figure 7 for each run as a horizontal line segment, that extends from $t = 0$ to the contraction time t^{cont} found from this fit. In all cases, t^{cont} coincides with the rapid decay of $\mathrm{v} \times \hat{\mathrm{p}}$ to zero, providing independent confirmation that rotation ceases when the system contracts to a semi-aster.

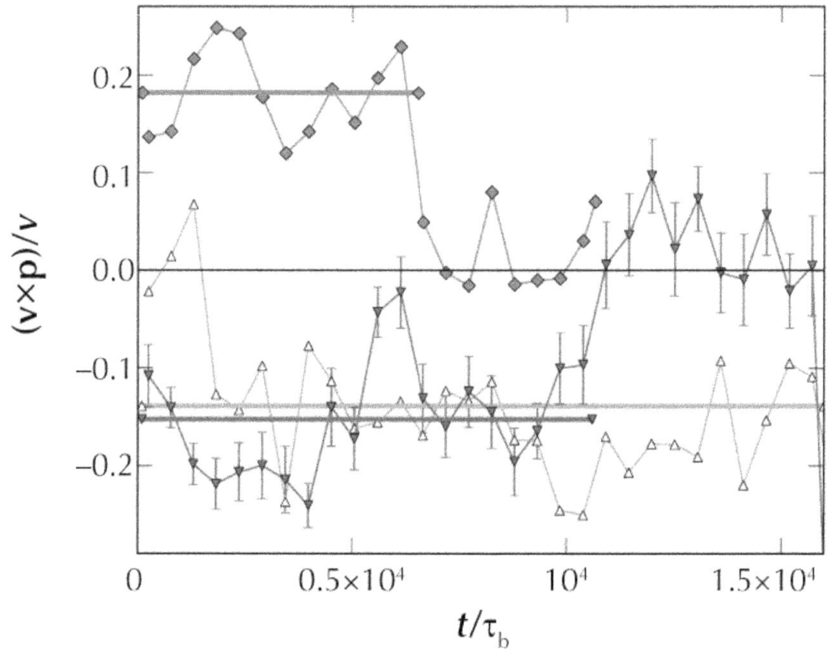

FIGURE 7 Example of vorticity. Examples of the vorticity $(\mathrm{v} \times \hat{\mathrm{p}})|_z$ scaled by a characteristic velocity v for $k_A/k_D = 35$, $k_{Mtb} = 7.5 \times 10^{-3}$ and $P/P_0 = 0.024$. For clarity error bars are only given for a single run. The thick horizontal line segments denote the mean value up to the time when the vortex contracts to a semi-aster, and are plotted up to this time.

It is now possible to define a vorticity order parameter for each point in parameter space. For each run α, the mean $\mu\alpha$ and standard deviation $\sigma\alpha$ of $(\mathrm{v} \times \hat{\mathrm{p}})|_z$ is calculated starting from $t = 0$ up to the time that the system contracts. This is regarded as significant if the mean is comparable to or larger than the standard deviation, but since the sign is arbitrary we also take the absolute value to give the *vorticity* for a single run,

$$\mathcal{V}_\alpha = \left| \frac{\mu_\alpha}{\sigma_\alpha} \right| \geq 0 \tag{3}$$

This is then averaged over all runs with the same parameters to give the mean vorticity $\mathcal{V} = \overline{\mathcal{V}_\alpha}$. A given point in parameter space is then regarded as exhibiting a (transient) vortex if $\mathcal{V}$ exceeds some arbitrarily-chosen value of order unity. The corresponding region of parameter space for $\mathcal{V} > 0.7$ is plotted in Figure 3 and arises for higher densities of motors that are not so fast that they aggregate at [+]-ends, which would stabilize an aster relative to a vortex. Independently varying the fraction of [+]-ended motors by increasing the end-detachment rate k_E supports the existence of a critical fraction for vortex formation.

The reciprocal relationship between vorticity and contraction time is clearly evident when both quantities are plotted together; see Figure 8. Stronger vortices have a shorter lifetime than weaker vortices. The distribution of contraction times is presented in Figure 9(a) for a single pressure. As the data is noisy we do not attempt to extract an arbitrary distribution, but instead make two statistical tests for the most basic possibilities, *i.e.* that contraction happens at a fixed *time*, which would give a Gaussian distribution, or that it happens at a fixed *rate* corresponding to an exponential distribution. To give some measure of the goodness-of-fit, the AndersonDarling statistics for an exponential distribution with an unknown mean has a significance level of $P \approx 0.2$, whereas that for a normal distribution of unknown mean and variance has a significance level of $P \approx 0.025$ [31, 32]. This clearly favours the exponential over the Gaussian distribution. Attaining even this noisy data consumed considerable computing resources and were unable to repeat this procedure for other parameter values.

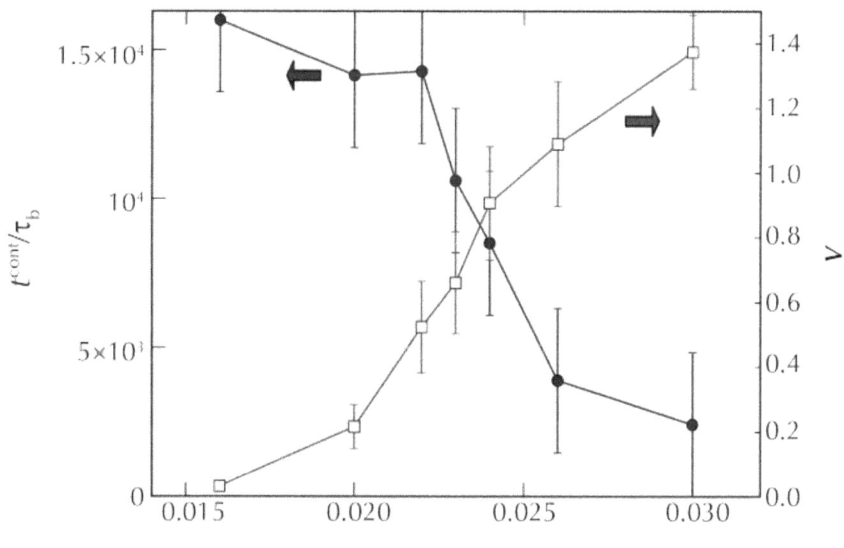

FIGURE 8 **Vorticity and contraction times.** Contraction time to a semi-aster $t^{cont}/\tau b$ (left axis; solid circles) and vorticity $\mathcal{V}$ (right axis; open squares) versus pressure P/P_0 for $k_A/k_D = 35$ and $k_{M\tau b} = 7.5 \times 10^{-3}$. The contraction time was determined by the fit of the radius to a hyperbolic tangent, or assigned the maximum value of $t^{cont} = 1.6 \times 10^{4\tau b}$ if no contraction had occurred within this time. Each point represents five independent runs.

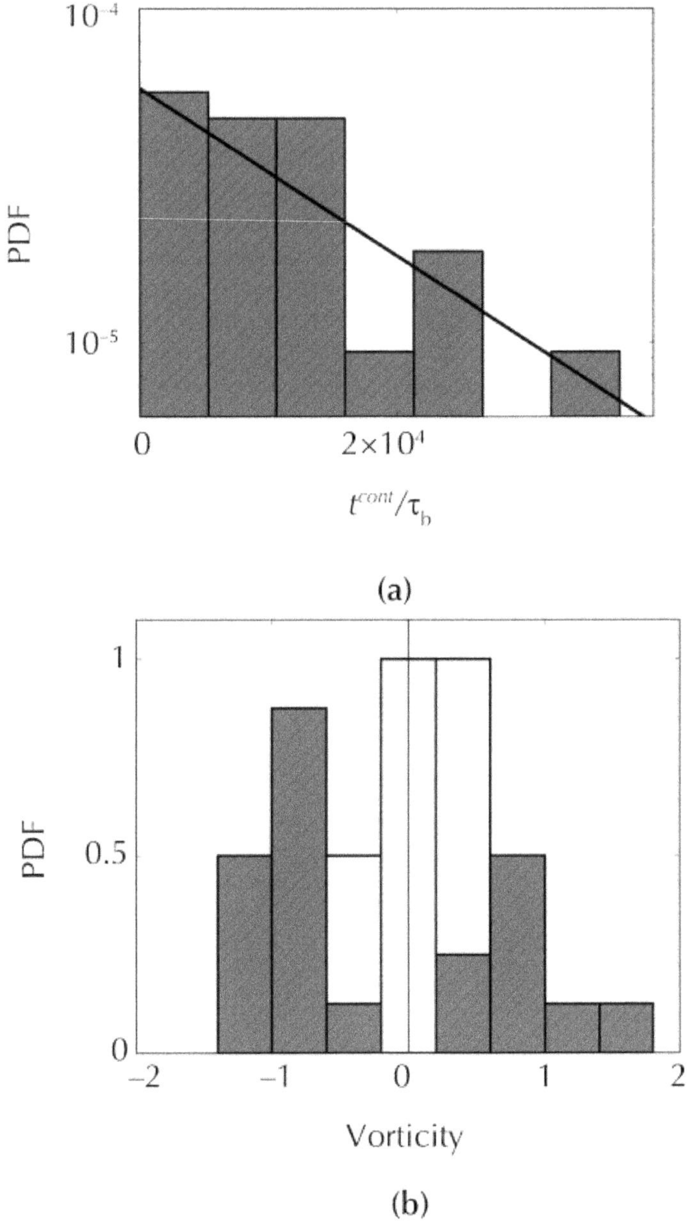

FIGURE 9 Contraction time and vorticity statistics. (a) Probability distribution of contraction times on log-linear axes for $P/P_0 = 0.024, k_A/k_D = 35$ and $k_{Mrb} = 7.5 \times 10^{-3}$. The thick line gives the best fit to an exponential distribution which has a mean $\approx 1.7 \times 10^4$, corresponding to ≈ 0.75 full rotations (the longest vortex survived for ≈ 2.6 rotations). Data corresponds to 20 independent runs. (b) Normalised probability histogram of signed vorticity for $P/P_0 = 0.020$ (white bars in the background; 5 runs) and $P/P_0 = 0.024$ (shaded bars in the foreground; 20 runs).

Assuming the true distribution is exponential, this would suggest that contraction is triggered by spontaneous fluctuations that occur at a constant rate in time. From observation of movies of filament arrangements, a likely candidate is the transient void formation frequently observed near the outer wall, where nearby filaments are attached purely by motors at their [+]-ends and not along their length. Such voids, when large enough, lead to a 'hinge'-like mechanism in which the void expands and one section of the polarity field inverts, leading to the semi-aster.

The onset of vorticity is also evident in the histogram of the *signed* vorticity, *i.e.* the $\mathcal{V}_\alpha$ before taking the modulus in eqn. (3), which can be positive or negative depending on the direction of rotation. For low pressures with $\mathcal{V} \ll 1$ this distribution is unimodal around the origin, but becomes bimodal when vorticity is more evident as demonstrated in Figure 9(b). Of the 20 runs presented here, 12 rotated in one direction and 8 in the other, which has a significance interval of $P \approx 0.5$ as determined from a Binomial test with equal probabilities for both directions. This is to be expected given the use of stochastic initial condition that does not predispose the system to any preferred rotational direction.

Independent confirmation of vorticity can be inferred from the mean-squared angular deviation $\langle(\Delta\theta)^2\rangle$ already defined. This is plotted in Figure 10 for the same parameters as above as a function of the lag time Δt, averaged over all waiting times t up until t^{cont}. There is a crossover from linear behavior $(\Delta\theta)^2 \sim \Delta t$ for low pressures with low vorticity, to a more rapid scaling $(\Delta\theta)^2 \sim (\Delta t)^2$ for pressures well into the vortex regime. Since, this quantity is the angular analogue of the mean squared displacement for translation degrees of freedom, these two limits can be regarded as *diffusive* and *ballistic*, respectively. Microscopically the diffusive limit corresponds to fluctuations with no net drift, whereas the ballistic limit arises when all filaments are rotating around the system with a constant angular velocity in the same direction. Therefore, the vortex state should correlate with ballistic motion, and comparison of Figures 8 and 10 confirms this. This figure also demonstrates that the integrated angular rotation of the vortex before contraction is typically larger than $\pi/2$, much larger than the diffusive drift $\approx \pi/10$ over the same time frame.

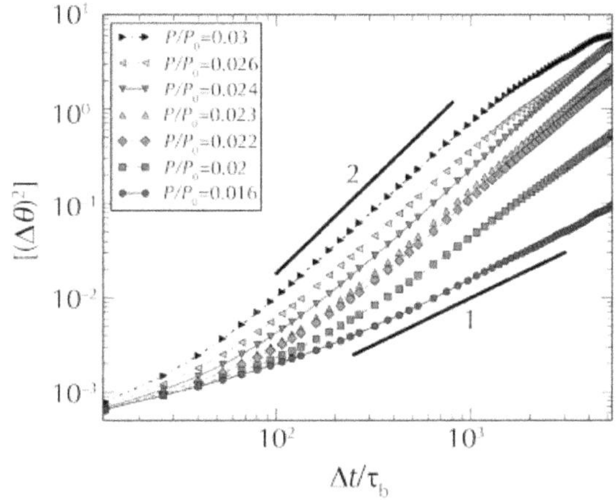

FIGURE 10 **Mean-squared angular changes**. Mean-squared changes in angle $\langle(\Delta\theta)^2\rangle$ versus lag time Δt for the pressures given in the legend, $k_A/k_D = 35$ and $k_{\text{Mrb}} = 7.5 \times 10^{-3}$. The short thick line segments have the slope given.

2.4.3 Enhanced Detachment from Ends

In the simulations that accompanied the microtubule experiments, it was claimed that the residence time at the microtubule [+]-ends played a crucial role in determining the vortex stability, with an enhanced end-detachment rate required to form vortices [16]. By contrast, for our model it is the *fraction* of motors at filament [+]-ends that determines vortex stability relative to an aster. The vorticity, motor density and fraction of motors at [+]-ends are plotted in Figure 11against end-detachment rates $k_E \geq k_D$ for two sets of k_A, k_M and P. It is clear that increasing k_E can both destroy a vortex that existed when $k_E = k_D$, and create a vortex when $k_E = k_D$ gave an aster. In order of increasing k_E, the sequence aster → vortex → semi-aster (where any vortex is either absent or too short lived to be discerned) is typically observed, although we do not claim this sequence is followed by all points in parameter space. There is a slight increase in motor density in the semi-aster state as evident from the figure, resulting from an increase in potential attachment points due to the increased density.

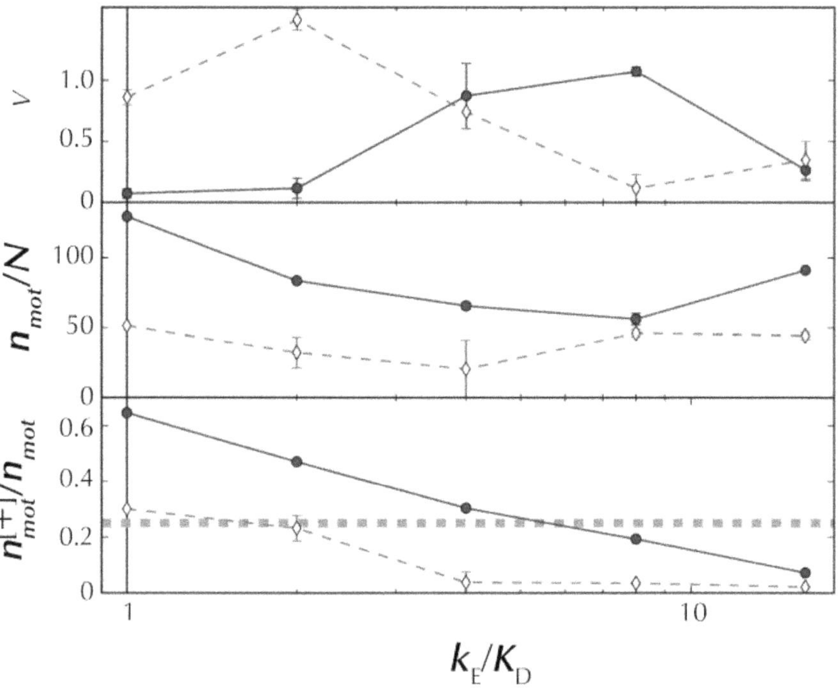

FIGURE 11 Varying end-detach rates. Variation of vorticity, motor density and fraction of [+]-end motors with k_E for $k_A/k_D = 35$ and $k_{mrb} = 7.5 \times 10^{-3}$ (solid lines, filled squares) and $k_A/k_D = 60$ and $k_{Mrb} = 37.5 \times 10^{-3}$ (dashed lines, open diamonds). $P/P_0 = 0.024$ in both cases. The thick dashed line in the lower plot corresponds to 25%. Quantities were measured just prior to contraction, or in steady-state if there was no contraction or it happened too rapidly to discern.

Thus residence time at [+]-ends, which is ∝ k_E^{-1}, is *not* the determining factor with regards vortex stability here. Rather, vortices coincide with around 25% of motors at [+]-ends as highlighted in the figure. There is no critical dependency on the motor density, although we speculate that below some minimum value spindles or nematic states would be observed instead. The critical fraction 25% will likely depend on parameters that were not varied in this work, such as filament length M and the motor spring stiffness. A systematic survey of these parameters, or of $k_E \neq k_D$ for all k_A,k_M and P.

2.4.4 Controlled Volume

One message from the previous sections is that the observed steady-state is predominately determined by the density of motors and the fraction at [+]-ends. It may appear that the primary role of motor motion, which would be the source of any non-equilibrium effects in this model, is merely to select the fraction of [+]-ended motors, faster motors giving a higher fraction. It might even be speculated that even the transient vortex state is driven, not by motor motion, but rather as a protracted buckling event powered by the pressurised walls.

It is straightforward to show that motor motion can drive vortex motion, however. Plotted in Figure12 is the rotational velocity $\langle \mathbf{v} \times \hat{\mathbf{p}} \rangle / v$ for two independent runs in a box with fixed radius, where v is a characteristic filament velocity. The system is initially in an aster configuration, but when the radius is suddenly reduced by $b/2$ at a time $t/\tau b \approx 1.6 \times 10^4$, the system switches to a rotating vortex state that appears to be long-lived; the total time window in this figure is an order of magnitude longer than the longest vortex (which has the same parameters). Since, there is no energy input from the walls, the only possible cause for this rotation is the motor motion. Thus, the pressure ensemble is important to let the system adjust its density to the vortex state; however, the same pressure also destabilizes the vortex state, because it favors further contraction into the semi-aster.

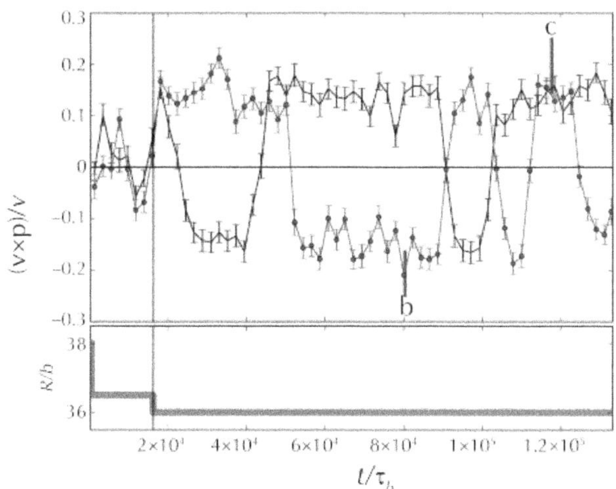

(a)

FIGURE 12 *(Continued)*

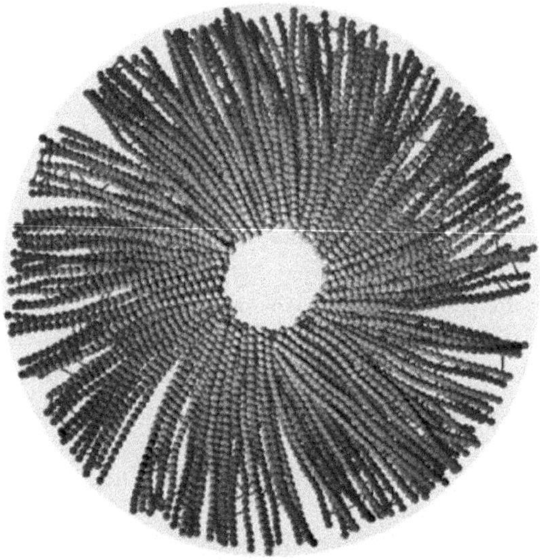

(b)

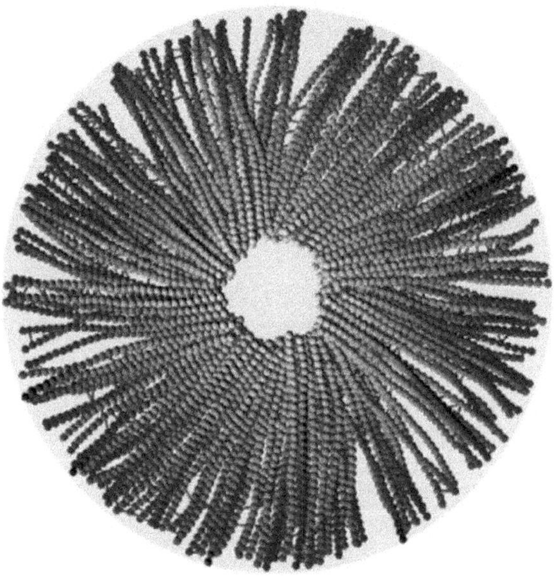

(c)

FIGURE 12 Fixed volume vortices. (a) Filament rotation for two independent runs at fixed volume. The imposed radius R is given in the lower panel. Snapshots for the run corresponding to the solid black line at points (b) and (c) are given in the lower figures. A short time at the initial radius $R/b = 38$ was required to avoid numerical instabilities.

Although, the magnitude of the rotational velocity remains fixed (note that the characteristic velocity v is the same for both runs and constant in time), the direction aperiodically reverses as evident in the changes of sign in the figure. The statistics of time intervals between direction switching suggest that the underlying mechanism may be the same as for the contraction to the semi-aster state in the constant pressure case. Specifically, the mean switching time $\Delta t^{switch}/\tau b = 21.6 \times 10^4 \pm 5.8 \times 10^4$ is consistent with the mean contraction time $\approx 1.7 \times 10^4$ measured earlier, and again is consistent with an exponential distribution (significance level $P \approx 0.2$ from $n = 7$ values using the AndersonDarling test [31,32]). This suggests that the same spontaneous fluctuation that permits contraction under constant pressure instead promotes rotational reversal under constant volume.

2.5 CONCLUSION

Systematically varied motor speed and density in filaments confined to a pressurised cylindrical cell, and have uncovered four qualitatively different types of steady state, namely aster, semi-aster, spindle and nematic. The corresponding regions of parameter space for each state were delineated by modal analysis of the filament polarities. Furthermore, in one region of parameter space we found a vortex state in which filaments rotated about the system centre for a finite time before buckling to a semi-aster. Quantitative analysis of rotation speed and mean-squared angular displacement provided unambiguous evidence of coherent filament rotation in this state. The vortex state persisted for far longer times with fixed walls, albeit with stochastic changes of direction, demonstrating that motors and not pressure alone are necessary for the observed vortex rotation.

COMPETING INTERESTS

The authors declare that they have no competing interests.

AUTHORS' CONTRIBUTIONS

DAH, WJB, and GG designed research; DAH performed simulations and analyzed data; DAH, WJB, and GG wrote the paper. All authors read and approved the final manuscript.

KEYWORDS

- **Eukaryotic Cells**
- **Filamentous Proteins**
- ***Fission* Yeast Cells**
- **Spindles**
- **Steady-States**

ACKNOWLEDGEMENTS

Financial support of this project by the European Network of Excellence "SoftComp" through a joint postdoctoral fellowship for DAH is gratefully acknowledged.

REFERENCES

1. Alberts, B., Johnson, A., Lewis, J., Raff, M., Roberts, K., and Walter, P. *Molecular Biology of the Cell. Garland Science*. 5th edition (2007).
2. Bray, D. *Cell Movements: From Molecules to Motility. Garland Science*. 2nd edition. 2000.
3. Mizuno, D., Tardin, C., Schmidt, C. F., and Mackintosh, F. Nonequilibrium mechanics of active cytoskeletal networks. *Science* , **315**(5810)**:**370373 (2007).
4. Carazo-Salas, R. E. and Nurse, P. Self-organization of inter-phase microtubule arrays in fission yeast. *Nature Cell Biology* , **8**(10)**:**11021107 (2006).
5. Daga RR, Lee KG, Bratman S, Salas-Pino S, Chang F: Self-organization of microtubule bundles in anu-cleate fission yeast cells. *Nature Cell Biology*, **8**(10)**:**11081113 (2006).
6. Haase, S. B. and Lew, D. J. Microtubule Organization: Cell Shape Is Destiny. *Current Biology*, **17**(7)**:**R249R251 (2007).
7. Verde, F., Berrez, J., Antony, C., Karsenti, E. Taxol-induced microtubule asters in mitotic extracts of xenopus eggs - Requirement for phosphorylated factors and cytoplasmic dynein. *Journal of Cell Biology* , **112**(6)**:**11771187 (1991).
8. Heald, R., Tournebize, R., Blank, T., Sandaltzopoulos, R., Becker, P., Hyman, A., and Karsenti, E. Self-organization of microtubules into bipolar spindles around artificial chromosomes in Xenopus egg extracts. *Nature*, **382:**420 (1996).
9. Hentrich, C. and Surrey, T. Microtubule organization by the antagonistic mitotic motors kinesin-5 and kinesin-14. *The Journal of Cell Biology*, **189**(3)**:**465480 2010.
10. Terenna, C. R., Makushok, T., Velve-Casquillas, G., Baigl, D., Chen, Y., Bornens, M., Paoletti, A., Piel, M., and Tran, P. T. Physical Mechanisms Redirecting Cell Polarity and Cell Shape in Fission Yeast. *Current Biology*, **18**(22)**:**17481753 (2008).
11. Pinot, M., Chesnel, F., Kubiak, J., Arnal, I., Nedelec, F., Gueroui, Z. Effects of Confinement on the Self-Organization of Microtubules and Motors. *Current Biology*, **19:**954960 (2009).
12. Voituriez R, Joanny JF, Prost J: Spontaneous flow transition in active polar gels. *Europhysics Letters (EPL)*, **70**(3)**:**404410 (2005).
13. Cates, M. E., Fielding, S. M., Marenduzzo, D., Orlandini, E., and Yeomans, J. M. Shearing Active Gels Close to the Isotropic-Nematic Transition. *Physical Review Letters*, **101**(6)**:**068102 (2008).
14. Giomi, L., Liverpool, T. B., and Marchetti, M. C. Sheared active fluids: Thickening, thinning, and vanishing viscosity. *Physical Review E (Statistical, Nonlinear, and Soft Matter Physics)* , **81**(5)**:**051908 (2010).
15. N'ed'elec F, Surrey T, Maggs A, Leibler S: Self-organization of microtubules and motors. *Nature* , **389**(6648)**:**305308 (1997).
16. Surrey, T., Nédélec, F., Leibler, S., and Karsenti, E. Physical Properties Determining Self-Organization of Motors and Microtubules. *Science* , **292**(5519)**:**11671171 (2001).
17. Kruse, K., Joanny, J. F., Julicher, F., Prost, J., and Sekimoto, K. Asters, vortices, and rotating spirals in active gels of polar filaments. *Physical Review Letters* , **92**(7)**:**078101 (2004).
18. Elgeti J, Cates ME, Marenduzzo D: Defect hydrodynamics in 2D polar active fluids. *Soft Matter* , **7**(7)**:**3177 (2011).
19. Head, D. A., Gompper, G., and Briels, W. J. Microscopic basis for pattern formation and anomalous transport in two-dimensional active gels. *Soft Matter* , **7:**31163126 (2011).
20. Liverpool, T. B. and Marchetta, M. C. Instabilities of isotropic solutions of active polar filaments. *Phys Rev Lett* , **90:**138102 (2004).
21. Ziebert, F. and Zimmermann, W. Nonlinear competition between asters and stripes in filament-motor systems. *Euro Phys J E* , **18:**41 (2005)

22. Kraikivski, P., Lipowsky, R., and Kierfeld, J. Enhanced Ordering of Interacting Filaments by Molecular Motors. *Physical Review Letters*, **96**(25):258103 (2006).
23. Schaller, V., Weber, C., Semmrich, C., Frey, E., and Bausch, A. R. Polar patterns of driven filaments. *Nature* 2010, **467**(7311):7377.
24. Schaller, V., Weber, C., Frey, E., and Bausch, A. R. Polar pattern formation: hydrodynamic coupling of driven filaments. *Soft Matter* , **7**(7):32133218 (2011).
25. Head, D. A. and Mizuno, D. Nonlocal fluctuation correlations in active gels. *Physical Review E (Statistical, Nonlinear, and Soft Matter Physics)*, **81**(4):041910 (2010).
26. Czirok, A., BenJacob, E., Cohen, .I, and Vicsek, T. Formation of complex bacterial colonies via self-generated vortices. *Physical Review E (Statistical, Nonlinear, and Soft Matter Physics)*, **54**(2):17911801 (1996).
27. Ramaswamy, S. The Mechanics and Statistics of Active Matter. *Annual Review of Condensed Matter Physics* , **1**:323345 (2010).
28. Toner J, Tu Y: Flocks, herds, and schools: A quantitative theory of flocking. *Physical Review E (Statistical, Nonlinear, and Soft Matter Physics)*, **58**(4):48284858(1998).
29. Allen, M. P. and Tildesley, D. J. *Computer Simulation of Liquids.* Oxford University Press, USA; (1989).
30. Doi, M. and Edwards, S. F. *The Theory of Polymer Dynamics (International Series of Monographs on Physics)*. Oxford University Press, USA; (1988).
31. Spurrier, J. D. *Comm Stats - Thy Mech.*, **13**:1635(1984).
32. D'Agostino, R. B. and Stephens, M. A., (Eds): *Goodness-of-fit-techniques (Statistics: A Series of Textbooks and Monographs). Dekker*. 1st edition. **68** (1986).
33. Howard, J: *Mechanics of Motor Proteins and the Cy-toskeleton*. Sinauer, USA; (2001).
34. Soares e Silva, M., Alvarado, J., Nguyen, J., Georgoulia, N., Mulder, B., and Koenderink, G. H. Self-organized patterns of actin filaments in cell-sized confinement. *Soft Matter* , **7**(22):1063110641 (2011).

3 Actomyosin-Dependent Cortical Dynamics

Patrizia Sommi, Dhanya Cheerambathur, Ingrid Brust-Mascher, and Alex Mogilner

CONTENTS

3.1 Introduction 49
3.1.1 Background 49
3.1.2 Principal Findings 50
3.1.3 Conclusions 50
3.2 Materials And Methods 52
3.2.1 Fly Stocks 52
3.2.2 Drosophila Embryo Injection 53
3.2.3 Image Acquisition 53
3.3 Discussion 53
3.4 Results 58
3.4.1 Myosin Inhibition in a Drosophila Syncytial Embryo Results in Greater Centrosome Separation During Prophase 58
3.4.2 Myosin II Inhibition Changes the Actin Distribution at the Cortex 59
3.4.3 Astral MTs Reach the Increased Actin Cap Area in Myosin-Inhibited Embryos 62
3.4.4 A Change in Actin Distribution is not Sufficient to Impair Early Spindle Development but Shows an Additive Effect in Combination with Myosin Inhibition 62
Keywords 67
Acknowledgments 67
References 67

3.1 INTRODUCTION

3.1.1 Background

The assembly of the *Drosophila* embryo mitotic spindle during prophase depends upon a balance of outward forces generated by cortical dynein and inward forces generated by kinesin-14 and nuclear elasticity. Myosin II is known to contribute to the dynamics of the cell cortex but how this influences the prophase force- balance is unclear.

3.1.2 Principal Findings

Here this question is investigated by injecting the myosin II inhibitor, Y27632, into early *Drosophila* embryos. A significant increase in both the area of the dense cortical actin caps and in the spacing of the spindle poles is observed. Tracking of microtubule plus ends marked by EB1-GFP and of actin at the cortex revealed that astral microtubules can interact with all regions of these expanded caps, presumably via their interaction with cortical dynein. In *Scrambled* mutants displaying abnormally small actin caps but normal prophase spindle length in late prophase, myosin II inhibition produced very short spindles.

3.1.3 Conclusions

These results suggest that two complementary outward forces are exerted on the prophase spindle by the overlying cortex. Specifically, dynein localized on the mechanically firm actin caps and the actomyosin-driven contraction of the deformable soft patches of the actin cortex, cooperate to pull astral microtubules outward. Thus, myosin II controls the size and dynamic properties of the actin-based cortex to influence the spacing of the poles of the underlying spindle during prophase.

Microtubule (MTs) and actin-myosin arrays interact and cooperate in many mechanochemical modules of cell motility and cell division [1] but the functional implications of such interactions are not well understood. In particular, interactions of mitotic spindles with the F-actin cortex are crucial for spindle positioning and orientation [2–4] as well as the regulation of cytokinesis [5], yet whether the actin-myosin network affects internal processes of mitotic spindle assembly and maintenance, or only external phenomena involving the spindle's interactions with other regions of the cell such as the cortex, is still a controversial question [4]. Some evidence suggests that myosin II is needed only for cytokinesis: inhibition of myosin II in echinoderm blastomeres blocks cytokinesis but not mitosis [6]; similarly, RNAi depletion of myosin II in S2 cells blocks cytokinesis but metaphase and anaphase spindles are morphologically normal [7].

On the other hand, myosin II has recently been reported to exert force on the spindle poles during prophase, presumably via a drag on cortex-anchored astral microtubules subsequent to nuclear envelope breakdown (NEB) through myosin-powered cortical flow [2]. In locust spermatocytes, there is evidence that actin and myosin are involved in anaphase chromosome movement [8]. Curiously, it was recently reported that F-actin promotes spindle lengthening, perhaps by interactions with astral MTs, while Myosin-10 works antagonistically to shorten the spindle [9].

The early *Drosophila* embryo is a very convenient system for investigating the coupling between the spindle and the actomyosin cortex because of this organism's amenability to genetic analysis, inhibitor microinjection and microscopy [10]. In early *Drosophila* embryogenesis, some morphogenetic events such as cellularization [11] and nuclear migration [12] indicate interactions between the actomyosin cytoskeleton and microtubule arrays; myosin II is thought to have additional as yet unidentified functions [13]. Following our earlier efforts [14], here it is focused on the syncytial blastoderm divisions that occur at the cortex of the embryo, just beneath the plasma membrane, where dramatic redistribution of the cortical actin accompanies spindle morphogenesis [15]. In these cycles, actin concentrates into ‘caps’ centered above each nucleus and centrosomes. As the nuclei progress into prophase, the centrosomes migrate toward opposite poles and the caps expand in synchrony with the centrosomes [14]. After NEB, a transient steady state is reached in prophase, after which the centrosomes separate and the spindle elongates further.

Here, it is focused on the early stage of mitosis–prophase–because myosin-dependent contraction of the cortex has been reported at this stage, while at prometaphase myosin concentration starts to decline rapidly throughout the cortex [16]. The role of myosin II in cellularization [11] and the influence of astral MT arrays on the rapid spatial reorganizations of the actomyosin cortex [15, 17] are well documented. Actin dynamics must play an important role in centrosome separation based on the observations that separation is incomplete in *Drosophila* embryos treated with cytochalasin D [18] and that actin polymerization is crucial for the centrosome separation before NEB [19], but details of this cortex-to-spindle feedback and myosin II involvement were not studied.

The question about the nature of the spindle-cortex interaction is intimately linked to another outstanding question–the relation between the internal and external forces shaping the spindle [20]. Recent work points to a principal role for a molecular motor-generated force balance in spindle assembly and pole-pole separation [14, 21, 22] (Figure 1). Recent data and modeling suggest that cooperative interactions between multiple spindle associated internal mitotic motors must be complemented by external cortical dynein motors that pull out astral MTs to coordinate prophase spindle assembly. It is plausible that dynein anchored on the actin cortex can slide astral MTs against the cortex [23] and separate spindle poles. Based on time-lapse imaging, functional perturbation experiments and quantitative modeling, it is hypothesized that the force-balance for spindle pole separation depends on outward forces generated by dynein associated with the actin caps being antagonized by inward forces exerted by the kinesin-14, Ncd and nuclear elasticity [14, 21] (Figure 1).

Here it is set out to further explore the molecular basis of this force-balance, asking whether myosin II contributes to the early spindle assembly, perhaps by generating a cortical flow that pulls actin radially outward, and with it the cross-linked astral MTs [2]. Time lapse microscopy of *Drosophila* embryos expressing fluorescently tagged myosin light chain, microinjected with rhodamine-tubulin was used to study and compare the dynamics and distribution of myosin II and microtubules. To examine the role of myosin II on spindle dynamics a myosin II inhibitor (the Rho kinase inhibitor, Y-27632 [16, 24]) is used. In order to assess the role of F-actin in the early mitotic spindle, *Scrambled* (*Sced*) mutants which display defects in furrow formation and early centrosome separation are used [18, 25]. It is observed, that myosin inhibition, rather than decreasing the spindle length, as expected from an outward myosin-dependent force, led to spindle lengthening. However, microscopy revealed that the actin caps expand when myosin II is inhibited, so dynein can exert a greater outward force in this case. It is further reported that in *Sced* mutants with disorganized actin caps, the pole-pole distance changes very little, and that it shortens significantly when myosin II is inhibited in these mutants. It is proposed that dynein and myosin II synergize in elongating the early mitotic spindle, albeit not directly. Rather, under circumstances when relatively rigid actin caps are present, myosin-II has little effect and dynein alone generates the outward force on spindle poles. However, if the actin cortex weakens, myosin contracts it locally and augments dynein in pulling out the astral MTs.

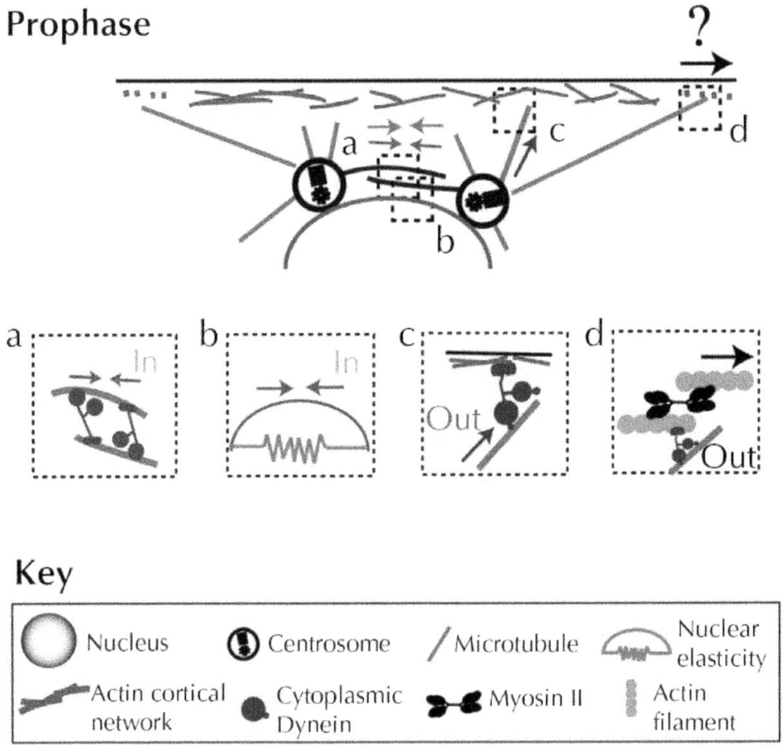

FIGURE 1 Force balance in prophase. The previous work supports the hypothesis that in prophase the balance between the inward forces generated by kinesin-14 sliding interpolar MTs together (a) and nuclear elasticity (b) and the outward forces generated by dynein pulling on astral MTs (c) governs the spindle length. Here, the idea that an additional outward force is exerted by myosin-II (d) indirectly – by contracting the actin cortex and pulling outward on astral MTs cross-linked with the deforming actin network is tested.

3.2 MATERIALS AND METHODS

3.2.1 Fly Stocks

Flies were maintained and 0 to 2 h embryos were collected as described [26]). Experiments were performed using embryos expressing GFP-Tubulin (provided by Dr. Allan

Spradling, Carnegie Institute, Washington), EB1-GFP (provided by Dr Steve Rogers, University of North Carolina at Chapel Hill, Chapel Hill, NC), GFP-tagged myosin II light chain and *Sced* embryos (provided by Dr. Wiliam Sullivan, UCSC, Santa Cruz, California).

3.2.2 *Drosophila* Embryo Injection

Microinjection of 0–2 h embryos was performed as described [27]. Briefly, as indicated for each experiment, rhodamine conjugated monomeric actin (Cytoskeleton, Denver, CO) was injected into GFP-tubulin expressing embryos but in the case of Sced, FITC-tubulin and rhodamine-actin were coinjected. The myosin inhibitor, Y27632 was used at the concentration of 50 mM in the needle.

3.2.3 Image Acquisition

Time lapse confocal images were acquired on an Olympus (Melville, NY) microscope equipped with an Ultra View spinning disk confocal head (Perkin Elmer-Cetus Wallac, Inc., Gaithersburg, MD) and a 100× 1.35NA objective. Images were analyzed using MetaMorph Imaging software (Molecular Devices, Downintown, PA). Quantitation of spindle pole positions was carried out using MetaMorph. The positions of the two poles were logged, and the distance between them was calculated and plotted as a function of time using Excel.

3.3 DISCUSSION

Earlier studies suggested that the myosin-powered contraction of the cortex can generate a force on the centrosomes by pulling on the astral MTs cross-linked with the actin cortex [2]. For example, MTs were observed to undergo directed movement along cortical actin filaments during wound healing experiments in Xenopus oocytes [28] and along polarized actin cables in yeast [29] (in the latter case, the Myosin-II-Kar9-Bim1 complex is responsible for this transport). However, the nature of specific actin-myosin-MT interactions in various systems and organisms remains an open question.

The observations support the hypothesis that in the early *Drosophila* embryo during prophase, dense actin patches provide a scaffold from which dynein can pull astral MTs outward promoting pole-pole separation, while myosin regulates the actin cap size (Figure 2, WT and WT+Y27632). However, when the actin distribution is compromised, myosin can exert local mechanical action on the loose weakened actin patches and generate contraction that pulls astral MTs cross-linked with actin filaments (Figure 2, *Sced*). This myosin pulling from the loose actin patches synergizes with the dynein pulling from the remaining dense actin patches, so that the pole-pole separation in *Sced* embryos is almost normal. However, when both myosin and actin are compromised, more 'weakened actin areas' are formed at the cortex, thus, the dynein-generated pulling force from the remaining firm actin patches (Figure 2,*Sced*+Y27632) is decreased resulting in an incomplete pole-pole separation.

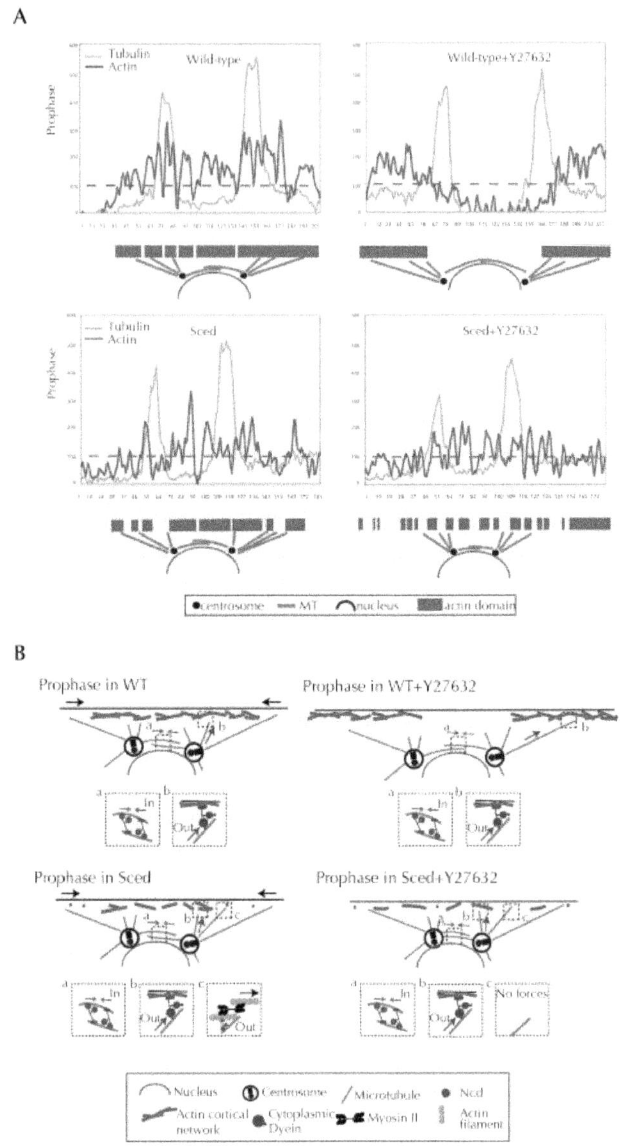

FIGURE 2 Synergistic actions of actin, myosin and dynein in spindle development. Hypothesis for each condition (B) is based on the actin distribution as shown in the corresponding line scans (A). The intensity value of 100 has been chosen as the arbitrary cut-off value to discriminate between the 'soft' (below 100) and 'firm' (above 100) actin patches. In the WT embryo, most of the actin cap is firm, and dynein pulls from the large firm actin cortex (inset b) while kinesin-14 exerts an inward force (inset a). When myosin is inhibited, the area of the firm actin is shifted outward focusing the dynein pulling in the outward direction. In theSced embryo, many soft actin patches appear, but myosin contracts actin filaments there (inset c) complementing the pulling action of dynein (inset b). When myosin is inhibited in the Sced embryo, dynein pulls from the small infrequent firm patches of actin and does not separate the poles effectively.

This hypothesis suggests that myosin II participates in regulating the actin cap geometry and dynamics in two ways. First, myosin antagonizes the spreading of the cap outward, but does not disrupt the actin assembly within the cap. This effect is unlikely to be based on centripetally contracting the cap because myosin and actin in the early WT embryo are largely separated in space [11, 15] and no significant deformations of the actin caps are observed in WT embryos. More likely, myosin facilitates disassembly of actin filaments [30] at the rim of the cap or acts by a combination of this and other recently uncovered mechanisms in different systems [31, 32].

Earlier it was reported that dynein co-localizes with the actin cortex and that dynein inhibition leads to a severe spindle shortening in prophase [14, 21]. It is proposed that when the actin assembly is normal within the cap, then dynein alone is responsible for the generation of the outward force in WT embryos: dynein motors pull on astral MTs effectively from the dense and robust undeformed cap. When myosin II is inhibited, the dense actin cap area expands outward, which probably focuses the dynein-mediated pulling in the outward direction increasing the outward force and spindle length, as observed.

It is hypothesized that when the actin dynamics in the cap is disrupted, the F-actin assembles throughout the cap with an uneven density, so that high density (and likely mechanically firm) patches of actin network are interspersed with low density (likely mechanically soft) patches. We call such a cap 'loose'. It is propose that dynein can effectively pull from the dense actin patches, but not from the soft ones, because the result of such pulling would be local deformations of the cortex, rather than 'reeling in' the astral MTs (Figure 2B). However, myosin II motors, which cannot deform the dense actin patches, are able to locally deform the soft patches [33], and astral MT tips reaching and associating with the F-actin in these soft patches can be pulled outward by these myosin-caused upward deformations (Figure 2B). Therefore, on the loose actin cortex dynein and myosin synergize in the generation of outward force. This model predicts that the outward force and thus the spindle length remains unchanged if the actin cap is loose, because diminished dynein action is supplemented by the local myosin contraction. If myosin is inhibited in this situation, then the force becomes weaker and the spindle–shorter. To test this idea, It was investigated *Sced* mutants that have a loosely organized actin layer at the cortex (Figure 2A) with functional myosin II, and discovered that the spindle pole-pole distance in these cortical mutants was similar to that in WT embryos. Myosin inhibition in these mutants indeed led to spindle shortening.

The model which is being proposed (Figure 2B) does not address the very complex question about the mechanical and biochemical regulatory interactions of astral MTs with the actin-myosin cortex in the early *Drosophila* embryo. There are too many possible interactions [1] and not enough data to discriminate between various possibilities. For example, MTs can transport either F-actin itself or actin-nucleating agents to the cap [17], as well as molecules regulating myosin contraction [1] and myosin itself [34]. Myosin II could be concentrated at the cap's margins because both actin and myosin are transported along astral MTs growing along the cortex towards their plus ends, with actin and myosin exhibiting different affinities for the cell's cortex [15]. There is also the possibility of a feedback loop from actin-myosin to MTs [35]. The nature of these MT-actin-myosin interactions will have to be addressed in future studies. Recently, it was observed [36] that the membrane and actin-myosin cortex deformed significantly by MT-mediated pulling forces when the cortex was weakened in one-cell C. elegans embryos, consistent with the hypothesis that the MTs need to be anchored to a stiff actomyosin network to be able to transduce forces. Such a large scale deformations are not detected clearly, but further, higher resolution, microscopy in *Drosophila* embryos will be needed to investigate a possibility of a similar phenomenon. We did, however, noticed patches of actin at the side of the nucleus below and near the centrosomes in the myosin-inhibited embryo (see side views in Figure 3A). These patches could indicate large-scale cortex deformations when greater pulling forces are applied to the cortex.

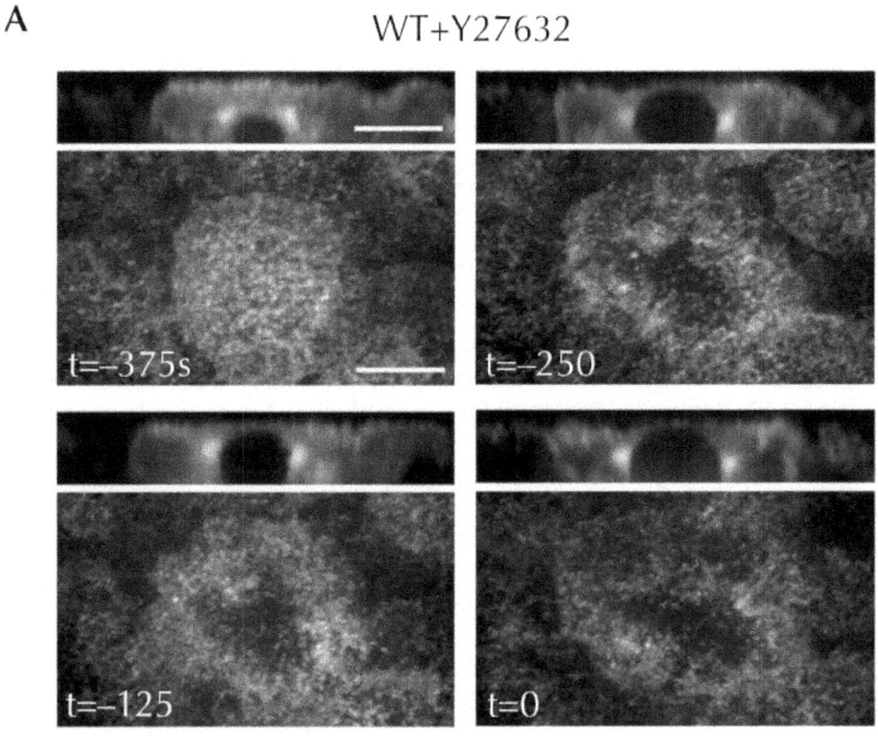

FIGURE 3 *(Continued)*

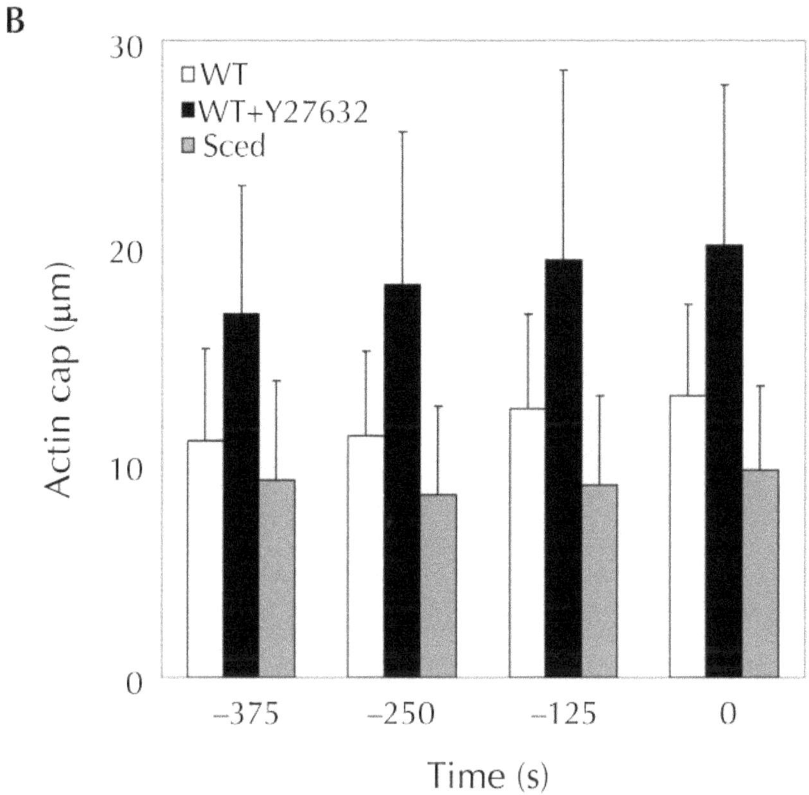

FIGURE 3 Quantification of the actin cap size. A) Side views of actin, nucleus and centrosomes (rhodamine-actin, red; GFP-Tubulin, green) and corresponding top views of actin in prophase WT embryos in the presence of the myosin inhibitor Y27632 reveal that the centrosomes segregate around the nucleus, likely being pulled outward by the increasing forces. The resulting vertical component of the force pushes the nucleus upward against the actin cap thinning the cortex and creating the 'hole'. NEB is t = 0 s. Bars, 10 μm. B) The square root of the actin cap areas as functions of time (NEB is t = 0 s) in WT embryos in the absence and presence of the myosin inhibitor Y27632 and in the Sced embryos.

To conclude, it is suggested that there is a complex synergy between dynein and myosin II actions before NEB (Figure 2B): dynein pulling on astral MTs from the dense and firm parts of the actin cortex is responsible for most of the outward force elongating the spindles, while myosin II regulates the cap size and supplements the dynein-generated force when actin organization is compromised locally by contracting the weakened cortical patches and pulling on the actin-cross-linked astral MTs. Further analysis is required to understand the precise molecular mechanism by which myosin-II contributes to other stages of spindle development in the *Drosophila* embryo and other organisms.

3.4 RESULTS

3.4.1 Myosin Inhibition in a *Drosophila* Syncytial Embryo Results in Greater Centrosome Separation During Prophase

To test the hypothesis that myosin II plays a role in spindle dynamics, a specific Rho kinase inhibitor was microinjected, Y27632, into *Drosophila* syncytial embryos expressing GFP-tubulin. This drug disrupts myosin II activity by inhibiting Rho kinase which is responsible for myosin II activation via the serine and threonine phosphorylation [24]. Y27632 has been successfully used in previous studies to inhibit myosin function [16, 17] and its specificity for myosin II has been documented: depletion of DROK (Rho kinase in *Drosophila*) by either using RNAi or Y27632 gave identical results in S2 cells in preventing cell shape change, a process which depends on functional myosin [37]. In human cells, the use of Y27632 prevented myosin II localization at the cleavage furrow and generated defects in cytokinesis [38].

Injection of Y27632 into Sqh-GFP expressing embryos disrupted myosin II localization, as shown by the disappearance of the myosin II regulatory light chain, spaghetti-squash, from the cortex (Figure 4A), indicating that Y27632 is an appropriate inhibitor to study the function of myosin II during spindle morphogenesis in early embryogenesis. After microinjection of Y27632 in GFP-tubulin expressing embryos an increase in the distance between the two centrosomes was found during prophase compared to control embryo (Figure 4B).

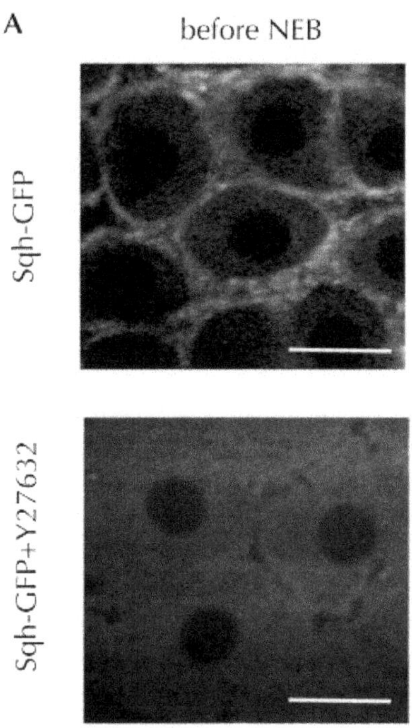

FIGURE 4 *(Continued)*

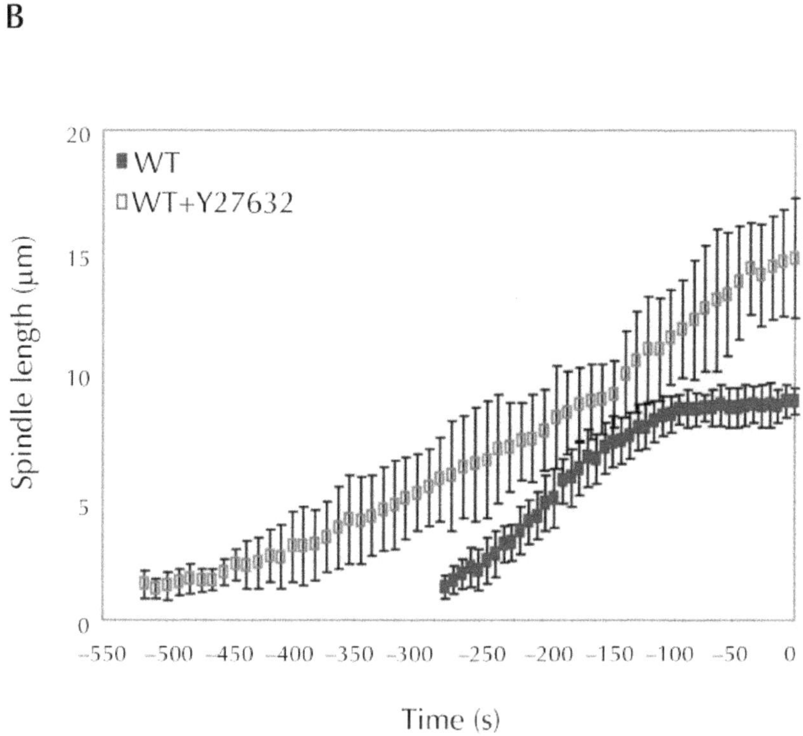

FIGURE 4 Effect of myosin inhibition on spindle development in WT embryos. A) Injection of the myosin inhibitor Y27632 causes myosin to lose its cortical localization. In Sqh-GFP expressing embryos the spaghetti-squash signal largely disappears from the cortex and becomes diffuse after Y27632 is injected. Bars, 10 μm. B) Graph of pole-pole distance as a function of time (NEB is t = 0 s) for GFP-Tubulin expressing embryos (WT) and myosin inhibited embryos (WT+Y27632). After myosin inhibition pole separation during prophase is greater than in WT.

3.4.2 Myosin II Inhibition Changes the Actin Distribution at the Cortex

To see whether myosin perturbation affected actin structure and distribution, rhodamine-actin was co-injected with the myosin inhibitor. In addition to the increased pole-pole separation described before, a dramatic change in F-actin distribution and organization before NEB was observed (Figure 5A). In control embryos, F-actin is organized into a compact actin cap right above the nucleus/centrosomes in prophase. The most striking effect of the myosin inhibition is the increment of the actin cap area suggesting that myosin II is playing a role in restricting the actin expansion (see Discussion).

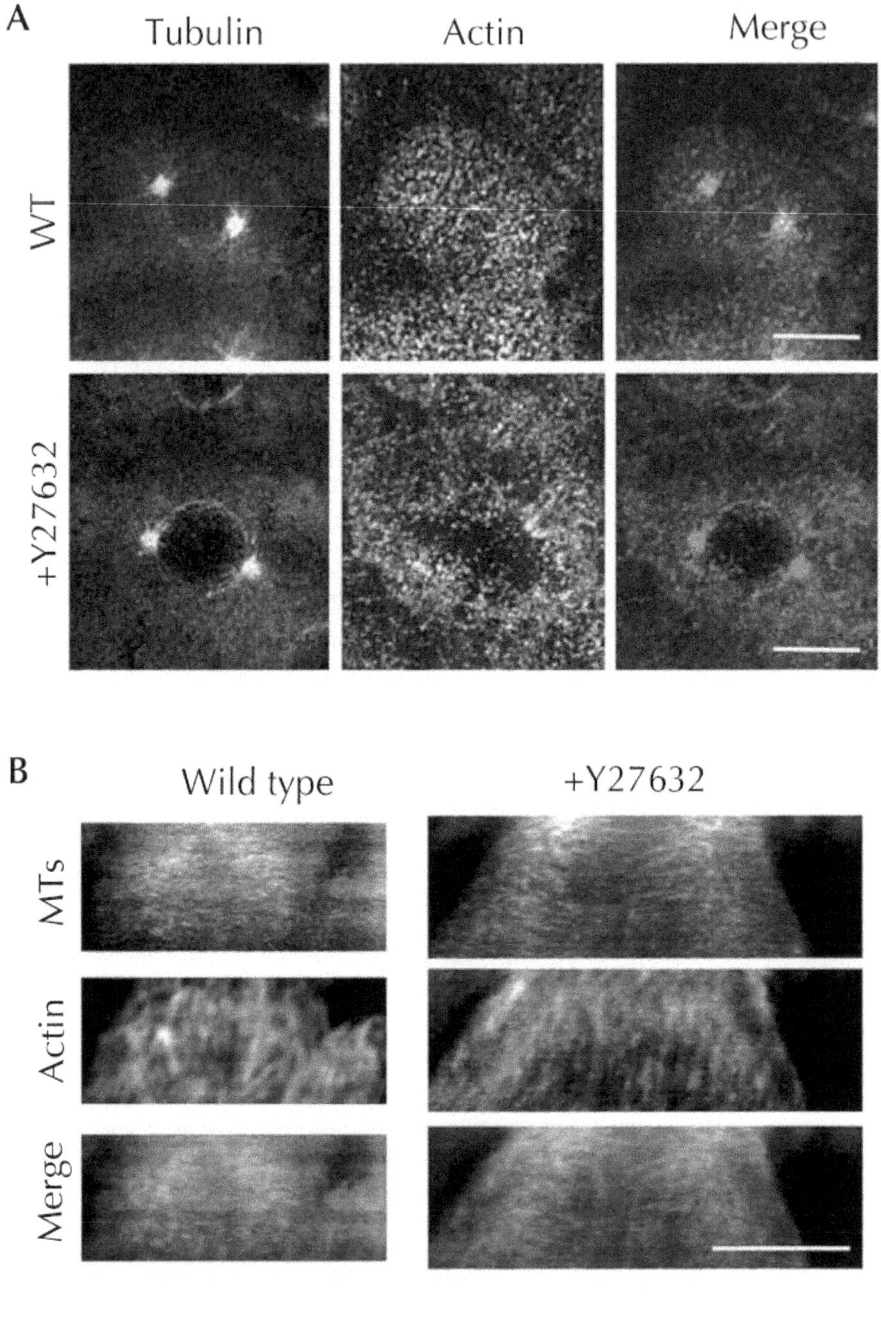

FIGURE 5 *(Continued)*

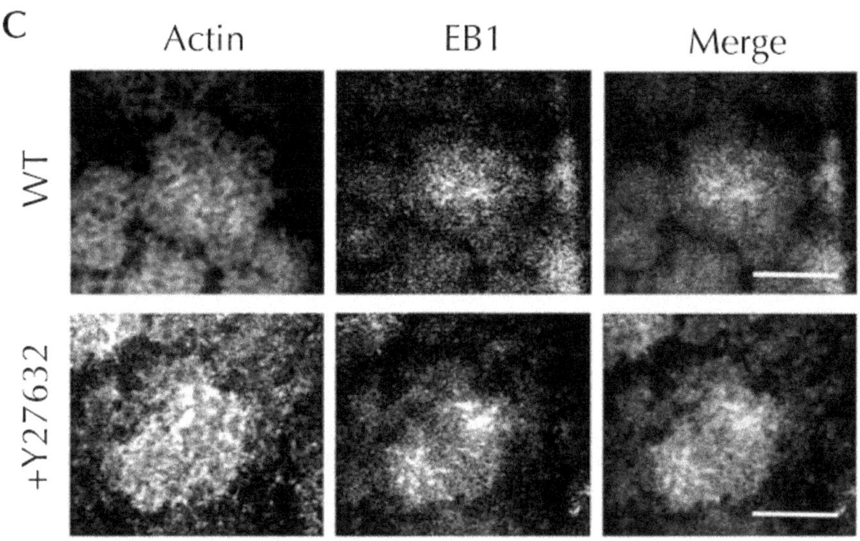

FIGURE 5 Effect of myosin inhibition on the dynamics of actin and MTs. A) Images from a time-lapse movie of a WT embryo expressing GFP-tubulin injected with rhodamine-actin and the myosin inhibitor Y27632. When myosin is inhibited, actin distribution in prophase changes compared to control showing a less compact and wider distributed actin structure above the nucleus (note the void area). Images are projections of 8–10 confocal planes taken 0.5 μm apart. B) Kymographs (from −250 to −100 seconds before NEB) of prophase embryos injected with rhodamine actin with or without Y27632 confirm that in the absence of myosin the outer edge of the actin cap expands at a faster rate and to a greater width than in control. Same confocal planes taken at the very top of the actin cap were imaged for both actin and MTs. Red, actin; green, tubulin. C) Images from a time-lapse movie of GFP-EB1 expressing embryos injected with rhodamine-actin with and without Y27632 during prophase. Tracking of MT tips shows that both in WT and myosin-inhibited embryos, MT tips uniformly reach the surfaces of both narrower and wider actin caps (EB1-GFP, green; rhodamine-actin, red). Bars, 10 μm.

To gain more insight into the actin expansion process at the cortex in both control and myosin-inhibited embryos, kymographs of cortical actin along the pole-pole axis was analyzed (Figure 5B). It was seen that the actin cap expands faster and to a greater extent in myosin-inhibited embryos, and observed 'streaks of actin' in the kymographs expanding outward with respect to the position of the centrosomes in both WT and myosin-inhibited embryos. These streaks had a much higher slope in myosin-inhibited embryos indicating a faster expansion of the cortical actin array. Note that it cannot be distinguish at this point between physical centrifugal F-actin movements and outward polymerization waves near the cap edge. Indeed, in some cases it was also possible to see nascent spots of F-actin polymerized along the cap edge in the direction of the actin expansion, perhaps suggesting that MTs reaching that area promote actin polymerization which is responsible for the growth of the actin cap.

It was also noticed the formation of a "hole" in the center of the actin cap in myosin-inhibited embryos (Figure 5A). Its origin could be the steric hindrance of the nucleus being pushed up into the cortex with a greater force coming from a stronger outward pull on the centrosomes (Figure 3A). The average thickness of the actin cap was also measured right above the nuclear top in control and myosin-inhibited embryos and found that the thickness decreased from 2.2±0.3 μm in control to 1.25±0.4 μm in myosin-inhibited embryos. To quantify the expansion of the actin cap in the myosin-inhibited embryos, the areas of the actin caps were measured as functions of time in control and myosin-inhibited embryos. The square root of the areas characterizes the cap sizes: for almost circular caps the square root of the area is close to the cap diameter. Both cap size and spindle length increase with myosin inhibition (Figure 3B, 4B). The cap size also increases with time in myosin inhibited embryos in sync with spindle length growth (Figure 3B).

3.4.3 Astral MTs Reach the Increased Actin Cap Area in Myosin-Inhibited Embryos

Dynein co-localizes with the actin cortex [14, 21, 39], and its inhibition decreases spindle elongation during prophase and prometaphase [21]. Modeling of the available data from perturbation experiments strongly suggests that during prophase dynein is the outward force generator [14, 22]. Thus, it is possible that the cause of the increment in pole-pole distance after myosin inhibition is due to astral MTs reaching the increased actin cap area and interacting with a greater number of dynein molecules, resulting in a stronger outward force on the spindle poles. To see if this is plausible, the distribution of actin and MTs at the cortex (Figure 5C) were examined by checking whether dynamic MT plus ends were able to reach the whole cortex after myosin II inhibition.

To do this, MT plus end localization was followed with respect to actin at the cortex by injecting rhodamine-actin into embryos expressing the plus-end tip-tracker protein, EB1-GFP (Figure 5C). In WT embryos, we could see spots of EB1-GFP transiently contacting the cortex almost evenly wherever actin was present. Even when myosin was inhibited, these EB1-GFP spots were able to reach uniformly the wider distributed cortical actin and showed a distribution similar to WT embryo, over a greater area of the expanded actin cap. This observation suggests that myosin II inhibition does not affect the ability of MT plus ends to reach the cortex and possibly generate pulling forces. In addition, no significant difference in the density of the EB1 spots were observed between wild type and myosin inhibited caps (# tips/actin area: WT, 159.6±32.3; WT+Y27632, 136.9±54.9). It is hypothesized that myosin inhibition does not change the density of the dynein force generators (number of motors pulling on the astral MTs per unit area of the cap). Thus, the total outward pulling force is greater in the myosin-inhibited embryos because the cap area is greater.

3.4.4 A Change in Actin Distribution is not Sufficient to Impair Early Spindle Development but Shows an Additive Effect in Combination with Myosin Inhibition

To test further whether the cortical area had any effect on the force balance required for centrosome separation, a mutant was used for *Sced* protein, which shows defective actin organization at the cortex: in *Sced* mutants, smaller actin caps and absence of actin furrows were reported [18, 25]. Although the *Sced* protein has not been shown to be closely related to sequences currently in the non-redundant or expressed sequence tag (EST) databases, it has been proposed, based on its localization to centrosomes and mitotic furrows, that *Sced* could be involved in actin reorganization via a MTs-independent process either by direct recruitment of actin or by promoting the localization of actin polymerization factors [18].

The *Sced* mutant, which shows a strong phenotype in the early syncytial divisions in *Drosophila* [18], allowed us to test the effect of a reduced actin cap area, from where dynein could pull on MTs, without the use of any drug. *Sced* mutants have smaller actin caps (Figure 3B). Therefore, it was expected to see a decrease in pole-pole separation as a consequence of a decrease in forces acting on the poles from the cap. The spindle pole dynamics were followed in *Sced* embryos before NEB, and not a significant difference was found in the pole-pole distance between WT and *Sced* mutant embryos in late prophase (Figure 6A). The average cap thickness above the nucleus also changed very little (2.2±0.3 μm in control versus 2.0±0.3 μm in *Sced* embryos). Note though that the centrosomes start to separate earlier in mutants. However, when myosin II was inhibited in *Sced* mutants, the pole-pole distance was reduced compared to both WT and Sced embryos (Figure 6A). Actin structures were also affected in *Sced* mutants in the absence of myosin: the actin cap became smaller and less defined (Figure 6B).

A

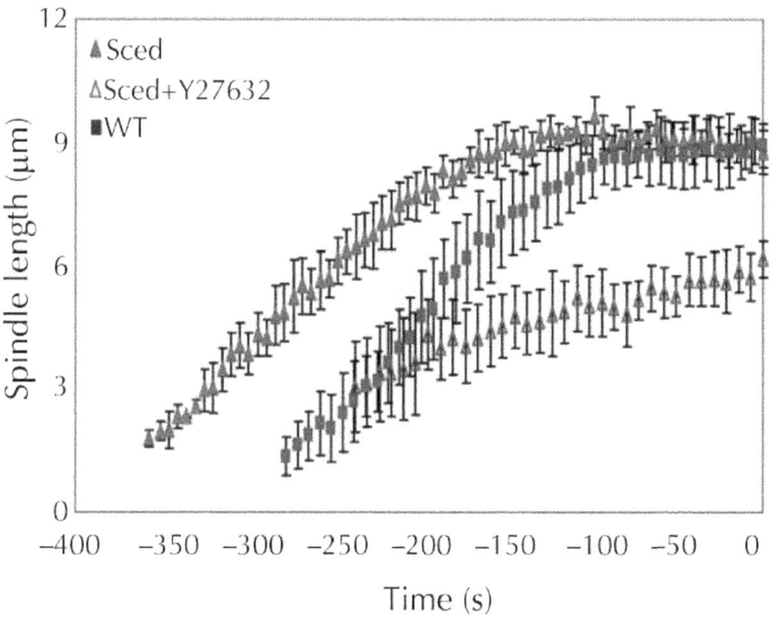

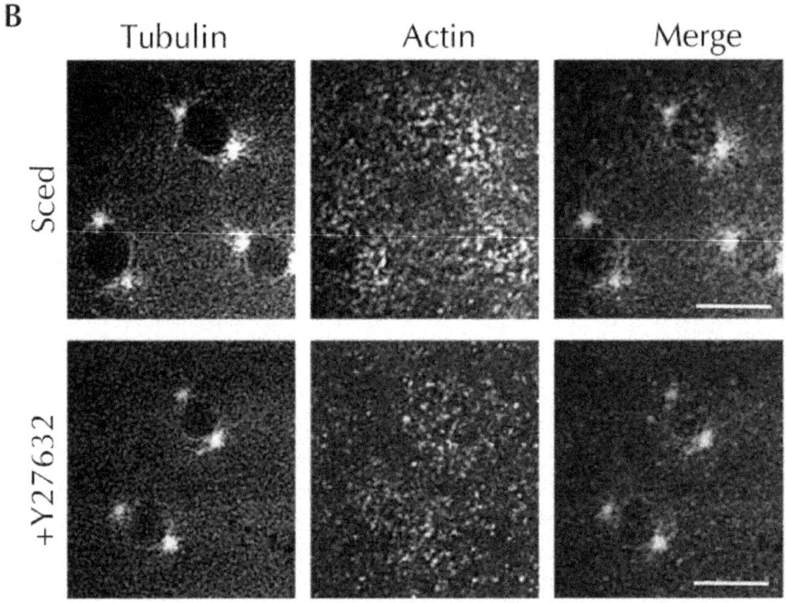

FIGURE 6 Effect of myosin inhibition on spindle development in Sced embryos. A) Pole-pole distances in prophase as functions of time (NEB is t = 0 s) show a different degree of pole separation in Sced embryos in the presence and absence of the myosin inhibitor Y27632. The pole-pole separation in Sced embryos is comparable to that in control, but when myosin is inhibited, centrosomes separate less than in control and Sced embryos. B) Images from a time-lapse movie of a Sced embryo in prophase co-injected with FITC-tubulin and rhodamine-actin in the presence or absence of Y27632. Scedembryos have smaller and less organized actin caps which become even less compact when myosin is inhibited. Images are projections of 8–10 confocal planes taken 0.5 microns apart. Red, actin; green, tubulin. Bars, 10 μm.

To probe deeper into the apparent puzzle as to why the spindle length remained the same in *Sced* embryos, where actin is disorganized, a detailed analysis of actin organization was undertaken in WT, *Sced* mutant and myosin-inhibited embryos by looking at the distribution of actin in the cap with respect to the centrosomes. Figure 7A shows side views of the actin cap during prophase in WT and *Sced* mutants with and without myosin inhibition. In *Sced* mutant embryos actin covers a smaller area than in control. In WT myosin-inhibited embryos, actin accumulates at the periphery and expands forming small furrows. In *Sced* embryos in the absence of myosin action, the cap expands outward while the actin density fluctuates in space becoming patchy (small, micron-size patches of lower density are intermittent with patches of higher density). Combined line scans of actin signal taken along the cortical actin layer and tubulin intensity taken across the two centrosomes quantitatively confirm these qualitative differences in actin cap distributions.

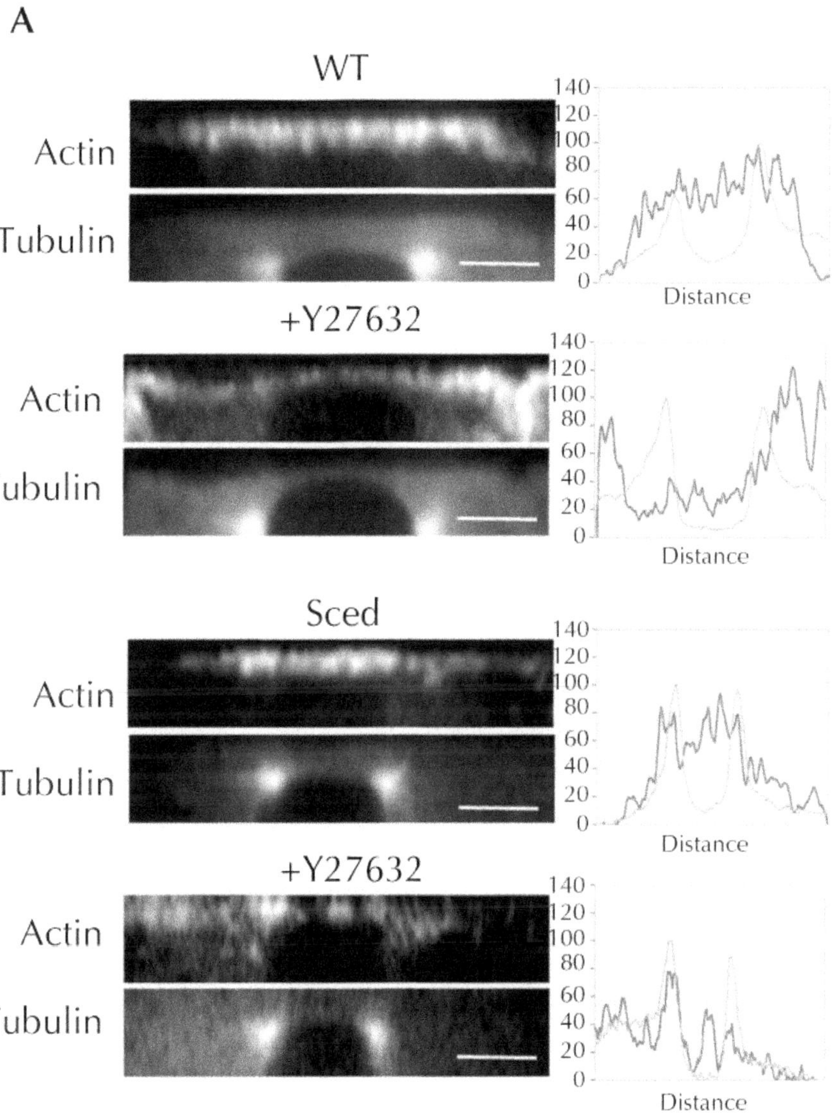

FIGURE 7 *(Continued)*

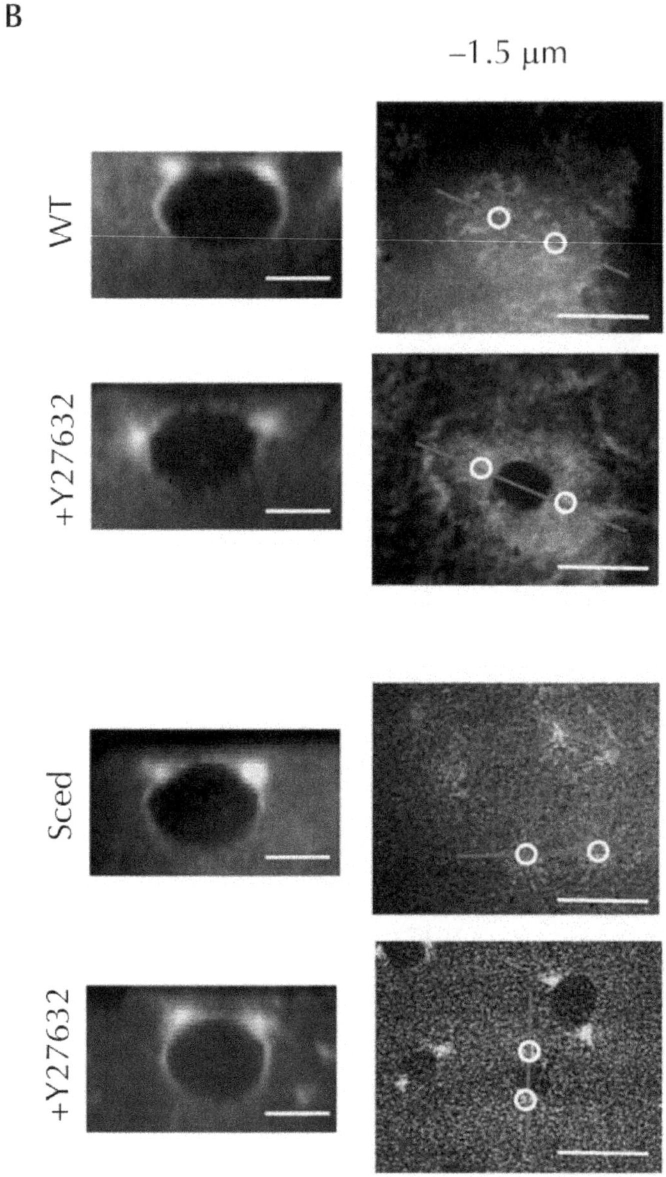

FIGURE 7 Effect of myosin inhibition on actin distribution in WT and Sced embryos. A) Side views of actin at the cortex in WT and Scedembryos in the absence and presence of the myosin inhibitor Y27632 reveal different actin distributions and cap sizes. The WT actin cap is compact with an evenly distributed thick layer of actin. When myosin is inhibited, the cap expands remaining thick at the margins and becoming thinner in the center. Note that the nucleus ascends 'indenting' the cap. In Sced embryos the cap becomes smaller compared to WT embryos, and when myosin function is inhibited, the actin in the cap becomes disorganized. Corresponding line scans of fluorescence intensity (tubulin, green; actin, red) taken along the actin cap in relation to the centrosomes' position (taken along the pole-pole axis), confirm the qualitative

description of the changes in actin distribution. Bars, 5 μm. B) Side view of the centrosomes' position with respect to the nucleus. The nuclear membrane was stained in vivo with fluorescent conjugated WGA in WT embryos expressing GFP-tubulin or inSced embryos injected with FITC-tubulin. When myosin is inhibited in WT, the centrosomes move further down and around the nucleus than in control. In Sced, the centrosomes are positioned slightly more apically, and when myosin in inhibited, this apical shift and lesser separation becomes more evident. Bars, 10 μm. Next to the side views are corresponding images of a wild-type embryo expressing GFP-tubulin (green) injected with rhodamine-actin (red) and a Sced embryo injected with FITC-tub (green) and rhodamine-actin (red) in the presence and absence of the myosin inhibitor Y27632. The prophase actin cap taken 1.5 μm below the top shows the difference in actin density and distribution between wild-type and Sced embryos in the presence or absence of myosin. Circles mark the position of the centrosomes which in some cases are not clearly visible. Bars, 10 μm.

Finally, the position of the centrosomes was examined relatively to the nucleus in prophase and found that in WT the centrosomes separate further around the nucleus in the absence of myosin function, while the nucleus itself appears to be pulled up indenting the actin cap (Figure 7B). This could be an indication of a greater pulling outward force applied by the astral MTs on the centrosomes [14]. Sometimes, detached centrosomes could also be seen in myosin-inhibited WT embryos (data not shown and [19]) lending further support to the hypothesis of a stronger pulling force. Note that the top view sometimes shows a larger nucleus in a myosin-inhibited embryo, however, the side view clearly indicates that the centrosomes have moved further down along the nuclear surface, so that the nuclear cross-section in the centrosomal plane is greater. In fact, the nuclear volume does not seem to change in the myosin-inhibited embryos. This observation is in contrast with the result reported in [19], where the increased inter-centrosomal distance in myosin-inhibited embryos was observed and interpreted as nuclear swelling.

KEYWORDS

- **Centrosomes**
- ***Drosophila***
- **Spindle Poles**
- **Myosin-II**
- **Rhodamine-Actin**

ACKNOWLEDGMENTS

We thank Dr. Jonathan Scholey for hosting all experimental work in his laboratory and for fruitful discussions.

REFERENCES

1. Rodriguez, O. C., Schaefer, A. W., Mandato, C. A., Forscher, P., Bement, W. M. et al. Conserved microtubule-actin interactions in cell movement and morphogenesis. *Nat Cell Biol* **5**, 599–609 (2003).

2. Rosenblatt, J., Cramer, L. P., Baum, B., and McGee, K. M. Myosin II-dependent cortical movement is required for centrosome separation and positioning during mitotic spindle assembly. *Cell* **117,** 361–372 (2004).
3. Théry, M., Jiménez-Dalmaroni, A., Racine, V., Bornens, M., and Jülicher, F. Experimental and theoretical study of mitotic spindle orientation. *Nature* **447,** 493–496 (2007).
4. Kunda, P. and Baum, B. The actin cytoskeleton in spindle assembly and positioning. *Trends Cell Biol* **19,** 174–179 (2009).
5. Glotzer, M. The molecular requirements for cytokinesis. *Science* **307,** 1735–1739 (2005).
6. Kiehart, D. P., Mabuchi, I., and Inoué, S. Evidence that myosin does not contribute to force production in chromosome movement. *J Cell Biol* **94,** 165–178 (1982).
7. Somma, M. P., Fasulo, B., Cenci, G., Cundari, E., and Gatti, M. Molecular dissection of cytokinesis by RNA interference in Drosophila cultured cells. *Mol Biol Cell* **13,** 2448–2460 (2002).
8. Fabian, L. and Forer, A. Possible roles of actin and myosin during anaphase chromosome movements in locust spermatocytes. *Protoplasma* **231,** 201–213 (2007).
9. Woolner, S., O'Brien, L. L., Wiese, C., and Bement, W. M. Myosin-10 and actin filaments are essential for mitotic spindle function. *J Cell Biol* **182,** 77–88 (2008).
10. Glover, D. M. Polo kinase and progression through M phase in Drosophila: a perspective from the spindle poles. *Oncogene* **24,** 230–237 (2005).
11. Royou, A., Field, C., Sisson, J. C., Sullivan, W., and Karess, R. Reassessing the role and dynamics of nonmuscle myosin II during furrow formation in early Drosophila embryos. *Mol Biol Cell* **15,** 838–850 (2004).
12. Wheatley, S., Kulkarni, S., and Karess, R. Drosophila nonmuscle myosin II is required for rapid cytoplasmic transport during oogenesis and for axial nuclear migration in early embryos. *Development* **121,** 1937–1946 (1995).
13. Halsell, S. R. and Kiehart, D. P. Second-site noncomplementation identifies genomic regions required for Drosophila nonmuscle myosin function during morphogenesis. *Genetics* **148,** 1845–1863 (1998).
14. Cytrynbaum, E. N., Sommi, P, Brust-Mascher, I., Scholey, J. M., and Mogilner, A. Early spindle assembly in Drosophila embryos: role of a force balance involving cytoskeletal dynamics and nuclear mechanics. *Mol Biol Cell* **16,** 4967–4981 (2005).
15. Foe, V. E., Field, C. M., and Odell, G. M. Microtubules and mitotic cycle phase modulate spatiotemporal distributions of F-actin and myosin II in Drosophila syncytial blastoderm embryos. *Development* **127,** 1767–1787 (2000).
16. Royou, A., Sullivan, W., and Karess, R. Cortical recruitment of nonmuscle myosin II in early syncytial Drosophila embryos: its role in nuclear axial expansion and its regulation by Cdc2 activity. *J Cell Biol* **158,** 127–137 (2002).
17. Riggs, B., Rothwell, W., Mische, S., Hickson, G. R., Matheson, J. et al. Actin cytoskeleton remodeling during early Drosophila furrow formation requires recycling endosomal components Nuclear-fallout and Rab11. *J Cell Biol* **163,** 143–154 (2003).
18. Stevenson, V. A., Kramer, J., Kuhn, J., and Theurkauf, W. E. Centrosomes and the Scrambled protein coordinate microtubule-independent actin reorganization. *Nat Cell Biol* **3,** 68–75 (2001).
19. Cao, J., Crest, J., Fasulo, B., and Sullivan, W. Cortical Actin Dynamics facilitate early-stage centrosome separation. *Curr Biol* **20,** 770–776 (2010).
20. Mitchison, T. J., Maddox, P., Gaetz, J., Groen, A., Shirasu, M. et al. Roles of polymerization dynamics, opposed motors, and a tensile element in governing the length of Xenopus extract meiotic spindles. *Mol Biol Cell* **16,** 3064–3076 (2005).
21. Sharp, D. J., Brown, H. M., Kwon, M., Rogers, G. C., Holland, G. et al. Functional coordination of three mitotic motors in Drosophila embryos. *Mol Biol Cell* **11,** 241–253 (2000).
22. Wollman, R., Civelekoglu-Scholey, G., Scholey, J. M., and Mogilner, A. Reverse engineering of force integration during mitosis in the Drosophila embryo. *Mol Syst Biol* **4,** 195 (2008).
23. Dujardin, D. L. and Vallee, R. B. Dynein at the cortex. *Curr Opin Cell Biol* **14,** 44–49 (2002).
24. Narumiya, S., Ishizaki, T., and Uehata, M. Use and properties of ROCK-specific inhibitor Y-27632. *Methods Enzymol* **325,** 273–284 (2000).

25. Sullivan, W., Fogarty, P., and Theurkauf, W. Mutations affecting the cytoskeletal organization of syncytial Drosophila embryos. *Development* **118,** 1245–1254 (1993).
26. Sharp, D. J., McDonald, K. L., Brown, H. M., Matthies, H. J., Walczak, C. et al. The bipolar kinesin, KLP61F, cross-links microtubules within interpolar microtubule bundles of Drosophila embryonic mitotic spindles. *J Cell Biol* **144,** 125–138 (1999).
27. Brust-Mascher, I. and Scholey, J. M. Microinjection techniques for studying mitosis in the Drosophila melanogaster syncytial embryo. *J Vis Exp* **31,** doi:10.3791/1382 (2009).
28. Mandato, C. A. and Bement, W. M. Actomyosin transports microtubules and microtubules control actomyosin recruitment during Xenopus oocyte wound healing. *Curr Biol* **13,** 1096–1105 (2003).
29. Hwang, E., Kusch, J., Barral, Y., and Huffaker, T. C. Spindle orientation in Saccharomyces cerevisiae depends on the transport of microtubule ends along polarized actin cables. *J Cell Biol* **161,** 483–488 (2003).
30. Wilson, C. A., Tsuchida, M. A., Allen, G. M., Barnhart, E. L., Applegate, K. T. et al. Myosin II contributes to cell-scale actin network treadmilling through network disassembly. *Nature* **465,** 373–377 (2010).
31. Backouche, F., Haviv, L., Groswasser, D., and Bernheim-Groswasser, A. Active gels: dynamics of patterning and self-organization. *Phys Biol* **3,** 264–273 (2006).
32. Paluch, E., Piel, M., Prost, J., Bornens, M., and Sykes, C. Cortical actomyosin breakage triggers shape oscillations in cells and cell fragments. *Biophys J* **89,** 724–733 (2005).
33. Chaudhuri, O., Parekh, S. H., and Fletcher, D. A. Reversible stress softening of actin networks. *Nature* **445,** 295–298 (2007).
34. Dean, S. O., Rogers, S. L., Stuurman, N., Vale, R. D., and Spudich, J. A. Distinct pathways control recruitment and maintenance of myosin II at the cleavage furrow during cytokinesis. Proc *Natl Acad Sci USA* **102,** 13473–13478 (2005).
35. Hu, C. K., Coughlin, M., Field, C. M., and Mitchison, T. J. Cell polarization during monopolar cytokinesis. *J Cell Biol* **181,** 195–202 (2008).
36. Redemann, S., Pecreaux, J., Goehring, N. W., Khairy, K., Stelzer, E. H. et al. Membrane invaginations reveal cortical sites that pull on mitotic spindles in one-cell C. elegans embryos. *PLoS One* **5,** e12301 (2010).
37. Rogers, S. L., Wiedemann, U., Häcker, U., Turck, C., and Vale, R. D. Drosophila RhoGEF2 associates with microtubule plus ends in an EB1-dependent manner. *Curr Biol* **14,** 1827–1833 (2004).
38. Kosako, H., Yoshida, T., Matsumura, F., Ishizaki, T., Narumiya, S. et al. Rho-kinase/ROCK is involved in cytokinesis through the phosphorylation of myosin light chain and not ezrin/radixin/moesin proteins at the cleavage furrow. *Oncogene* **19,** 6059–6064 (2000).
39. Sharp, D. J., Rogers, G. C., and Scholey, J. M. Cytoplasmic dynein is required for poleward chromosome movement during mitosis in Drosophila embryos. *Nat Cell Biol* **2,** 922–930 (2000).

4 Filament Depolymerization During Bacterial Mitosis

Edward J. Banigan, Michael A. Gelbart, Zemer Gitai, Ned S. Wingreen, and Andrea J. Liu

CONTENTS

4.1 Introduction 72
4.2 Methods 75
4.2.1 Biochemistry 75
4.2.2 Units 75
4.2.3 Interactions 75
4.2.4 Equations of motion 77
4.3 Discussion 78
4.3.1 Self-Diffusiophoresis can Explain Para Pulling 78
4.3.2 Translocation is most Robust for Side-Binding of Parb to Para with Disassembly only from the Tip 80
4.3.3 Comparisons with Experiments on Par-Mediated Chromosome Pulling ... 80
4.3.4 Implications for Other Phenomena 81
4.4 Results 83
4.4.1 Simulating ParB Translocation 83
4.4.2 Translocation is most Robust when ParB Binds Side-on to ParA 83
4.4.3 Tip-Only Binding 84
4.4.4 Side-Binding with Filament Severing 84
4.4.5 Side-Binding with Tip-Only Disassembly 84
4.4.6 Weak Binding 84
4.4.7 Fast Hydrolysis and Depolymerization 84
4.4.8 The Rate of Disassembly Controls the ParB Translocation Velocity 84
4.4.9 Three Regimes of Translocation Velocity 88
4.4.10 Dependence of the Translocation Velocity on Binding Energy, Binding Sites, Applied Load, and Other Physical Variables 89
4.4.11 Detachment Force for the Parb Polymer 89

4.4.12 The Parb Polymer Translocates even When the Para Bundle is not Anchored 91
Keywords 91
Acknowledgments 91
References 91

4.1 INTRODUCTION

Chromosome segregation is fundamental to all cells, but the force-generating mechanisms underlying chromosome translocation in bacteria remain mysterious. *Caulobacter crescentus* utilizes a depolymerization-driven process in which a ParA protein structure elongates from the new cell pole, binds to a ParB-decorated chromosome, and then retracts via disassembly, pulling the chromosome across the cell. This poses the question of how a depolymerizing structure can robustly pull the chromosome that disassembles it. Brownian dynamics simulations is performed with a simple, physically consistent model of the ParABS system. The simulations suggest that the mechanism of translocation is "self-diffusiophoretic": by disassembling ParA, ParB generates a ParA concentration gradient so that the ParA concentration is higher in front of the chromosome than behind it. Since the chromosome is attracted to ParA via ParB, it moves up the ParA gradient and across the cell. It is found that translocation is most robust when ParB binds side-on to ParA filaments. In this case, robust translocation occurs over a wide parameter range and is controlled by a single dimensionless quantity: the product of the rate of ParA disassembly and a characteristic relaxation time of the chromosome. This time scale measures the time it takes for the chromosome to recover its average shape after it is has been pulled. Our results suggest explanations for observed phenomena such as segregation failure, filament-length-dependent translocation velocity, and chromosomal compaction.

Reliable chromosome segregation is crucial to all dividing cells. In some bacteria, segregation has been found to occur in a rather counterintuitive way: the chromosome attaches to a filament bundle and erodes it by causing depolymerization of the filaments. Moreover, unlike eukaryotic cells, bacteria do not use molecular motors and/or macromolecular tethers to position their chromosomes. This raises the general question of how depolymerizing filaments alone can continuously and robustly pull cargo as the filaments themselves are falling apart. In this work, the first quantitative physical model is introduced for depolymerization-driven translocation in a many-filament system. Our simulations of this model suggest a novel underlying mechanism for robust translocation, namely self-diffusiophoresis, motion of an object in a self-generated concentration gradient in a viscous environment. In this case, the cargo generates and sustains a concentration gradient of filaments by inducing them to depolymerize. It is demonstrated that this model agrees well with existing experimental observations such as segregation failure, filament-length-dependent translocation velocity, and chromosomal compaction. In addition, several predictions–including predictions are made for the specific modes by which the chromosome binds to the filament structure and triggers its disassembly–that can be tested experimentally.

Several processes involved in DNA partitioning rely on depolymerization of filaments for translocation. In eukaryotes, depolymerizing microtubules [1] position chromosomes before cell division via macromolecular couplers and/or molecular motors bound to the microtubules [2, 3]. In prokaryotes, however, no such coupler or motor has been identified. Instead, proteins bound to the chromosome or plasmid bind directly to filaments and trigger their depolymerization [4, 5]. This poses the question of whether in the absence of a coupler, DNA can be pulled in a robust fashion, without becoming detached from the filaments as they disassemble.

Type I low-copy-number-plasmids [6, 7], chromosome I of *Vibrio cholerae* [8], and the chromosome of *Caulobacter crescentus* [9–12] all share a common segregation mechanism that relies on pulling mediated by filament depolymerization. This conserved system relies on three central components: the ATPase ParA, the DNA-binding protein ParB, and a centromere-like DNA locus. ParA is a deviant Walker-type ATPase that upon binding ATP forms dimers that can polymerize and associate with DNA [10, 13]. ParB interacts with ParA directly and stimulates ATP hydrolysis, causing ParA to dissociate into free monomers [13]. The spatial and temporal organization of ParA and the ParB-binding *parS*chromosomal locus can lead to robust chromosome segregation *in vivo*. For example, in *C. crescentus*, the chromosomal origin (*ori*) is initially localized at a single cell pole (the "stalked" pole) [14], and must translocate to the opposite "swarmer" pole before cell division. In predivisional cells, approximately one thousand ParB are bound via *parS* near the origin of the chromosome (*ori*) [9, 15]. There appear to be several distinct stages of ParB-*parS-ori* complex translocation [11]; our focus is on the final, most rapid stage in which the complex binds to filaments of ParA and translates from partway across the cell to the swarmer pole at a velocity of $v \approx 0.3\ \mu\text{m}/\text{min}$ [9, 11, 16, 17]. As the ParA bundle depolymerizes, presumably due to ParB-induced ATP hydrolysis or nucleotide exchange [7–11, 13, 15, 18, 19], the ParB-*parS-ori* complex remains localized near the edge of the ParA structure [8, 10–12].

For eukaryotic chromosome segregation driven by depolymerization of microtubules [2, 3], models generally assume the existence of a "coupler" that attaches the chromosome to the depolymerizing microtubules. This coupler moves along the microtubule ahead of the depolymerizing end, either because it slides along it diffusively [20–24], because it is pushed by conformational changes near the tip of the microtubule [23–26], or because it has a complex internal structure that causes it to process [3, 27]. Of the existing models of bacterial chromosome segregation [28–33], only a few address the question of how depolymerizing proteins can cause translocation. Typically, these models attempt to explain ParAB partitioning systems with reaction-diffusion models or general thermodynamic arguments, but do not address the conditions required for robust translocation [31, 32].

Here it is asked whether depolymerization of ParA by ParB without a coupler is sufficient to explain the observed translocation in prokaryotic DNA partitioning. Brownian dynamics simulations are performed that explicitly incorporate the biochemistry of the primary constituents of the ParABS segregation system. In this simulations, a polymer representing the ParB-*parS-ori* complex (henceforth referred to as the "ParB polymer"), binds to a filamentous ParA bundle and initiates disassembly of ParA. It is found that the ParB polymer can indeed exhibit robust, depolymerization-driven translocation via a novel mechanism (Figure 1), provided certain conditions are met.

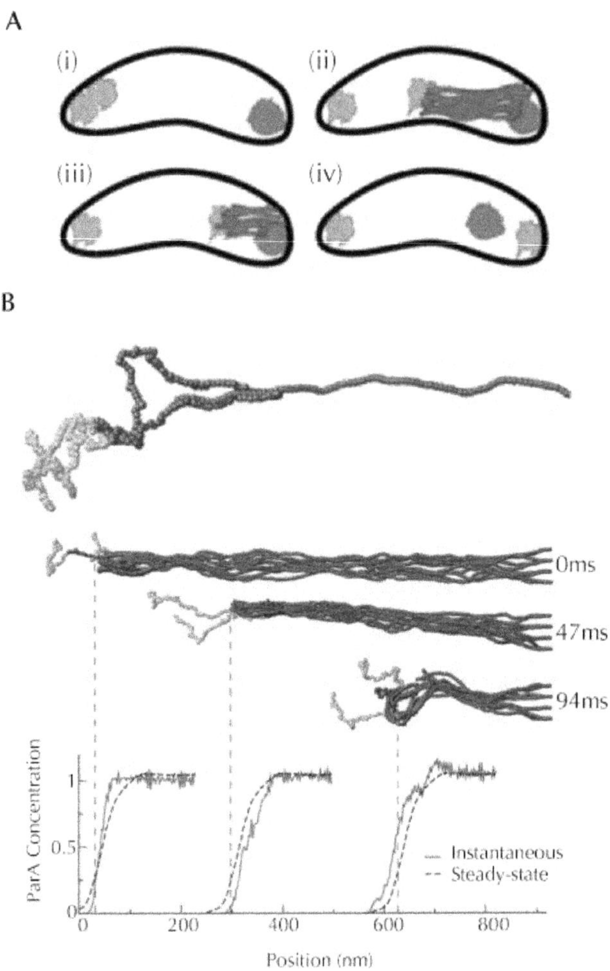

FIGURE 1 Schematic model for chromosome segregation and simulation snapshots. (A) Model of chromosome segregation in Caulobacter crescentus. (i) Initially, the two copies of the origin of replication (ori - green) and the terminus (ter - blue) of the chromosome are localized at the stalked and swarmer poles, respectively. (ii) The two origins separate and a structure of ParA protein (red) emanating from the swarmer pole comes into contact with the medial origin; ParB, polymerized on the chromosome near the origin, binds to ParA. (iii) ParB and the origin localize with the end of the ParA and move across the cell as ParA depolymerizes. (iv) The origin localizes near the swarmer pole; the terminus moves towards mid-cell. (B) Snapshot of ParB-ParA binding in simulation. The central strip of the ParB polymer (dark green) binds side-on to ParA filaments, whereas the peripheral segments of the ParB polymer (light green) cannot bind to ParA. (C) Snapshots of the full simulation and corresponding ParA filament concentration profiles (red). The dashed green lines indicate the center of mass of the ParB polymer. ParB binds to ParA and disassembles the ParA bundle (for clarity, depolymerized ParA monomers are not displayed). This interaction creates a steady-state ParA filament concentration gradient (black), which moves with and transports the ParB across the cell, providing a mechanism for chromosome segregation.

4.2 METHODS

At the start of each simulation, ParA monomeric subunits form a cross-linked bundle of filaments. The ParB-decorated chromosome is represented by a semi-flexible chain of monomeric subunits, typically of length 100 subunits, divided into three sections. The center section, typically of length 50 subunits, represents the part of the chromosome bound to ParB; these subunits can bind specifically to ParA subunits. The two end sections of the ParB polymer flanking the ParB section do not bind to ParA.

4.2.1 Biochemistry

The process of ParA disassembly begins when a ParB subunit binds to a ParA-ATP subunit. If the interaction energy, U_{AB}, exceeds a certain threshold, 0.75ϵ, the ParA-ATP hydrolyzes at rate k_h. Once the ParA subunit hydrolyzes, it may detach from the ParA filament by depolymerization at rate k_d(after which it continues to interact with other subunits by the interaction U_R). In this standard model, ParB binds to the sides of ParA filaments, and a hydrolyzed ParA subunit can only depolymerize if it is located at the tip of a ParA filament.

4.2.2 Units

Simulation units are converted into physical units by taking the subunit length to be a=5nm. The typical subunit diffusion coefficient is taken to be D=7.7 μm^2 D=7.7 μm^2/s, as measured in [34], and the diffusion coefficient for a particular subunit is ζ D_i= D/ζ_i (typically $\zeta_i = 1$ or 5, see below), giving a cell viscosity η=11.4cP and a characteristic time scale $\tau = a^2/D$ = 3.3 μm. Typical runs are approximately 100ms and simulation steps are 0.81ns.

4.2.3 Interactions

Several interactions are included in the model; their specific forms are given below. All subunits are spheres with diameter a that repel each other if they overlap:

$$U_R(r_{ij}) = \begin{cases} \frac{1}{2}K_R(r_{ij}-a)^2, & \text{for } r_{ij} < a \\ 0 & \text{for } r_{ij} \geq a, \end{cases} \tag{1}$$

where r_{ij} is the center-to-center distance between subunits i and j and $KR = 100k_BT/a^2$. Within a ParA or ParB polymer chain, neighboring subunits are held together through an attractive harmonic potential:

$$U_B(r_{ij}) = \frac{1}{2}K_B(r_{ij}-a)^2, \text{ for } r_{ij} > a \tag{2}$$

with $K_B = 100k_BT/a^2$. In order to hold the ParA bundle together, typically 40% of ParA subunits are taken for cross-linking to a subunit in a nearby filament through an attractive potential:

$$U_C(r_{ij}) = \frac{1}{2}K_C(r_{ij}-b)^2, \text{ for } r_{ij} > b \tag{3}$$

where $b = 5a$ is the initial spacing of filaments in the ParA bundle and $K_C = K_B/2$. ParA filaments are stiffened by a bending potential [35]:

$$U_S(\theta_i)=\frac{1}{2}K_S(\cos\theta_i - \cos\theta_0)^2 \tag{4}$$

where θ_i is the angle between the bond vector, $\vec{b}_{i-1,i}$, between ParA subunits $i - 1$ and i, and the bond vector, $\vec{b}_{i,i+1}$, between subunits i and $i + 1$. Thus, $\cos\theta_i=\hat{b}_{i-1,i}\cdot\hat{b}_{i,i+1}$, where $\hat{b}=\vec{b}/|\vec{b}|$. $K_S=500k_BT$ and $\theta_0=0$ is taken. Similarly, the stiffness of the ParB polymer can be controlled by an interaction potential of form of Eq. 4 (however, in this standard model, $K_S=0k_BT$ in the ParB polymer).

In addition, interactions are introduced so that binding between ParA and ParB occurs in specific spatial locations on the spheres representing the subunits. Each subunit i has a unit polarization vector, $\hat{p}_i$, that determines the location of the binding site for the ParB–ParA interaction, and the following interaction potential aligns it to be at an angle θ_p to the bond vectors $\vec{b}$ connecting adjacent subunits:

$$U_P(\theta_i)=\frac{1}{2}K_P(\hat{p}_i\cdot\hat{b}_{i,i+1} - \cos\theta_p)^2 + \frac{1}{2}K_P(\hat{p}_i\cdot\hat{b}_{i-1,i} - \cos\theta_p)^2 \tag{5}$$

$\theta_p=\pi/2$ is chosen so that $\hat{p}_i$ tends to be perpendicular to the bond vectors, and fix $K_P=100k_BT$ for ParA filaments and $K_P=25k_BT$ in the ParB polymer, which is relatively more flexible. Binding sites are arranged helically on the ParA filaments and the ParB polymer due to two additional interaction potentials. The first constrains polarization vectors on nearest-neighbor subunits on a given chain:

$$U_{H1}(\psi_{i,1})=\frac{1}{2}K_{H1}(\cos\psi_{i,1} - \cos\psi_{01})^2 \tag{6}$$

where $\cos\psi_{i,1}\equiv\hat{p}_i\cdot\hat{p}_{i+1}$ and $\psi_{i,1}\equiv\hat{p}_i\cdot\hat{p}_{i+1}$ sets the pitch of the helix. Here, $K_{H1}=200k_BT$ for ParA and $K_{H1}=50k_BT$ for ParB. The second potential has the same form,

$$U_{H2}(\psi_{i,2})=\frac{1}{2}K_{H2}(\cos\psi_{i,2} - \cos\psi_{02})^2 \tag{7}$$

but constrains polarization vectors on the next-nearest-neighbor subunits with $\cos\psi_{i,2}\equiv\hat{p}_i\cdot\hat{p}_{i+2}$ and $\psi_{02}=\pi/9$. Here, $K_{H2}=100k_BT$ in ParA and $K_{H2}=25k_BT$ for ParB. Note that in addition to regulating the locations of the binding sites, Eqs. 6 and 7 implicitly regulate torsion within the ParB polymer.

Finally, ParB binds to ParA with a site-specific, short-ranged interaction potential:

$$U_{AB}(r_{AB},\phi_1,\phi_2,\phi_3)=\begin{cases} \frac{\epsilon}{C}[\frac{\sigma^{14}}{r_{AB}^{14}} - \frac{\sigma^{12}}{r_{AB}^{12}}](\cos\phi_1\cos\phi_2\cos\phi_3)^2 \\ \quad \text{for } \cos\phi_1>0,\ \cos\phi_2,\cos\phi_3<0 \\ \quad 0 \quad \text{otherwise.} \end{cases} \tag{8}$$

where $\vec{r}_{AB}=\vec{r}_B-\vec{r}_A$ is the vector distance between the ParA and ParB subunits and ϵ is the binding energy. In this standard model, $\epsilon=8k_BT$. The normalization factor $C=(\sigma/a)^{12}-(\sigma/a)^{14}$ ensures that ϵ is the relevant energy scale for binding. The distance $\sigma=a\sqrt{6/7}$ sets $r_{AB}=a$ as the minimum of the binding potential. Binding site specificity is implemented through regulation of the angles between the polarization vectors on the ParA and ParB subunits as well as $\hat{r}_{AB}=\vec{r}_{AB}/r_{AB}$. In Eq. 8, $\cos\phi_1=\hat{p}_A\cdot\hat{r}_{AB}$, $\cos\phi_2=\hat{p}_B\cdot\hat{r}_{AB}$, and $\cos\phi_3=\hat{p}_A\cdot\hat{p}_B$. Binding is strongest when the two polarization vectors point towards each other and along $\vec{r}_{AB}$.

Several variations have also been studied of these models. For example, in a separate set of simulations, $K_{H1}=0$ and $K_{H2}=0$ is set for both ParA and ParB, so that the binding sites were not arranged helically on the ParA filaments and ParB polymer. The orientation of the polarization vectors was set by Up, where $\theta_p=0$ for tip binding and $\theta_p=\pi/2$ for side-on binding. Cases were also studied in which monomeric ParB subunits did not possess specific orientations (polarization vectors). In these cases, ParA polarization vectors were set by Up, where $\theta_p=0$ for both tip-binding and side-binding. Binding only weakly depended on the orientation of the ParA-ParB bond through a modified version of U_{AB}, which is denoted as U^{*}_{AB} and U^{**}_{AB} for tip-binding and side-binding, respectively. For tip-binding without ParB polarization vectors:

$$U^{*}_{AB}(r_{AB},\phi_1)=\begin{cases}\frac{\epsilon}{C}\left[\frac{\sigma^{14}}{r_{AB}^{14}}-\frac{\sigma^{12}}{r_{AB}^{12}}\right]\cos^6\phi_1 \\ \quad \text{for } \cos\phi_1>0 \\ 0 \quad \text{otherwise}\end{cases} \tag{9}$$

For side-binding without ParB polarization vectors:

$$U^{**}_{AB}(r_{AB},\phi_1)=\frac{\epsilon}{C}\left[\frac{\sigma^{14}}{r_{AB}^{14}}-\frac{\sigma^{12}}{r_{AB}^{12}}\right]\sin^{10}\phi_1 \tag{10}$$

where r_{AB}, φ_1, σ, and C are as defined above.

4.2.4 Equations of motion

All subunits in the system translate and rotate according to Brownian dynamics [36]. Thus, a system of coupled Langevin equation is solved where the velocity of each subunit is governed by the forces exerted by other subunits in the system as well as thermal forces, $\vec{F}$ from the surrounding liquid medium:

$$\zeta\dot{\vec{r}}=-\vec{\nabla}_r(U_R+U_B+U_C+U_S+U_P+U_{AB})+\vec{F}(t) \tag{11}$$

$$\langle\vec{F}(t)\rangle=0,\ \langle\vec{F}(t)\cdot\vec{F}(t')\rangle=6k_BT\zeta\delta(t-t') \tag{12}$$

and

$$\zeta_p \dot{\hat{p}} = -\vec{\nabla}_p (U_P + U_{H1} + U_{H2} + U_{AB}) + \vec{G}(t) \tag{13}$$

$$\langle \vec{G}(t) \rangle = 0,\ \langle \vec{G}(t) \cdot \vec{G}(t') \rangle = \frac{a^2}{3} \langle \vec{F}(t) \cdot \vec{F}(t') \rangle = 6k_B T \zeta_p \delta(t - t') \tag{14}$$

The subunit friction constant is $\zeta = 3\pi\eta a\zeta_i$, where η is the viscosity, and ζ_i is a constant that determines the relative magnitude of the drag on subunit *i*. Typically, $\zeta_i = 1$ for ParA and normal ParB subunits, and $\zeta_i = 5$ for ParB subunits that cannot bind to ParA. $\zeta_p = \pi\eta a^3 \zeta_i$ is the rotational friction coefficient.

4.3 DISCUSSION

Based on recent experimental observations [7–12, 15], several simulation models have been tested (Figure 2) and discovered a robust mechanism for chromosome segregation in *C. crescentus* via the ParABS system.

4.3.1 Self-Diffusiophoresis can Explain Para Pulling

Our simulations point to a specific physical mechanism underlying translocation in the ParABS system. It is found that disassembly of ParA generates a steady-state ParA filament concentration gradient that remains fixed in the center-of-mass frame of the translocating ParB polymer (Figure 1c). In other words, disassembly of ParA allows the ParA filament concentration gradient to translocate with the particle across the cell so that at all times the ParB polymer is moving up the concentration gradient of ParA to satisfy its attraction to ParA. Our simulations do not include fluid flow, but it is known that external concentration gradients can also drive motion of a particle in a fluid environment; the latter phenomenon is known as "diffusiophoresis." If the particle (in this case, the ParB-*parS-ori* complex) is attracted to the solute (the ParA filament bundle), it will translocate up the concentration gradient towards high solute concentrations [37]. In "self-diffusiophoresis," the particle itself (the ParB-*parS-ori*complex) generates and sustains the solute concentration gradient [38, 39] via disassembly of ParA. It is emphasized that ParB-induced depolymerization (particle-induced destruction of solute) is central to this process. Without depolymerization, the ParA bundle would remain intact and the concentration of ParA filaments would not change with time. As a result, the ParA concentration profile would not be able to move with the particle and translocation would not occur.

This intrinsically many-body mechanism is distinct from biased diffusion. In contrast to biased diffusion mechanisms which apply to a coupler that attaches a load to a single filament or fiber [20–22, 26], self-diffusiophoretic translocation can occur even if the ParB polymer does not diffuse, as long as the ParB-ParA interaction range is finite. In self-diffusiophoresis, "diffusio" refers not to diffusion of a coupler, but to the key role of the solute gradient, just as the prefix in "electrophoresis" refers to an electric potential gradient [37]. The self-diffusiophoretic mechanism also differs from ones involving motion of a coupler [3, 20–27]; in this case, the load is not attached to a coupler that cannot detach from the depolymerizing filaments. Instead, the load is attached directly to the depolymerizing filaments via many non-permanent bonds.

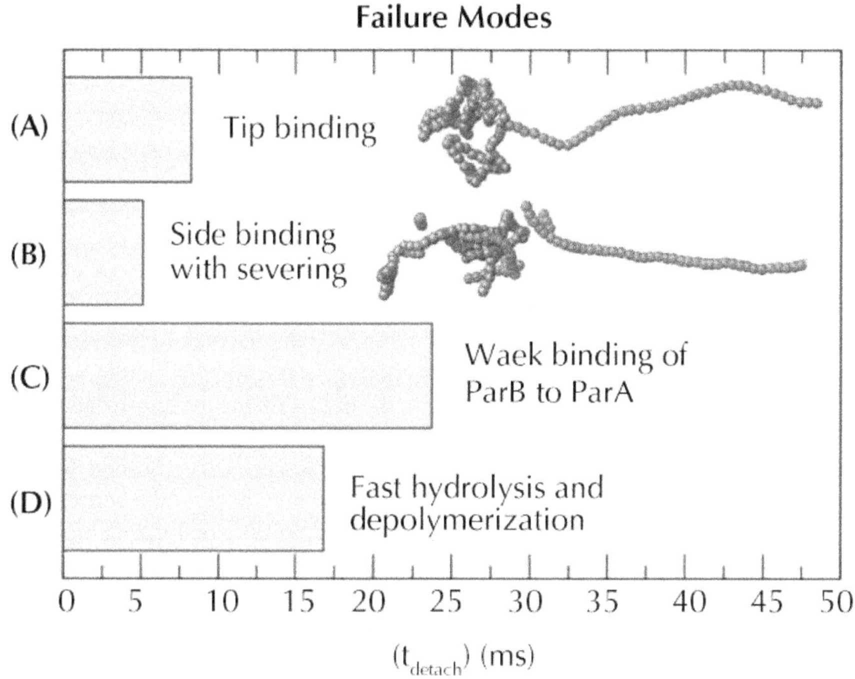

FIGURE 2 Mean time, $\langle t_{detach} \rangle$, to first detachment of ParB polymer from ParA for various failure modes. In a standard simulation, ParB binds ParA filaments side-on, and hydrolyzes individual ParA subunits. Hydrolyzed ParA disassembles from the tip of each filament in the bundle. In a typical simulation it takes about 200ms for the ParB polymer to translocate across a distance of 1μm. However, ParB completely detaches in a short time if (A) the ParB polymer binds to only the tips of the ParA filaments or (B) if ParA filaments disassemble via mid-filament severing. In addition, the ParB polymer detaches if (C) ParB binds too weakly to ParA or (D) ParA disassembles too quickly. Measurements for (A) and (B) are taken from simulations with side-binding with severing or tip binding, respectively, with standard parameters. Measurements for (C) and (D) are taken from simulations with the slowest disassembly rate or highest binding energy, respectively, for which the mean time to first detachment is shorter than the time required for the ParB polymer to translocate across the cell.

It has been suggested that polymerization-driven motility, as in the case of F-actin in the lamellipodium of eukaryotic cells, also constitutes an example of self-diffusiophoretic motility [40, 41]. In that case, the object to be moved (*e.g.*, the cell membrane) is repelled by the structure (the branched actin network) that it builds in order to move. In depolymerization-driven translocation, on the other hand, the object to be moved (the ParB-DNA complex) is attracted to the structure (ParA) that it destroys in order to move.

The self-diffusiophoretic mechanism suggests modes of failure for translocation. For example, overexpression of ParA leads to segregation defects, and it has been suggested that these defects arise due to the increase in the quantity of delocalized ParA [12, 15]. This effect may be analogous to what it is observed in this simulations with severing (Video S2), where instead of binding to the ParA bundle, ParB can bind to severed ParA filaments. This disrupts the steady-state generation of a translating ParA concentration gradient so that it does not support steady-state ParB polymer translocation. Similarly, when ParA is overexpressed, extra ParA monomers or protofilaments may diminish or erase the ParA concentration gradient created by depolymerization. Alternatively, the extra ParA could saturate ParB, preventing translation of the ParA gradient.

4.3.2 Translocation is most Robust for Side-Binding of ParB to ParA with Disassembly only from the Tip

Robust translocation over a wide range of physical parameters are observed only if ParB binds to the sides of ParA filaments, triggering disassembly only from the tips of filaments (Figure 1b–c). If ParB binds only to the tips of filaments, translocation is far less robust for two reasons. First, there are many fewer ParA subunits to which ParB can bind so the overall attraction between ParB and ParA is weaker. Second, the ParB polymer is localized near the tip of the bundle, at the very edge of the concentration gradient of ParA that drives translocation. In contrast, in the side-binding model, the ParB polymer penetrates further into the bundle so that it is localized near the steepest, central section of the concentration gradient (Figure S6). Thus, in the tip-binding-only model, the ParB polymer is much more likely to detach from the ParA bundle due to thermal noise (Figure 2a). This failure mode can only be averted by greatly increasing the binding energy or the number of filaments, and thus tips, in the ParA bundle.

It is also found that ParA disassembly via severing does not provide robust translocation (Figure 2b) because severed protofilaments can bind to ParB, reducing the attraction between the ParB polymer and the main ParA bundle, leading to detachment.

It is therefore predicted that ParB binds to the sides of ParA filaments and ParA filaments disassemble primarily from the tip. This prediction can be tested with *in vitro* experiments.

4.3.3 Comparisons with Experiments on Par-Mediated Chromosome Pulling

Our model is sufficiently versatile to account for a range of experimental observations. For example, by varying the initial density and cross-linking of the ParA filament bundle in this simulations, cases are found in which some ParA filaments remain partially assembled even though the ParB polymer has translocated across the cell (Figure S7). This is in agreement with the observations of Ptacin *et al.* [10], who found that in some cases, a fiber of ParA extended across the predivisional cell after *ori* had translocated.

It is found that the robustness of translocation is primarily controlled by the quantity $\tau_a k_0$, the product of an effective relaxation time (Eq. 16) and the maximum rate of disassembly of ParA (Eq. 17). The underlying details of the ParB polymer are only important insofar as they affect quantitative results such as the precise value of the relaxation time; they do not affect the qualitative physical principles described above.

Specifically, if $\tau_a k_0$ is too high, the ParB polymer stretches out and can detach from the ParA bundle. This finding suggests a possible role for chromosome organizing factors such as the SMC protein [14, 42]. In order to translocate reliably and efficiently, the chromosome of four million base pairs [14, 16] must be organized such that it does not overload the pulling mechanism. It is proposed that one important physical function of chromosomal organization and condensation is to minimize the effective relaxation time, τ_a, so that the chromosome can keep up with the retracting ParA bundle, to ensure robust translocation.

In addition, it is found that the velocity is simply the product of the ParA subunit length and the maximum disassembly rate, k_0, provided disassembly is slow enough to guarantee that $\tau_a k_0 \leq 1$ (Eq. 18). From the observed *ori* translocation velocity, $v = 0.3$ μm/min [9, 11, 16, 17], it is estimated that the *in vivo* ParA disassembly rate to be k $\approx 0.9 s^{-1}$, which is slower than the measured disassembly rate of dynamically unstable ParM filaments [43], but comparable to the disassembly rate of actin filaments [44].

The translocation velocity in this simulations is considerably higher, typically several *μm/s*, because high disassembly rates are used. Translocation is robust in this simulations at these high values of k_0 because the effective relaxation time, τ_a, of this ParB polymer is fairly short. In the real system, where the effective relaxation time of the chromosome is likely to be considerably longer, it could be a biological necessity that both ParA disassembly and *ori* translocation proceed at slower than the simulated rates.

Likewise, in this simulations the ParB polymer detaches when it is pulled with a force f^* of order tens of pN, but this detachment force is likely to be much higher in the real system. The most important difference between this simulations and the actual bacterium lies in the number of ParB binding sites N. To estimate the detachment force, f^*, under realistic conditions, first N is estimated, the maximum possible binding energy $N\epsilon$, the extent of the chromosome R, and the diffusion coefficient of the chromosome D. First $N \approx 1000$ is estimated by assuming that ParB decorates the approximately 10 kilobase segment of the chromosome that was found to be the site of force exertion during translocation in [9]. For $\epsilon \approx 10 k_B T$ therefore a maximum binding energy of $N\epsilon \approx 10^4 k_B T$ is obtained. For ideal polymer chains [45], $Rz \propto \sqrt{N}$ and $D \propto 1/N$. Thus, it is estimated $R_z \approx 100$nm and $D \approx 10^{-3}$ $\mu m^2/s$. This crude estimate of R actually agrees well with experimental snapshots of *C. crescentus* during chromosome segregation [10, 11]. The estimate of D falls within the range 10^{-1} $\mu m^2/s$ $< D < 10^{-5}$ $\mu m^2/s$, which is measured in *E. coli* for DNA segments of varying sizes [46, 47]. It is noted that f^* is insensitive to D, and varies by less than 1pN over that range.

According to experiments [9, 11, 16, 17], the ParB-*parS*-*ori* complex translocates across the cell in about 10 minutes. Using Eq. 20, it is found that the detachment force is $f^* \approx 200$pN. This value is of the same order of magnitude as the 700pN stall force for chromosome segregation along kinetochore fibers in eukaryotes [48, 49]. Thus, this estimate suggests that the mechanism that is proposed is both physically reasonable and biologically relevant.

4.3.4 Implications for Other Phenomena

Insights from the results may extend to plasmid segregation by ParAB. In *Escherichia coli*, the ParA concentration profile is known to oscillate as plasmid pB171 is partitioned [6, 19, 50]. This dynamic behavior appears to be required for proper plasmid partitioning [7, 19]. It suggests that ParB creates a moving ParA filament concentration gradient that pulls the plasmid along as ParA disassembles.

In addition, this findings suggest an alternative explanation for observations that the distance that plasmid pB171 translocates in a given time interval increases approximately linearly with the initial ParA filament length [7]. Ringgaard *et al.* [7] suggest that this effect arises from a ParA filament-length-dependent plasmid detachment rate. However, it is shown that the relative velocities of the ParB polymer and the ParA bundle depend on the ratio of the viscous drags on ParA and ParB, ζParA/ζParB (Figure 3). Thus, the observed dependence of plasmid translocation distances and velocities on ParA filament length may simply be a result of Newton's third law, due to the variation of ζParA/ζParB with ParA filament length.

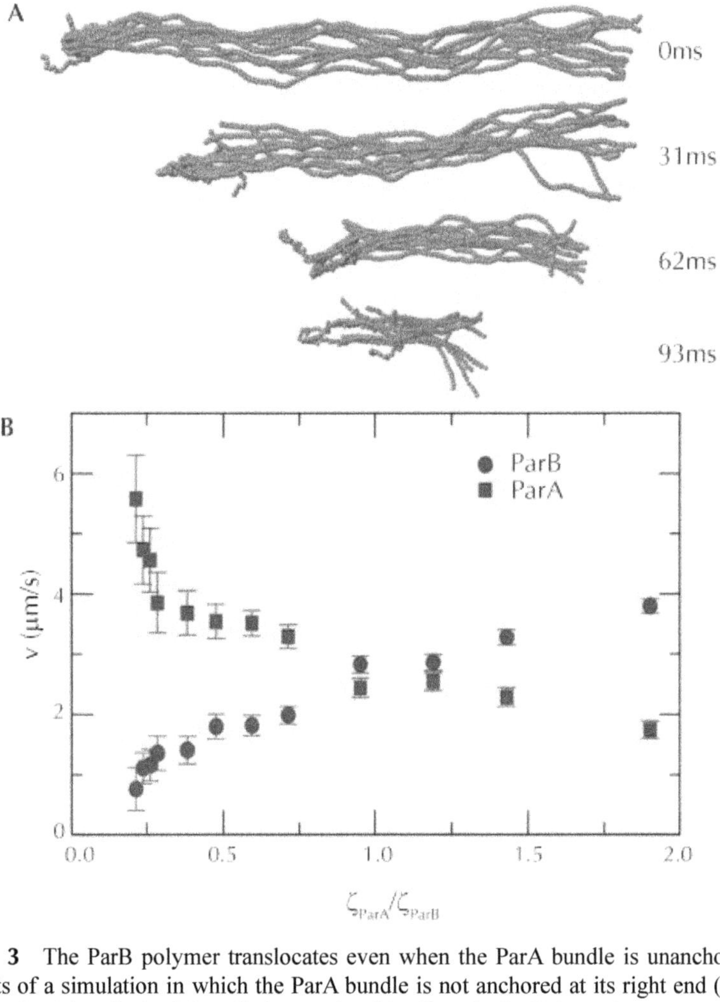

FIGURE 3 The ParB polymer translocates even when the ParA bundle is unanchored. (A) Snapshots of a simulation in which the ParA bundle is not anchored at its right end (swarmer pole). The ParA bundle (red) is pulled towards mid-cell as the ParB (green) moves towards the swarmer pole. (B) Dependence of speeds of ParA (red) and ParB (green) on the ratio of drags, ζParA/ζParB. In these simulations, ζParB = 1.1×10^{-3}g/s and $L = 1000$nm.

where ζParB and ζParA are the drag coefficients of the ParB polymer and ParA filament bundle, respectively.

Our simulations with unanchored ParA filaments suggest a new possibility for the mechanism of terminus segregation in *C. crescentus*. As translocation begins, the ParA filaments are long, so ζParA > ζParB and the ParB polymer is pulled rapidly towards the swarmer pole. However, as the ParB polymer nears the swarmer pole the ParA filaments are much shorter and ζParA < ζParB may be satisfied, so that the ParA bundle is pulled toward mid-cell. Experiments have indicated that ParA binds non-specifically to DNA [7, 10, 18]. Thus, it is proposed that DNA near the terminus is non-specifically bound to ParA and translocates away from the swarmer pole as ParA filaments are pulled toward mid-cell by the ParB-*parS-ori* complex. In contrast to previously suggested passive mechanisms [16, 30, 33], this is an active process, directly linked to *ori* translocation.

Our results provide a new paradigm for understanding depolymerization-driven translocation in prokaryotic DNA segregation systems. Since self-assembly and disassembly are ubiquitous in cellular systems, the creation of concentration gradients by these processes provides a general and robust mechanism for translocation.

4.4 RESULTS

4.4.1 Simulating ParB Translocation

To understand the mechanism by which ParA translocates ParB, Brownian dynamics simulations of a ParB polymer is performed interacting with an anchored ParA filament bundle (Figure 1c). The ParB polymer, shown in Figure 1b–c, corresponds to the ParB-*parS-ori* complex. It is represented by a semi-flexible chain of monomeric subunits, typically of length 100 subunits. The center section (dark green in Figure 1b), typically of length 50 subunits, represents the part of the chromosome that binds to ParA via ParB, while the two peripheral segments (light green in Figure 1b only) cannot bind to ParA.

During robust translocation, the ParB polymer remains localized near the tip of the ParA bundle and moves across the cell (see snapshots in Figure 1c and Video S1). By inducing disassembly, ParB creates a concentration gradient of ParA filaments that remains fixed with respect to the center of mass of the ParB polymer. Thus, the ParA concentration profile translocates with the ParB, and exhibits only small, short-lived fluctuations around a well-defined steady-state mean (Figure 1c).

4.4.2 Translocation is most Robust when ParB Binds Side-on to ParA

Since the precise nature of the ParB–ParA interaction is unknown, simulations are used to identify the modes of binding and disassembly that provide robust translocation. In this model (see Methods), ParB binds to ParA subunits in the filament bundle (Figure 1c). The ParB polymer hydrolyzes ParA subunits that it binds to; once a subunit at the tip of a ParA filament is hydrolyzed, it can depolymerize from the filament. Monomers rapidly diffuse away once they have depolymerized. Some interaction/disassembly mechanisms or parameter ranges lead to robust translocation of the ParB polymer, while others lead to failure by rapid detachment:

4.4.3 Tip-Only Binding

In this model, ParB binds only to the tips of ParA filaments (Figure 2a). Since the number of ParA filament tips is limited, the ParB polymer is held only weakly to the ParA bundle, and small fluctuations can cause it to detach (Figure 2a). In principle, this failure mode could be suppressed by increasing the number of ParA filaments within the bundle, but translocation is intrinsically fragile for this model.

4.4.4 Side-Binding with Filament Severing

As an alternative, ParB is allowed to bind to the sides of the ParA filaments (Figure 1b). In this model, ParA filaments can disassemble by severing in addition to disassembling from the filament tips (Figure 2b). Severing may occur at the location of any ParA subunit that has been hydrolyzed by ParB. Typically, it is found that the ParB polymer binds to multiple severed ParA segments, preventing ParB from binding to the remaining filaments in the ParA bundle (Video S2). As a result, the ParB polymer rapidly detaches from the anchored bundle. ParA severing therefore does not lead to reliable ParB translocation.

4.4.5 Side-Binding with Tip-Only Disassembly

In this model, ParB binds side-on to ParA filaments (Figure 1b–c) and ParA disassembles only at the filament tips. In this case, the ParB polymer translocates across the cell without detaching from the ParA bundle for a wide range of parameters. However, under certain extreme conditions, translocation fails:

4.4.6 Weak Binding

If the ParB–ParA binding energy, ϵ, is too small, ParB quickly detaches from ParA due to thermal noise and the force from the rest of the ParB polymer (Figure 2c).

4.4.7 Fast Hydrolysis and Depolymerization

Rapid detachment occurs if the ParA hydrolysis rate, k_h, and ParA depolymerization rate, k_d, are both too large (Figure 2d).

Our major result is that translocation is most robust in the side-binding model with disassembly only from the tips of ParA filaments. The rest of the simulations use this robust mode of disassembly and translocation, and henceforth, side-binding is referred with tip-only disassembly as this standard model.

4.4.8 The Rate of Disassembly Controls the ParB Translocation Velocity

To understand how ParA translocates ParB, variables controlling the translocation velocity were identified, *v*ParB. In all cases, it is found that *v*ParB is given by the mean rate, *k*, of disassembly of a ParA filament, so that $v\text{ParB} = ak$, where a is the length of a ParA subunit. In order for a subunit to disassemble from the tip of a ParA filament, the subunit must bind to ParB, its ATP must hydrolyze, and the subunit must fall off. k therefore depends on the distance, l, that the ParB polymer typically penetrates into the ParA bundle and causes ParA-ATP hydrolysis, the rate, k_h, of ParA-ATP hydrolysis, and the rate, k_d, at which a ParA subunit depolymerizes once hydrolyzed.

In turn, the penetration length, l, depends on the shape of the ParB polymer. In this simulations, the freely diffusing ParB polymer adopts an isotropic, globular equilibrium shape. The maximum value, l_{max}, of the penetration length, l, is achieved if the ParB polymer is able to maintain this equilibrium shape as it is pulled by ParA. If the disassembly rate, k, is too high, the ParB polymer is pulled along so rapidly that it does not have time to relax to its equilibrium shape. In this case, the ParA bundle pulls the leading region of the ParB polymer faster than the rear of the polymer can respond to the perturbation and the ParB polymer stretches out. Because the part of ParB polymer does not keep pace with the retraction of the depolymerizing ParA bundle, the ParB polymer does penetrate as deeply into the ParA bundle, so $l < l_{max}$.

It is now estimated that the time for the ParB polymer to relax to its equilibrium size. In this simulations, since ParB decorates the center section of the polymer and binds to ParA, the undecorated peripheral segments of the chain are the first ones to stretch out when the ParB polymer is pulled too rapidly (Video S3). The stretching of the peripheral segments is governed by the equation:

$$\langle \dot{z} \rangle = v_{\text{ParB}} - D_s\langle z \rangle/(R_z^0)^2 = v_{\text{ParB}} - \langle z \rangle/\tau_r \qquad (15)$$

where $\langle z \rangle$ is the ensemble-averaged z-distance between the ends of a peripheral segment pulled by one end in the z-direction, D_s is the diffusion coefficient of the segment, R_z^0 is the Z-component of its equilibrium radius of gyration, and the relaxation time, $\tau_r = (R_z^0)^2/D_s$, is the ratio of its internal drag, k_BT/D_s, to the effective spring constant, $k_BT/(R_z^0)^2$ (see Text S1). Stretching is appreciable if $\langle z \rangle \gtrsim R_z^0$, so for translocation in steady state ($\langle \dot{z} \rangle = 0$), stretching becomes appreciable for vParB $\gtrsim v^* = D_s/R_z^0$, or, equivalently, $ak R_z^0/Ds \gtrsim 1$ (inset to Figure 4a). The shape of the pulled ParB polymer is therefore governed by the product $\tau_a k$, where it is defined

$$\tau_a = aR_z^0/D_s = (a/R_z^0)\tau_r \qquad (16)$$

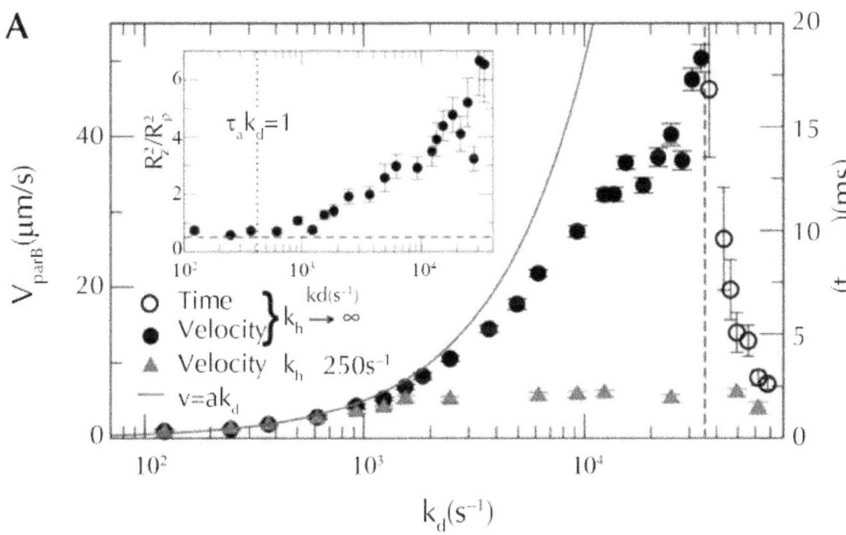

FIGURE 4 *(Continued)*

B

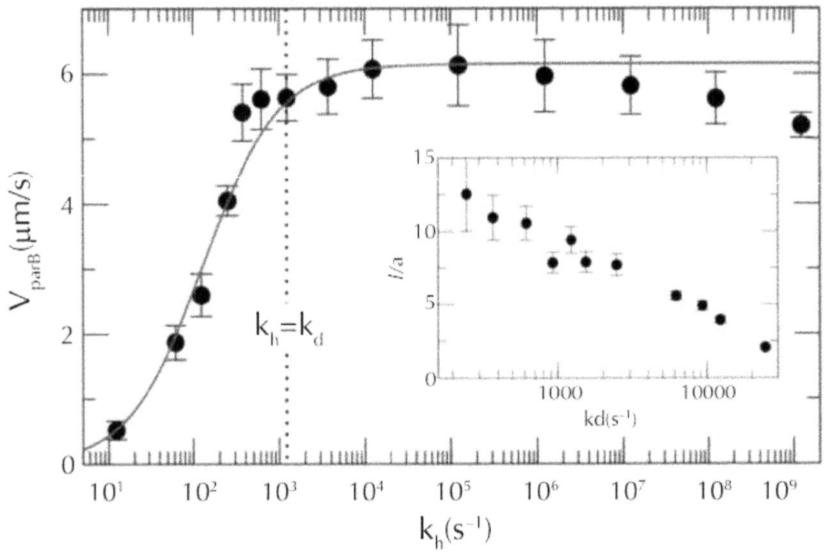

C

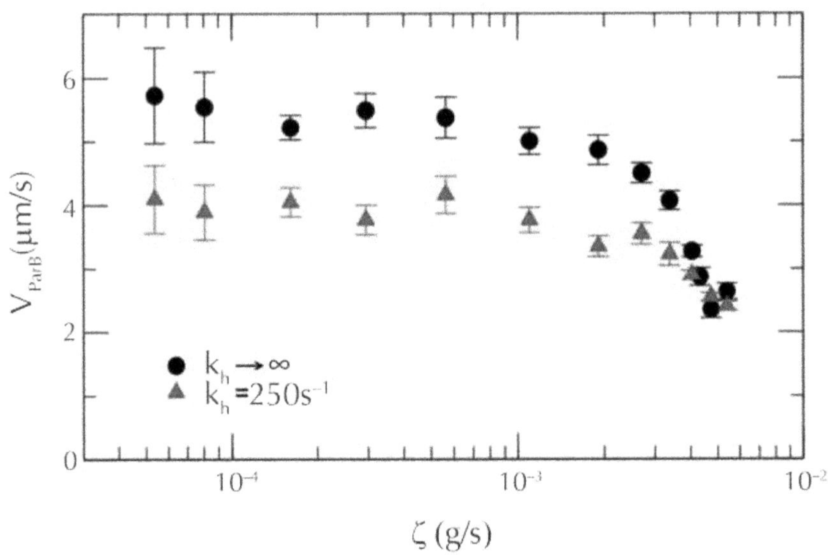

FIGURE 4 *(Continued)*

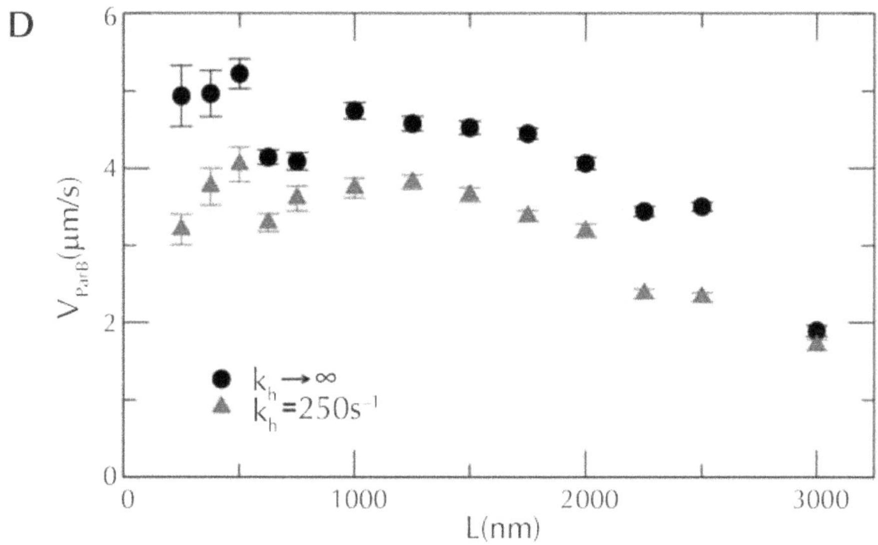

FIGURE 4 Dependence of translocation velocity on disassembly rate and relaxation time. (A) Translocation velocity, *v*ParB (solid symbols), increases with depolymerization rate, k_d. At low k_d, *v*ParB is linear in k_d, scaling as ak_d (green curve), where *a* is the diameter of a ParA subunit. At large k_d, with an arbitrarily fast hydrolysis rate, k_h, the ParB polymer detaches from the ParA bundle in an observably short time, $\langle t_{detach}\rangle$ (open symbols). The dashed line separates the regime of translocation from the regime of detachment. For small k_h (red triangles), translocation velocity saturates at intermediate values of k_d and *v*ParB. Inset: Ratio of the *Z*-component of the radius of gyration of the ParB polymer squared to the *ρ*-component squared (R_z^2 / R_ρ^2). At large k_d, the polymer stretches along the axis of motility. The black dotted line marks the k_d at which the depolymerization time, $1/k_d$, exceeds the effective relaxation time, τ_a (Eq. 16), of the ParB polymer. The green dashed line indicates $R_z^2 / R_\rho^2 = 1/2$, which is expected for an isotropic polymer coil. (B) *v*ParB grows with hydrolysis rate for small k_h and saturates at $k_h \approx k_d$ (indicated by dotted line). This behavior can be fit by $1/$ *v*ParB $= 1/lk_h + 1/ak_d$ (green, see Eq. 17). Inset: Variation of the best-fit length scale, *l*, over ParA subunit diameter, *a*, with k_d. (C) *v*ParB is insensitive to the total drag, ζParB, on the ParB polymer over several orders of magnitude for both fast k_h (black) and slow k_h (red). For very large ζParB, the ParB polymer translocates more slowly. (D) For a fixed quantity of ParB as one component of the polymer, longer polymers move more slowly than shorter polymers for both fast k_h (black circles) and slow k_h (red triangles). Unless noted to be varying, variables have the following values: k_d = 1230s^{-1}, $k_h = \infty$ (black circles) or k_h = 250s^{-1} (red triangles), $\epsilon = 8k_BT$, ζParB = 300ζ_0 = 1.6 × 10^{-4}g/s, L = 500nm, and there are 50 subunits that can bind to ParA in the ParB polymer. In (D), ζParB = (5(L/a–50) + 50)ζ_0.

The penetration length, *l*, depends directly on the shape of the ParB polymer. For large $\tau_a k$ the ParB polymer is pulled rapidly and *l* is small. This is because the ParB polymer is pulled away from the ParA bundle, leading to less overlap of the volume of the ParB polymer with the volume of the ParA bundle. As a result, there is less binding between individual ParB subunits with ParA subunits. As $\tau_a k$ decreases, *l* increases and saturates at l for $\tau_a k \gtrsim 1$ (inset to Figure 4b). In the latter regime, the disassembly rate is $k = k_0$, where

$$ak_0 = (1/\ell_{\max} k_h + 1/ak_d)^{-1} \tag{17}$$

Thus, the translocation velocity is controlled by the effective relaxation time, τ_a, and the maximum disassembly rate k_0.

4.4.9 Three Regimes of Translocation Velocity

It is found that the translocation velocity, vParB, falls into three regimes, depending on $\tau_a k_0$:

$$v_{\mathrm{ParB}} \begin{cases} \approx ak_0 \text{ for } \tau_a k_0 \lesssim 1 \text{ (regime I)} \\ \lesssim ak_0 \text{ for } \tau_a k_0 \gtrsim 1 \text{ (regime II)} \\ = 0 \text{ for } \tau_a k_0 \gg 1 \text{ (regime III)} \end{cases} \tag{18}$$

For $\tau_a k_0 \lesssim 1$ (regime I), the ParB polymer retains its equilibrium shape as it is pulled across the cell at the velocity vParB = ak_0. For $\tau_a k_0 \lesssim 1$ (regime II), the ParB polymer stretches as it is pulled and does not penetrate deeply into the ParA bundle. Since fewer ParA subunits bind to ParB, fewer are hydrolyzed and vParB drops below ak_0. For $\tau_a k_0 \gg$ 1 (regime III), the ParB polymer is so elongated that ParB binds to very few ParA subunits and the ParB polymer quickly detaches from the ParA bundle, leading to vParB = 0.

This physical picture explains the results shown in Figure 4, where both the disassembly rate are varied, k_0 (Figure 4a–b) and the effective relaxation time, τ_a (Figure 4c–d). Specifically, Figure 4a shows how vParB depends on the depolymerization rate, k_d. For the black circles in Figure 4a, the hydrolysis rate, k_h, is effectively infinite so that $k_0 = k_d$ (Eq. 17). In this case, for sufficiently small k_d, the system is in regime I and vParB $\approx ak_0 = ak_d$. As k_d increases, $\tau_a k_0 = \tau_a k_d$ also increases; as a result, the ParB polymer stretches (inset to Figure 4a) and the system crosses into regime II, where vParB drops below $ak_0 = ak_d$. At very large k_d, the system reaches regime III, and vParB = 0.

In contrast, if k_h is small (red triangles in Figure 4a), then k_0 cannot exceed $l_{\max} k_h$ as k_d increases (Eq. 17). Therefore, for small k_h, the ParB polymer remains in regime I, $\tau_a k_0 \lesssim 1$, for all k_d, so that vParB $\approx ak_0$ and translocation is robust for any k_d. Thus, by decreasing the overall rate of disassembly by lowering k_h, the system can achieve robust translocation, albeit at a cost to velocity.

Figure 4b shows how vParB varies as k_h increases. In this case, k_0 saturates to k_d at large k_h (Eq. 17). Since k_d is chosen to be small, $\tau_a k_0 \approx 1$ is found over the entire range of k_h, meaning the system is in regime I and vParB $\approx ak_0$.

The different velocity regimes can also be explored by varying τa instead of k_0. Figure 4c shows that vParB is insensitive to the total drag, ζParB = $k_B T/D$ParB, on the polymer when ζParB and thus τa are small. In this case, $\tau a k_0$ is small, and the system is in regime I. As ζParB increases, $\tau_a k_0$ increases, causing vParB to drop below ak_0 as the system crosses into regime II.

Figure 4d shows the effect of the total contour length, L, of the ParB polymer. For small L, vParB $\approx ak_0$ is constant since the system is in regime I. As L increases, τ_a increases, and when $\tau_a k_0 \gtrsim 1$, vParB crosses into regime II and vParB drops below ak_0.

4.4.10 Dependence of the Translocation Velocity on Binding Energy, Binding Sites, Applied Load, and Other Physical Variables

Figure 5 shows that vParB has a threshold dependence on the ParB–ParA binding energy, ϵ. As shown in Figure 2c, 5, ParB rapidly detaches from the ParA bundle if ϵ is too small. However, as long as ϵ is sufficiently large, the ParB polymer remains attached to the bundle throughout the simulation and translocates with a velocity that is insensitive to ϵ and is set by $\tau_a k_0$ (Eq. 18). Similar behavior is observed as the number of binding sites on the ParB polymer is varied. If there are too few binding sites, the ParB polymer quickly detaches from ParA. Above a threshold value, however, vParB does not sensitively depend on the length of the binding strip (Figure S1). The translocation velocity is also insensitive to the filament density within the ParA bundle, the arrangement of filaments in the bundle, and stiffness of the ParB polymer (Figure S2, S3, S4). Finally, it is also verified that the main results hold when the form of the ParB–ParA binding potential is altered to allow binding by multiple points on ParB and/or ParA subunits.

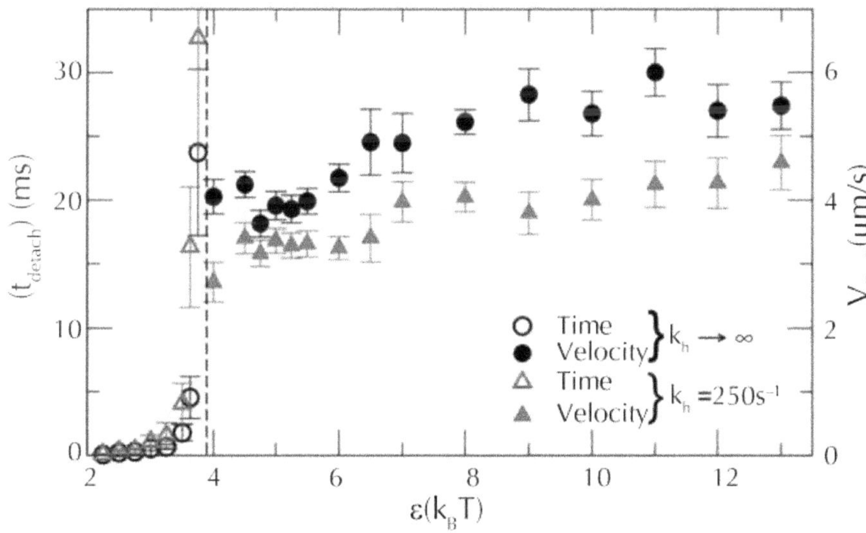

FIGURE 5 Dependence of translocation velocity on ParB–ParA binding energy. ParB detaches in an observably short time, $\langle \tau_{detach} \rangle$, when the binding energy, ϵ, is too small (open symbols). When ϵ is large enough, vParB(solid symbols) is non-zero, and is insensitive to ϵ over the observed range for both fast (black) and slow (red) k_h. The dashed line separates regimes of detachment and translocation.

4.4.11 Detachment Force for the Parb Polymer

It is next investigated that the extent to which motility is robust to an external force on the ParB polymer that opposes translocation. The external force, fext/2, opposes translocation by pulling on each end of the ParB polymer. In this simulations, it is found that vParB is unperturbed for $f_{ext} \gtrsim f^* \approx 10$pN (Figure S5). For $f_{ext} > f^*$, however, the ParB polymer rapidly detaches from the ParA bundle and translocation stalls.

In order to understand this behavior, the "detachment force," f^* is analytically estimated, required to pull the ParB polymer off of the ParA bundle in a time, τ^*, that is approximately equal to the time required for the ParB polymer to translocate across the cell (see Text S1 for details).

In this simulations, we model the ParB-*parS-ori* complex as a polymer chain comprised of N monomeric subunits. Each subunit in the central strip of the ParB polymer binds with a binding energy, $\in$, to a subunit in the ParA bundle. Thus, the total strength of the attraction between the ParB polymer and the ParA bundle is approximately proportional to $n\in$, where n is the number of ParB subunits actually bound to ParA. Since ParB subunits lie in approximately a Gaussian distribution about the center of mass of the ParB polymer [45], $n=n(z_{cm})$, depends on the location, z_{cm}, of the center of mass of the ParB polymer.

Now consider the effect of a force $-f\hat{z}$ on the ParB polymer that opposes translocation in the $\hat{z}$ direction. At the simplest level, based on the above analysis, the ParB polymer may be replaced by a point particle at the center of mass of the ParB polymer, z_{cm}, in an effective potential given by

$$U(z_{\text{cm}}) = -\epsilon n(z_{\text{cm}}) + f z_{\text{cm}} \tag{19}$$

The first term is due to ParB binding to ParA and the second term is the work done by the external pulling force, f. As f increases, the minimum of U shifts to lower values of z_{cm} and the number of bound ParB sites decreases, eventually leading to unbinding of the ParB polymer from the ParA bundle.

The mean time for the particle to escape from the potential well (to detach from the ParA bundle) is well approximated by the Kramers escape time, τ_K for this potential [51, 52]:

$$\tau_{\text{K}} = \frac{2\pi k_B T}{D} \left| U''(z_{\min}) U''(z_{\max}) \right|^{-1/2} e^{(U(z_{\max}) - U(z_{\min}))/k_B T} \tag{20}$$

Given these expressions, the detachment force f^* is calculated to be the force f for which the escape time, τ_K, is equal to τ^*, the time required for the ParB polymer to translocate across the cell.

In simulations with the standard model, the central binding strip has $R = 16$nm and $D = 0.054\ \mu m^2$. There are $N = 50$ ParB subunits that bind to ParA with energy, $\in = 8k_BT$, so the maximum total binding energy is $N\in = 400k_BT$. The ParB polymer translocates at vParB = 5μm/s, so that the time to translocate 1μm is $\tau^* = 200$ms. With these parameters, it is estimated that the detachment force is $f^* \approx 40$pN. An estimate for the detachment force under more realistic conditions (*in vivo*) is given in the Discussion section.

This order of magnitude estimate agrees with our simulations at high depolymerization rates, k_d (Figure 4a), large drag coefficients, ζParB (Figure 4c), and large external pulling forces, f_{ext} (Figure S5). In the first case, the mean time to first detachment is shorter than the translocation time for kd $\gtrsim 3.4 \times 10^5 s^{-1}$; this suggests that the force, *f*, required for rapid detachment is $f^* \approx \zeta_{ParB}\, v_{ParB} \approx 10$ pN. Similarly, it is found that the ParB polymer fails to translocate for $\zeta_{ParB} \gtrsim 5.4 \times 10^{-3}$g/s, giving a detachment force of $f^* \approx 15$pN. In addition, simulations are conducted in which an external force, $f_{ext}/2$ is applied to each of the ends of the polymer. For these simulations, robust translocation up to a detachment force of $f^* \approx 10\pi$N is found.

4.4.12 The ParB Polymer Translocates even When the Para Bundle is not Anchored

So far, it is assumed that the ParA bundle is anchored to the pole. Recent result suggest that in *C. crescentus*, ParA is localized to the swarmer pole by TipN [10, 12], but it is unclear if TipN actually anchors ParA. It is therefore examined whether ParB translocation could occur if the ParA bundle is localized but not anchored.

Figure 3 shows that the ParB polymer translocates even when the ParA bundle is unanchored. This is understood through Newton's third law, which dictates that the force, F_{BA}, that pulls ParB to ParA is equal in magnitude but opposite in direction to the force on ParA. Thus ParB is pulled towards the swarmer pole while ParA is simultaneously pulled away from it:

$$F_{BA} = \zeta_{ParB} v_{ParB} = -\zeta_{ParA} v_{ParA} \quad (21)$$

In the case of a long, unanchored ParA bundle, $\zeta_{ParA} >> \zeta_{ParB}$ and the ParB polymer translocates across the cell while the ParA bundle remains relatively stationary (Figure 3b). However, if the ParA bundle is sufficiently small (*e.g.*, when the ParB has nearly reached the swarmer pole), $\zeta_{ParA} >> \zeta_{ParB}$ is small, so the large ParB polymer remains relatively stationary while pulling the smaller, disassembling ParA bundle towards mid-cell (Figure 3b).

KEYWORDS

- ***Caulobacter crescentus***
- **Chromosome Segregation**
- **Detachment Force**
- **ParA filament**
- **Translocation Velocity**

ACKNOWLEDGMENTS

We thank Rob Phillips for helpful discussions.

REFERENCES

1. Mitchison, T. and Kirschner, M. Dynamic instability of microtubule growth. *Nature* **312,** 237–242 (1984).

2. Walczak, C. E., Cai, S., and Khodjakov, A. Mechanisms of chromosome behaviour during mitosis. *Nat Rev Mol Cell Biol* **11,** 91–102 (2010).
3. McIntosh, J. R., Volkov, V., Ataullakhanov, F. I., and Grishchuk, E. L. Tubulin depolymerization may be an ancient biological motor. *J Cell Sci* **123,** 3425–3434 (2010).
4. Gerdes, K., Howard, M., and Szardenings, F. Pushing and pulling in prokaryotic DNA segregation. *Cell* **141,** 927–942 (2010).
5. Kirkpatrick, C. L. and Viollier, P. H. A polarity factor takes the lead in chromosome segregation. *EMBO J* **29,** 3035–3036 (2010).
6. Ebersbach, G. and Gerdes, K. The double *par* locus of virulence factor pB171, DNA segregation is correlated with oscillation of ParA. *Proc Natl Acad Sci USA* **98,** 15078–15083 (2001).
7. Ringgaard, S., van Zon, J., Howard, M., and Gerdes, K. Movement and equipositioning of plasmids by ParA filament disassembly. *Proc Natl Acad Sci USA* **106,** 19369–19374 (2009).
8. Fogel, M. A. and Waldor, M. K. A dynamic, mitotic-like mechanism for bacterial chromosome segregation. *Genes Dev* **20,** 3269–3282 (2006).
9. Toro, E., Hong, S. H., McAdams, H. H., and Shapiro, L. *Caulobacter* requires a dedicated mechanism to initiate chromosome segregation. *Proc Natl Acad Sci USA* **105,** 15435–15440 (2008).
10. Ptacin, J. L., Lee, S. F., Garner, E. C., Toro, E., Eckart, M. et al. A spindle-like apparatus guides bacterial chromosome segregation. *Nat Cell Biol* **12,** 791–798 (2010).
11. Shebelut, C. W., Guberman, J. M., van Teeffelen, S., Yakhnina, A. A., and Gitai, Z. *Caulobacter* chromosome segregation is an ordered multistep process. *Proc Natl Acad Sci USA* **107,** 14194–14198 (2010).
12. Schofield, W. B., Lim, H. C., and Jacobs-Wagner, C. Cell cycle coordination and regulation of bacterial chromosome segregation dynamics by polarly localized proteins. *EMBO J* **29,** 3068–3081 (2010).
13. Leonard, T. A., Butler, P. J., and Löwe, J. Bacterial chromosome segregation: structure and DNA binding of the Soj dimer – a conserved biological switch. *EMBO J* **24,** 270–282 (2005).
14. Jensen, R. B. and Shapiro, L. The *Caulobacter crescentus smc* gene is required for cell cycle progression and chromosome segregation. *Proc Natl Acad Sci USA* **96,** 10661–10666 (1999).
15. Mohl, D. A. and Gober, J. W. Cell cycle-dependent polar localization of chromosome partitioning proteins in *Caulobacter crescentus*. *Cell* **88,** 675–684 (1997).
16. Viollier, P. H., Thanbichler, M., McGrath, P. T., West, L., Meewan, M. et al. Rapid and sequential movement of individual chromosomal loci to specific subcellular locations during bacterial DNA replication. *Proc Natl Acad Sci USA* **101,** 9257–9262 (2004).
17. Jensen, R. B. Coordination between chromosome replication, segregation, and cell division in *Caulobacter crescentus*. *J Bacteriol* **188,** 2244–2253 (2006).
18. Easter, J. Jr. and Gober, J. W. ParB-stimulated nucleotide exchange regulates a switch in functionally distinct ParA activities. *Mol Cell* **10,** 427–434 (2002).
19. Ebersbach, G., Ringgaard, S., Möller-Jensen, J., Wang, Q., Sherratt, D. J. et al. Regular cellular distribution of plasmids by oscillating and filament-forming ParA ATPase of plasmid pB171. *Mol Microbiol* **61,** 1428–1442 (2006).
20. Hill, T. L. Theoretical problems related to the attachment of microtubules to kinetochores. *Proc Natl Acad Sci USA* **82,** 4404–4408 (1985).
21. Peskin, C. S. and Oster, G. F. Force production by depolymerizing microtubules: load-velocity curves and run-pause statistics. *Biophys J* **69,** 2268–2276 (1995).
22. Raj, A. and Peskin, C. S. The inuence of chromosome exibility on chromosome transport during anaphase A. *Proc Natl Acad Sci USA* **103,** 5349–5354 (2006).
23. Efremov, A., Grishchuk, E. L., McIntosh, J. R., and Ataullakhanov, F. I. In search of an optimal ring to couple microtubule depolymerization to processive chromosome motions. *Proc Natl Acad Sci USA* **104,** 19017–19022 (2007).
24. Armond, J. W. and Turner, M. S. Force transduction by the microtubule-bound Dam1 ring. *Biophys J* **98,** 1598–1607 (2010).

25. Molodtsov, M. I., Grishchuk, E. L., Efremov, A. K., McIntosh, J. R., and Ataullakhanov, F. I. Force production by depolymerizing microtubules: a theoretical study. *Proc Natl Acad Sci USA* **102,** 4353–4358 (2005).
26. Liu, J. and Onuchic, J. N. A driving and coupling "Pac-Man" mechanism for chromosome poleward translocation in anaphase A. *Proc Natl Acad Sci USA* **103,** 18432–18437 (2006).
27. McIntosh, J. R., Grishchuk, E. L., Morphew, M. K., Efremov, A. K., Zhudenkov, K. et al. Fibrils connect microtubule tips with kintetochores: a mechanism to couple tubulin dynamics to chromosome motion. *Cell* **135,** 322–333 (2008).
28. Jacob, F., Brenner, S. and Cuzin, F. On regulation of DNA replication in bacteria. *Cold Spring Harb Symp Quant Biol* **28,** 329–348 (1963).
29. Norris, V. Hypothesis: Chromosome separation *Escherichia coli* involves autocatalytic gene expression, transertion and membrane-domain formation. *Mol Microbiol* **16,** 1051–1057 (1995).
30. Lemon, K. P. and Grossman, A. D. Localization of bacterial DNA polymerase: evidence for a factory model of replication. *Science* **282,** 1516–1519 (1998).
31. Hunding, A., Ebersbach, G., and Gerdes, K. A mechanism for ParB-dependent waves of ParA, a protein related to DNA segregation during cell division in prokaryotes. *J Mol Biol* **329,** 35–43 (2003).
32. Adachi, S., Hori, K., and Hiraga, S. Subcellular positioning of F plasmid mediated by dynamic localization of SopA and SopB. *J Mol Biol* **356,** 850–863 (2006).
33. Jun, S. and Mulder, B. Entropy-driven spatial organization of highly confined polymers: Lessons for the bacterial chromosome. *Proc Natl Acad Sci USA* **103,** 12388–12393 (2006).
34. Elowitz, M. B., Surette, M. G., Wolf, P. E., Stock, J. B, and Leibler, S. Protein mobility in the cytoplasm of *Escherichia coli*. *J Bacteriol* **181,** 197–203.
35. Rapaport, D. C. The art of molecular dynamics simulations. Cambridge, Cambridge University Press, p. 251 (1995).
36. Allen, M. P. and Tildesley, D. J. Computer simulation of liquids. Oxford University Press, pp. 259–264 (1989).
37. Anderson, J. L. Colloid transport by interfacial forces. *Annu Rev Fluid Mech* **21,** 61–99 (1989).
38. Golestanian, R., Liverpool, T. B., and Ajdari, A. Propulsion of a molecular machine by asymmetric distribution of reaction products. *Phys Rev Lett* **94,** 220801 (2005).
39. Golestanian, R., Liverpool, T. B., and Ajdari, A. Designing phoretic micro- and nano-swimmers. *New J Phys* **9,** 126–133 (2007).
40. Lee, K. C. and Liu, A. J. New proposed mechanism of actin-polymerization-driven motility. *Biophys J* **95,** 4529–4539 (2008).
41. Lee, K. C. and Liu, A. J. Force-velocity relation for actin-polymerization-driven motility from Brownian dynamics simulations. *Biophys J* **97,** 1295–1304 (2009).
42. Sullivan, N. L., Marquis, K. A., and Rudner, D. Z. Recruitment of SMC by ParB-parS organizes the origin region and promotes efficient chromosome segregation. *Cell* **137,** 697–707 (2009).
43. Campbell, C. S. and Mullins, R. D. *In vivo* visualization of type II plasmid segregation: Bacterial actin filaments pushing plasmids. *J Cell Biol* **179,** 1059–1066 (2007).
44. Pollard, T. D., Blanchoin, L., and Mullins, R. D. Molecular mechanims controlling actin filament dynamics in nonmuscle cells. *Annu Rev Biomol Struct* **29,** 545–576 (2000).
45. Doi, M. and Edwards, S. F. The theory of polymer dynamics. Oxford, Clarendon Press, pp. 14–16, 22–23, 95 (1986).
46. Cunha, S., Woldringh, C. L., and Odijk, T. Restricted diffusion of DNA segments within the isolated *Escherichia coli* nucleoid. *J Struct Biol* **150,** 226–232 (2005).
47. Elmore, S., Müller, M., Vischer, N., Odijk, T., and Woldringh, C. L. Single-particle tracking of *oriC* - GFP uorescent spots during chromosome segregation in *Escherichia coli*. *J Struct Biol* **151,** 275–287 (2005).
48. Nicklas, R. B. Measurements of the force produced by the mitotic spindle in anaphase. *J Cell Biol* **97,** 542–548 (1983).
49. Westermann, S., Drubin, D. G., and Barnes, G. Structures and functions of yeast kinetochore complexes. *Annu Rev Biohcem* **76,** 563–591 (2007).

50. Ebersbach, G. and Gerdes, K. Bacterial mitosis: Partitioning protein ParA oscillates in spiralshaped structures and positions plasmids at mid-cell. *Mol Microbiol* **52,** 385–398 (2004).
51. Kramers, H. A. Brownian motion in a field of force and the diffusion model of chemical reactions. *Physica (Utrecht)* **7,** 284–304 (1940).
52. Risken, H. The Fokker-Planck quation. Berlin, Springer-Verlag, pp. 96–99, 122–125 (1996).

5 Actin Ring Constriction

Alexander Zumdieck, Karsten Kruse, Henrik Bringmann, Anthony A. Hyman, and Frank Jülicher

CONTENTS

5.1 Introduction 95
5.2 Materials and Methods 97
5.2.1 End-on Imaging of C. Elegans 97
5.3 Discussion 97
5.4 Results 98
5.4.1 Ring Constriction in the C. Elegans Embryo 98
5.4.2 Force Balance and Material Exchange During Ring Contraction 100
5.4.3 Comparison with Experimental Observations 102
5.4.4 Force Generation in Filament Bundles by Molecular Motors 102
5.4.5 Force Generation in Filament Bundles by Filament Polymerization and Depolymerization 104
5.4.6 Continuum Description 105
Keywords 107
References 107

5.1 INTRODUCTION

A physical analysis of the dynamics and mechanics of contractile actin rings is presented. In particular, the dynamics of ring contraction during cytokinesis in the Caenorhabditis elegans embryo is analyzed. A general analysis of force balances and material exchange is presented and estimate the relevant parameter values. It is shown that on a microscopic level contractile stresses can result from both the action of motor proteins, which cross-link filaments, and from the polymerization and depolymerization of filaments in the presence of end-tracking cross-linkers.

During the division of eucaryotic cells, the cortical actin cytoskeleton plays a key role. At the end of mitosis, actin assembles at the site of cell division. As the assembly matures, a bundle of actin filaments and associated proteins is formed [1]. This bundle often forms a ring that encircles the cell. After maturation, the ring contracts and the cell is cleaved, a process called cytokinesis. The position of the cleavage furrow is determined by two signals induced by spindle microtubules [2]. It has been shown that forces are exerted at the cleavage furrow [3] suggesting that contractile stress in the bundle leads to normal forces in the membrane and consequently to the cleavage of the cell.

Cytokinesis is an active process that relies on the action of a number of proteins. Among these are myosin motor proteins that generate mechanical stress in the actin ring [4]. It has been suggested that sarcomere-like contractile elements play an important role in this process [5]. Other proteins involved in cytokinesis affect the nucleation, polymerization and depolymerization of actin filaments [1, 6]. These include capping proteins that stabilize polymerizing ends, formins that nucleate and polymerize actin filaments, the Arp2/3 complex that nucleates filament branches, and ADF/cofilin, which affects depolymerization. Finally, there are bundling and cross-linking proteins like α-actinin.

The kinetics of ring contraction has been observed in different cells. In *Schizosaccharomyces pombe*(fission yeast), the contraction velocity was found to be constant during cytokinesis [6]. In adherent *Dictyostelium* cells the observed contraction velocity decreased exponentially with time [7]. Myosin motors have been shown to play a key role for ring constriction [4]. The non-motor proteins also influence the contractile process. Measurements on fission yeast have shown that the velocity of ring contraction depends on factors that regulate the polymerization and nucleation of actin-filaments [6]. Remarkably, in fission yeast the turnover of actin and the associated proteins is rapid as compared to ring contraction [6].

It has been suggested that ring contraction can be described by a physical model that takes into account the balance of active contractile forces and hydrodynamic friction [7, 8]. In these calculations the contraction velocity was found to decrease with time. Computer simulations of such a model for the first division of sea urchin eggs that explicitly take into account flows of cytoplasmic material on the other hand lead to ring contraction with constant velocities [9].

Key aspects of cytokinesis can be discussed in a simplified geometry focusing on the properties of one-dimensional bundles forming contractile rings. The mechanical properties of filament bundles in the presence of motor proteins have been studied in vitro using purified systems. It has been shown that actin filament bundles in the presence of myosin motors contract spontaneously [10, 11]. Theoretical analysis of the relative filament sliding induced by active cross-linkers (e.g. motor aggregates) has revealed that contractile stresses can be generated even in bundles lacking a sarcomere structure [12–14]. In this case, motors and filaments self-organize to form a contractile filament configuration that is stable. Furthermore, as a result of motor action complex dynamic states can appear.

In this work, a multi-scale description of ring constriction is presented. It consists of a microscopic model and a macroscopic phenomenological description, which are connected through an intermediate continuum description. This phenomenological description is simple. Still, the processes that are considered are sufficient to account for the observations on ring contraction in *C. elegans* and in fission yeast. In particular, it can be predicted from this analysis that the observed constant contraction velocity depends essentially on a sufficiently fast actin turnover. This result is not trivial, as is, for example, demonstrated by ring contraction in *Dictyostelium* [7]. There, the contraction velocity is not constant and processes distinct from the ones we consider have to be taken into account for describing the observed time course. The microscopic model incorporates an essential additional feature compared to previous models for bundle dynamics as it takes into account effects of filament assembly and disassembly, in particular, treadmilling. This extension is important for the dynamics of contractile rings for two reasons: first because this phenomenological analysis shows that a high actin turnover is essential for constant contraction velocities. Secondly, it allows for studying mechanisms of stress generation in contractile rings that do not rely on the action of force generation by myosins. Such mechanisms have so far not been addressed from a theoretical point of view.

5.2 MATERIALS AND METHODS

5.2.1 End-on Imaging of *C. Elegans*

The worms were cultured as described [15]. NMY2::GFP [16] worms were maintained at 16° C and shifted to 25° C for 24h before an experiment. Then the worms were mounted on slides, covered with cell tac (BD Bioscience) in 10mM Tris Cl, pH 8.5. The embryos were put on their ends using a micromanipulator-controlled glass needle. Embryos were filmed using spinning disk microscopy at 23° C as described [2].

5.3 DISCUSSION

In summary, a phenomenological description of the dynamics of a contractile filament ring is presented. This chapter is motivated by the constriction of a contractile ring in eucaryotic cells during cytokinesis. In this description, the ring dynamics is driven by the contractile stress in the ring and accounts for effects of filament turnover. Furthermore, physical mechanisms of stress generation by active processes have been discussed and have shown that in addition to motor proteins the depolymerization of filaments also can contribute to contractile stress in the presence of end-tracking cross-linkers.

This results are compared to the observed ring constriction during the first division of the *C. elegans* embryo. This calculations can account for the observed contraction with constant velocity if filament turnover is sufficiently fast. A constant contraction velocity is also observed in fission yeast. There, the filament turnover rate is also known and sufficiently fast to lead to constant filament density as required for constant velocity.

This phenomenological description can be related to more microscopic models of force generation. This minimal model is based on the pair wise interaction of filaments and is valid if the density of cross-linkers is low. This assumption is most likely not satisfied in the cleavage furrow, where a high density of active elements is able to generate large forces of the order of tens of nano Newtons [17]. In a situation of high cross-linker density, individual filaments experience larger forces since they are transiently linked to clusters of other filaments via active and passive cross-linkers and the generated contractile stresses can increase. This more complex situation is not captured by the simple description. However, in the context of the model, such a situation corresponds to a case where individual filaments feel effective friction forces in the cross-linked filament network. Temporary cross-links keep filaments together only for a certain time. In the presence of forces, filaments will still slide with respect to each other but with significantly reduced velocity. This corresponds to an increase of the viscosity η in this description, which now becomes an effective viscosity that takes friction forces resulting from cross-linking into account. By choosing an appropriate effective viscosity in the bundle, which exceeds the viscosity of the solvent we can describe the contractile stress values relevant to the cleavage furrow. Note that this friction is different from the friction that characterizes ring contraction introduced in the phenomenological description. This latter friction depends on the viscosity of the cytoplasm and cortical protein networks that are formed during ring contraction.

In the minimal model, the contractile bundle stress is generated both by motor proteins as well as by filament depolymerization together with end-tracking cross-linkers. From this comparison to experiments, the material parameter A can be estimated that describes stress generation in the bundle and can be linked to the parameters of the minimal model by $A \simeq \eta \ell \gamma$. Here, n is the effective bundle viscosity, l denotes the typical filament length, and γ is an effective velocity of relative filament sliding, which can be related to the interaction strengths of the active processes as $\gamma = \alpha + \alpha' + \beta'/2$. Using $\gamma = 1$ μm/s, I≈1 μm, and $A \approx 1.2 \times 10^{-12}$Nμm^2 estimated for the *C. elegans* embryo, $\eta \approx 1.2$ Nsm^{-2} is found. This corresponds to about 1200 times the viscosity of water. To describe filament bundles that are highly cross linked and where large numbers of filaments interact simultaneously by active processes remains an important challenge. On a more coarse-grained level, the concept of active gels can provide a general description of the physics of highly cross-linked cytoskeletal systems [18–20].

5.4 RESULTS

5.4.1 Ring Constriction in the *C. Elegans* Embryo

The ring diameter is measured during the first cell division of a developing *C. elegans* embryo as a function of time, see Figure 1A. The radius as a function of time averaged over eight such experiments is displayed in Figure 1B (stars). Error bars indicate one standard deviation. The individual trajectories are displayed in the supporting text S1. The initial cell radius is $R_0 \approx 14.5$ μm and the ring contracts within $T \simeq 4$ min. During most of the contraction process, the ring constricts with a constant speed of $v_c \approx R_0/T \approx 60$ nm/s.

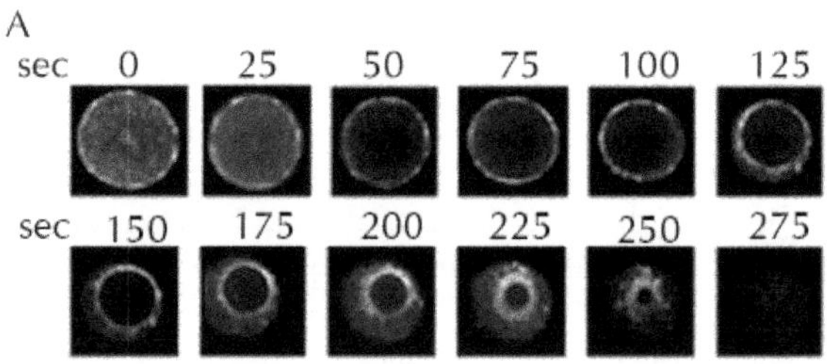

FIGURE 1 *(Continued)*

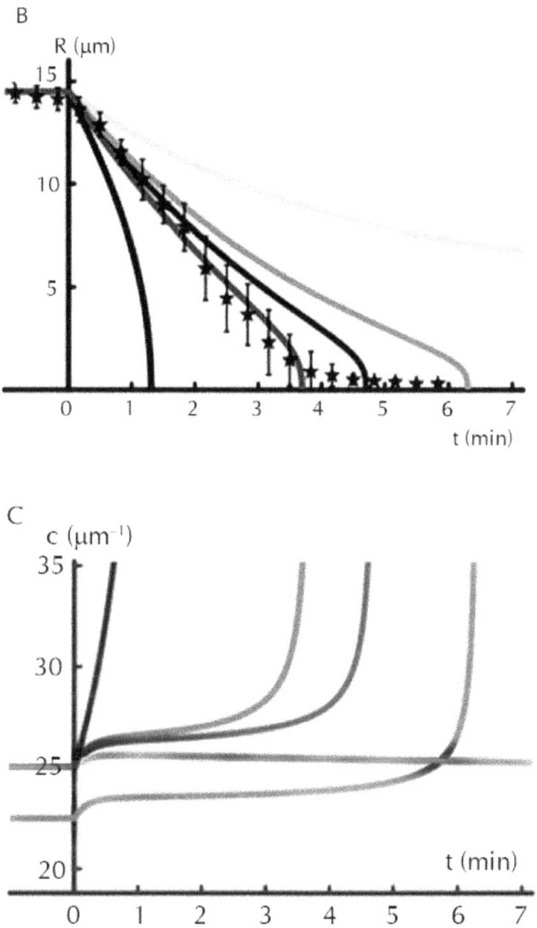

FIGURE 1 Dynamics of ring contraction. (A) Time lapse images of GFP tagged non-muscle myosin 2 during the first division using end-on imaging of a dividing *C. elegans* embryo at 21°C, which show the contracting ring structure. (B) Ring radius as a function of time. Stars indicate the average of eight experimentally observed ring radius traces as shown in (A). Bars mark one standard deviation. The colored lines represent solutions to Eqs. (1)–(4). For sufficiently large rate of filament turnover $k_p = 0.1 s^{-1}$, and using $A = 1.2 \times 10^{-3}$ nN μm^2, the calculated dynamics corresponds well to the observed one (red line). Reducing the polymerization rate to $k_p = 2.25$ s–1 μm^2, which implies a reduced initial filament concentration of $c_0 = 22.5\ \mu m^{-1}$ leads to slower ring contraction (green line). Using $k_d = 0.1\ s^{-1}$ and decreasing the generated contractile stress using A = 1.08 × 10–3 nN μm^2also leads to slower ring contraction (blue line). Using instead $A = 0.6 \times 10^{-3}$ nN μm^2 causes the ring to stop at a finite radius R≈5 μm (yellow line). In the absence of filament turnover, $k_d = 0 s^{-1}$ and with $A = 0.12 \times 10^{-3}$ nN μm^2, the contraction speed increases with time (black line). (C) Filament density c as a function of time shown for the same calculations as in (B). Parameter values are: initial ring radius R0 = 14.5 μm, initial filament density c0= 25 μm^{-1}, Nb = 40, K = 10 nN μm^{-1}, $\zeta = 3.75 \times 10^{-4}\ \mu m\ nN^{-1} s^{-1}$.

5.4.2 Force Balance and Material Exchange During Ring Contraction

Now a general theoretical analysis of the force is presented and material balances during the contraction process. Using a simplified geometry, a contractile ring is considered as wrapped around a cylindrical surface with radius R_0 representing the cell, see Figure 2A. The contractile stress $\sum$ in the ring is generated by active processes and will be specified below. It characterizes the mechanical work δW required to change the ring-diameter by δR, $\delta W = 2\pi\sum\delta R$, and has units of force. The contractile stress $\sum$ in the ring leads to a total force $2\pi\sum$ normal to the cylinder surface, which is balanced by viscous and elastic forces generated by the cell body. The viscous forces associated with the cell constriction is described by $-\zeta^{-1} R$, where R is the time derivative of the ring radius R and ζ^{-1} an effective friction coefficient. The elastic response of the cell is described by the energy $E(R)$. The work required to induce a deformation is thus $(\partial E/\partial R)\delta R$. A typical dependence of E on R is sketched in Figure 2B (black line). It is assumed that the initial radius, $R = R_0$, is locally stable. For simplicity, the elastic energy of cell deformations can be approximated by

$$E(R) = \frac{K}{2}(R - R_0)^2 \tag{1}$$

where K is an elastic modulus of the cell (It has been verified that different choices of $E(R)$ do not change the results). Balancing the active forces of the ring with the viscous and elastic forces exerted by the cell leads to a dynamic equation for the ring radius

$$\dot{R} = -\xi\left(\frac{\partial E}{\partial R} + 2\pi\sum\right) \tag{2}$$

The contractile stress $\sum$ depends on the internal dynamics of actin filaments and the associated proteins in the ring. Electron microscopy suggests that contractile rings consist of several distinct filament bundles [21]. Within each bundle, stress is generated by active processes, such as the action of motor protein aggregates, which link different filaments. Other possible contributions to active stress generation can result from filament polymerization in the presence of end-tracking cross-linkers. As will be shown below, the interaction of filament pairs leads to a quadratic dependence of the contractile stress on the filament density in a bundle. The total contractile stress in the ring can then be written as

$$\sum = A N_b c^2 \tag{3}$$

Here, A is an effective material coefficient that characterizes force generation within a bundle and depends on the density of motors and other associated proteins, as well as on the rates of filament polymerization and depolymerization. The sign of A is chosen so that positive stress is contractile. The number of distinct filament bundles in the ring is denoted N_b, and c denotes the number density of filaments per unit length along a bundle and has units of inverse length. In general, the expression for $\sum$ will involve also other powers in c. In particular, for large densities the generated stress could be proportional to c. For small filament densities, however, the quadratic term should dominate and the main results are not affected by the specific choice of $\sum$.

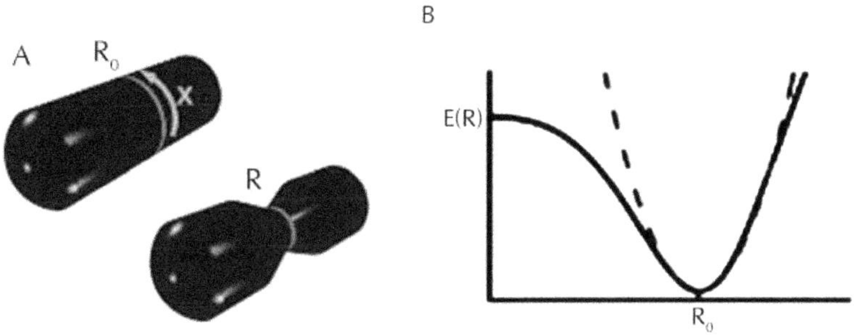

FIGURE 2 Illustration of key quantities used in the theoretical analysis. (A) Schematic representation of a cylindrical cell with a contractile ring of initial radius R0 (left) and contracted state with radius R<R0 (right). The position along the ring is denoted by x. (B) Schematic representation of the elastic energy E(R)as a function of ring radius (black). The dashed red line represents the approximation to E(R) given by Eq. (1), which are used in this calculations.

A filament bundle of length 2πR that consists of N filaments has a density c = N/(2πR). Therefore, the filament density increases with decreasing radius in the absence of polymerization and depolymerization of filaments as c~1/R. If filament turnover is taken into account the filament density obeys the material balance equation

$$\dot{c} = k_p - k_d c - \frac{\dot{R}}{R} c. \tag{4}$$

Here k_p and k_d describe rates of filament assembly and disassembly. Note, that here assembly includes also nucleation of new filaments. For simplicity, these parameters are assumed to be constant, which corresponds for example to a situation where the concentration of nucleators is constant. Equations (1)–(4) provide a physical description of the dynamics of ring contraction.

These equations are solved by numerical integration, starting at time $t = 0$ with $R = R_0$ and a steady state filament density $c_0 = k_p/k_d$, which corresponds to an inactive ring with $A = 0$. Assuming that at $t = 0$ the ring is activated and starts to generate contractile stress according to Eq. (3), A is set to a positive value. The ring then starts to contract as described by Eq. (2). Figure 1B shows the ring diameter as a function of time for different parameter values. In the absence of filament turnover, the contraction velocity increases as a function of time because an increasing filament density leads to higher stress. A finite filament turnover time k_d^{-1} reduces this effect. If this time is short compared to the contraction time, the filament density is controlled by the assembly and disassembly kinetics of filaments, and $c \simeq k_p / k_d$ is roughly constant. In this case, the generated contractile stress $\sum$ remains constant. If contractile stresses dominate over elastic stresses, that is, if $2\pi\sum > KR_0$ as is required for robust contraction, the contraction velocity $v_c \simeq 2\pi \sum \xi$ is constant during contraction, Figure 1B (red line). For reduced contractile stress, contraction can slow down for small radii (blue and green) or even become incomplete (yellow). In the latter case the system reaches a steady state at finite radius at which contractile and elastic stresses balance. The corresponding filament densities are displayed in Figure 1C.

5.4.3 Comparison with Experimental Observations

Using appropriate parameter values, the calculated contraction dynamics corresponds well to the observed one see Figure 1B (red line and stars). The ring in the *C. elegans* embryo contracts with constant velocity $v_c \simeq 0.060 \mu ms^{-1}$. The parameters used in the calculation were k_d= 0.1 s^{-1}, k_p= 2.5 s^{-1} μm^{-1}, N_b= 40, A= 1.2×10^{3} nNμm^{2}, ζ= 3.75×10^{-4} μm nN^{-1}s^{-1}, c_0= 25 μm^{-1}, K= 10 nNμm^{-1}, and R_0= 14.5 μm.

These parameter values can be related to structural, mechanical and kinetic properties of the actin cytoskeleton that have been observed in different systems. For filaments with a length $\ell \simeq 1\mu m$, the turnover rate k_d= 0.1 s^{-1} corresponds to a typical treadmilling velocity of 10 μm/min as is observed in reconstituted assays [22]. This value of k_d is indeed fast compared to the ring contraction and thus ensures constant contraction velocity. The value of the polymerization rate $k_{p=kd}c_0$ is then fixed by the initial filament density. The choice of c_0 and N_b is motivated by electron microscopy studies [21, 23].

The elastic modulus K is related to the whole cell elastic properties characterized by the Youngs modulus Y of the cell. A value of the order of 1 kPa has been found for dividing Ptk2 cells using AFM [24] and in the lamellipodium in fibroblasts using an optical stretcher [25]. Thus, the elastic constant of the cell $K \approx R_0 Y \approx 10^{-2} N/m$ is estimated. The contractile stress of the ring must exceed the elastic force, $\sum > KR_0/(2\pi) \approx 20$ nN. The stress in the contractile ring of echinoderm eggs has been measured by micromanipulation techniques. This suggest $\sum \approx 30$ nN [17]. Using this value $\xi = 3.75\times10^{-4} \mu$mnN$^{-1}s^{-1}$ and A= 1.2×10^{-3}nNμm^{2} is found.

A constant contraction velocity is also observed in fission yeast [6]. In this case the initial cell radius is $R_0 \approx 1.9$ μm. The total contraction time is T ≈ 30 min. The fluorescent signal of GFP-Cdc4, a myosin light chain, does not change remarkably during contraction, which is consistent with a constant ring density. The fluorescence recovery of GFP-Cdc8 (tropomyosin) is shorter than 30s, which suggests a filament depolymerization rate $k_d \approx 0.04$ s^{-1}. Experiments suggest that only one bundle of about 20 filaments exists [1, 26], corresponding to $N_{b=1}$ and fixing k_p= 0.8 μm^{-1}s^{-1}. The contraction speed is $v_c = R_0/T = 2\pi\xi\sum \approx 1\times10^{-3} \mu ms^{-1}$. Assuming the same mechanical properties of an individual filament bundle as discussed above, Σ~0.6 nN is estimated. From this estimate, $\xi = 3\times10^{-4} \mu$mnN$^{-1}s^{-1}$ is obtained. Note, that during cytokinesis in fission yeast a new cell wall is built while the ring contracts. Effects of cell elasticity are ignored and $K \simeq 0$ is assumed. The dynamics that results from Eqs. (1)–(4) using these parameter values is consistent with experimentally observed contraction kinetics (see supporting text S1).

5.4.4 Force Generation in Filament Bundles by Molecular Motors

Now the microscopic origin of stress generation in the contractile ring is discussed. Early experiments on sea urchin eggs have shown that myosin is involved in the generation of stresses in contractile rings [4]. As myosin minifilaments can displace actin filaments with respect to each other, a mechanism based on the sliding of actin filaments of opposite orientation has been proposed for stress generation in the contractile ring [3, 5]. Such a sliding filament mechanism is responsible for stress generation by sarcomeres in skeletal muscle cells [27].

In vitro experiments show that bundles of myosin motors and actin filaments lacking sarcomeric structure can also contract [10, 11]. Theoretical analysis has revealed that contractile stresses can be generated in such bundles if interactions of filaments of the same orientation are taken into account [13, 14]. For the analysis it is assumed that motors act as mobile cross-linkers that temporarily link filament pairs and induce relative sliding between the filaments, see Figure 3. Motor induced sliding of two filaments of opposite orientation will either shorten or lengthen the pair, depending on the relative filament position, see Figure 3B. As a consequence, for a homogenous filament distribution, on average no contractile stress is generated in the bundle by this process. Sliding filaments of the same orientation, however, tend to shorten the filament pair. It can be shown that this process on average generates contractile stresses in the bundle [14] (see Figure 3). Note that this implies that no organized sarcomere-like structure is required for the generation of contractile stresses in a bundle.

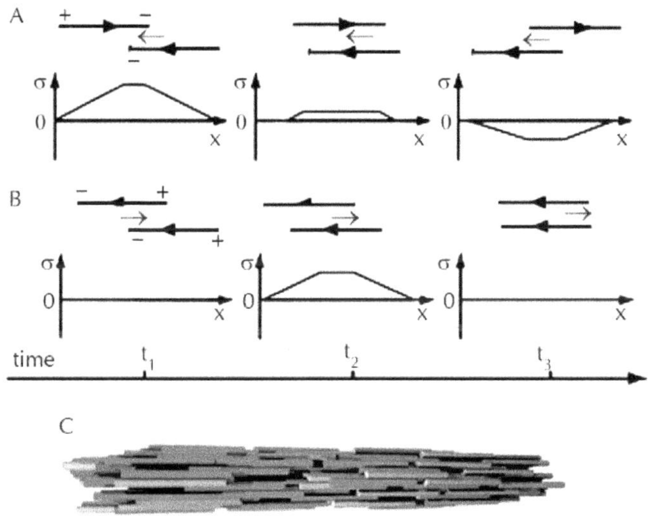

FIGURE 3 Stress generation in a filament bundle induced by plus end directed motor proteins. (A) Three stages of the relative sliding of two filaments of opposite orientation induced by a motor complex. During sliding, stresses result from the balance of motor forces, which are applied at the position where the motor is bound and friction forces that are distributed along the filaments. The combined stress profiles of both filaments σ(x) are displayed as a function of position x. When the centers of mass of the filaments slide towards each other (i.e. the pair contracts) the stress is positive (contractile) on average. When the filaments move apart, the stress is negative. Averaging over many filament pairs and motor position in the bundle, the net stress generated by this process is zero. (B) Filaments of equal orientation tend to align their plus ends if motors stay temporarily attached to the plus end. In this case, a positive (contractile) average stress is generated. Considering the combined effects of many filament pairs in a bundle, this process contributes to contractile bundle stress. (C) Snapshot of a stochastic simulation of a bundle consisting of many filaments that interact via the processes illustrated in (A) and (B). Displayed is a typical configuration in the steady state. Filaments oriented with their plus ends to the right and left are indicated by red and blue rods, respectively. This state has on average constant density as a function of position x and generates a net contractile stress along the bundle.

Motor-induced sliding of filaments of the same orientation occurs e.g. if an aggregate of plus-end directed motors that simultaneously binds to two filaments, reaches the end of one filament and stays attached for some time, see Figure 3B. Other microscopic mechanisms can amplify these effects [28].

This behavior is illustrated by a simple stochastic model for the dynamics of filaments along a common bundle axis. In this model, filaments can interact if they overlap. In a time-step of duration Δt pairs of overlapping filaments are randomly selected and displaced relative to each other in a direction, which depends on their relative orientation and a distance that depends on the interaction strength, see Figure 3 (see supporting text S1 for details). Finally, periodic boundary conditions are imposed, where the system length $L = 2\pi R$ is the ring circumference and assume that all filaments are of the same length l.

The simulations are started with an initially homogenous distribution of filaments. Depending on the value of the interaction strength of parallel filaments α, which has units of inverse time, the system reaches different dynamic states after long times. If α is smaller than a critical value α_c, the filament distribution remains homogenous. In this case, a contractile stress is generated in the bundle (see supporting text S1 for details). For $\alpha > \alpha_c$, the initially homogeneous bundle is unstable and develops into an inhomogenous distribution that is either a stationary distribution of segregated filaments or displays complex dynamics such as propagating density profiles. The critical value $\bar{\alpha}_c$ is positive and depends on the interaction strength β of filaments of opposite orientation.

5.4.5 Force Generation in Filament Bundles by Filament Polymerization and Depolymerization

The polymerization and depolymerization of filaments can also generate forces [29, 30]. In a filament bundle this can contribute to stress generation if end-tracking cross-linker are present [31]. Such proteins can bind along filaments and stay bound to depolymerizing filament ends for some time, Figure 4A. Such forces might be important for cytokinesis in particular in the absence of myosin II motors [32]. The description can be extended to include filament treadmilling and the role of end-tracking cross-linkers. Treadmilling leads to spontaneous motion of filaments with respect to the surrounding fluid, even in the absence of interactions with other filaments. In the presence of end-tracking cross-linkers, treadmilling can induce relative sliding between filaments of the same and opposite orientation, see Figure 4. Simulating this dynamics analogously to the case of motor-induced filament sliding, again the cases of stable homogenous bundles are found and cases where homogeneous bundles are unstable and complex dynamics emerges. The critical value of the interaction strength $\bar{\alpha}'$ mediated by end-tracking cross-linkers depends on the treadmilling velocity v and the interaction strength of filaments of opposite orientation. Remarkably, even in a homogenous bundle the interaction between anti-parallel filaments now leads to a net contractile stress, see Figure 4.

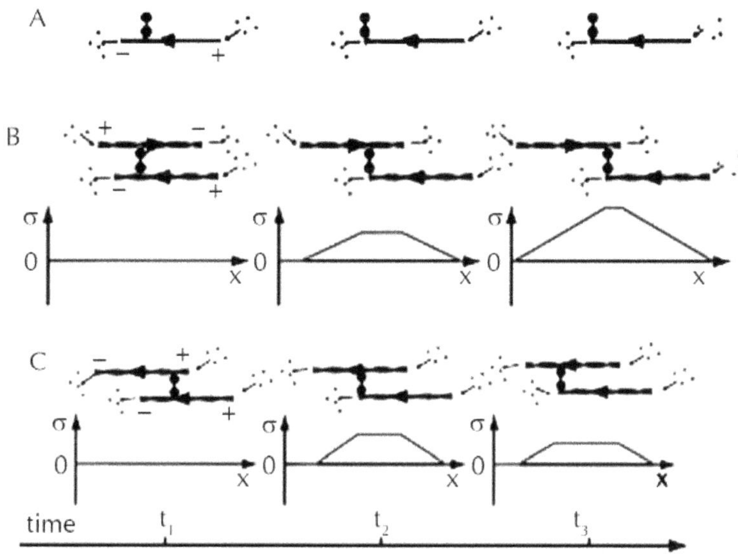

FIGURE 4 Stress generation in a bundle of treadmilling filaments and passive cross-linkers in the absence of motor proteins. (A) Three stages of the treadmiling of a pair of anti-parallel filaments. The arrows at the filament ends indicate that monomers are added at the plus ends and removed from the minus ends. As a consequence of treadmilling, in the absence of cross-linkers, the centers of mass of the filaments move relative to each other. At the same time, fixed subunits along the filaments do not move. Stresses can be generated if filaments are linked by passive cross-linkers, which have the ability to bind to filaments and also stay attached to depolymerizing filament ends (end-tracking cross-linkers). No stress is generated if monomers are cross-linked along filaments since these monomers do not move (left). As soon as one depolymerizing filament end is linked to the second filament by a cross-linker (middle), both filaments are physically moved relative to each other by depolymerization forces. This leads to a stress profile along the filament pair with positive (contractile) stress (middle and right). This process therefore contributes to contractile average bundle stress if many filament pairs are interacting in a filament bundle. Note that this is different from stress generation by motors, where anti-parallel filaments do not contribute to stress. (B) Stress profiles for a pair of treadmiling filaments that are arranged in parallel. Again, if a depolymerizing end of one filament is linked to the second filament, relative sliding occurs which is driven by filament depolymerization. The resulting stress is positive (contractile).

5.4.6 Continuum Description

The description of force and stress generation in filament bundles is stochastic in nature and based on microscopic interactions. In order to find simple expressions for the average stress and to connect the microscopic description with the phenomenological description of ring contraction, a coarse-grained continuum description is used, which can connect the different scales. By characterizing the system via the densities $c^{\pm}(x)$ of filaments pointing with their plus end in the positive and negative x-direction, respectively, a deterministic continuum description is obtained for the dynamics, which is given by equations of the form

$$\partial_t c^+ = D\partial_x^2 c^+ - \partial_x J^+ - k_d c^+ + k_p$$
$$\partial_t c^- = D\partial_x^2 c^- - \partial_x J^- - k_d c^- + k_p.$$

Average sliding of filaments induced by active processes is described by the currents $J^\pm$. The stochastic component of the same processes leads to diffusive motion characterized by an effective diffusion coefficient D. The currents can be expressed in terms of the densities. Interactions of filament pairs correspond to currents, which are quadratic in densities. For the case of motor induced sliding, these currents have been presented in [13, 14]. In the case of filament sliding induced by end-tracking cross-linkers, the currents can be decomposed as $J^\pm = J^{\pm\pm} + J^{\pm\mp}$, where

$$J^{\pm\pm} = \alpha' \int_0^\ell \mathrm{d}\xi \left[c^\pm(x+\xi) - c^\pm(x-\xi)\right] c^\pm(x)$$

describes interactions between filaments of the same orientation and

$$J^{\pm\mp} = \mp\beta' \int_{-\ell}^{\ell} \mathrm{d}\xi c^\mp(x+\xi) c^\pm(x)$$

interactions between filaments of opposite orientation. Here, α and β are effective interaction strengths with units of velocity. They are related to the rates α′ and β′ of the stochastic model by a coarse graining procedure. Similarly, α and β denote the corresponding interaction strengths for motor induced filament sliding. These coefficients capture the effective strength of the respective interactions including protein concentration as well as microscopic details of the interaction.

In this continuum description, stable homogeneous density profiles as well as profiles with propagating and oscillating density distribution are found. In the case of filament treadmilling, also extended regions with stationary filament patterns are found. Furthermore the coarse-grained profiles of the average stress in the bundle can be calculated [14]. In the special case of a homogeneous bundle, is found.

$$\sum_{\text{hom}} = N_b \frac{1}{2} \eta \ell^3 \left\{(\alpha+\alpha')\left[(c_{\text{hom}}^+)^2 + (c_{\text{hom}}^-)^2\right] + \beta' c_{\text{hom}}^+ c_{\text{hom}}^-\right\}$$

Here, n denotes an effective viscosity characterizing the dynamics of filaments in the bundle and l is the filament length. Note that on average interactions induced by motors (interaction strength α) and by end-tracking cross-linkers (strength α′) contribute to contractile bundle stress if filaments have the same orientation. For filaments pointing in opposite directions, end-tracking cross-linkers generate a contribution to stress proportional to β′, while motors do not contribute in this situation. This is due to the symmetry of this interaction that leads to positive (contractile) as well as negative (expansive) stresses in the filament pair, cf. Figure 3A. Note, that while the stress depends explicitly on the filament length l, its dependence on the myosin concentration is implicitly contained in the parameters α, α′, β, and β′.

KEYWORDS

- Actin Ring
- *C. Elegans* Embryo
- Cytokinesis
- Fission Yeast
- Force Balance

REFERENCES

1. Wu, J. Q. and Pollard, T. D. Counting cytokinesis proteins globally and locally in fission yeast. *Science* **310,** 310–314 (2005).
2. Bringmann, H. and Hyman, A. A. A cytokinesis furrow is positioned by two consecutive signals. *Nature* **436,** 731–734 (2005).
3. Rappaport, R. Cytokinesis in animal cells. Cambridge, Cambridge University Press, (1996).
4. Mabuchi, I. and Okuno, M. The effect of myosin antibody on the division of starfish blastomeres. *J Cell Biol* **74,** 251–263 (1977).
5. Schroeder, T. E. Dynamics of the contractile ring. *Soc Gen Physiol Ser* **30,** 305–334 (1975).
6. Pelham, R. J. and Chang, F. Actin dynamics in the contractile ring during cytokinesis in fission yeast. *Nature* **419,** 82–86 (2002).
7. Zhang, W. and Robinson, D. N. Balance of actively generated contractile and resistive forces controls cytokinesis dynamics. *Proc Natl Acad Sci USA* **102,** 7186–7191 (2005).
8. Biron, D., Alvarez-Lacalle, E., Tlusty, T., and Moses, E. Molecular model of the contractile ring. *Phys Rev Lett* **95,** 098102 (2005).
9. He, X. and Dembo, M. On the mechanics of the first cleavage division of the sea urchin egg. *Exp Cell Res* **233,** 252–273 (1997).
10. Takiguchi, K. Heavy meromyosin induces sliding movements between antiparallel actin filaments. *J Biochem (Tokyo)* **109,** 520–527 (1991).
11. Tanaka-Takiguchi, Y., Kakei, T., Tanimura, A., Takagi, A., Honda, M., Hotani, H., and Takiguchi, K. The elongation and contraction of actin bundles are induced by double-headed myosins in a motor concentration-dependent manner. *J Mol Biol* **341,** 467–476 (2004).
12. Kruse, K., Camalet, S., and Julicher, F. Self-propagating patterns in active filament bundles. *Phys Rev Lett* **87,** 138101 (2001).
13. Kruse, K. and Julicher, F. Actively contracting bundles of polar filaments. *Phys Rev Lett* **85,** 1778–1781 (2000).
14. Kruse, K. and Julicher, F. Self-organization and mechanical properties of active filament bundles. *Phys Rev E* **67,** 051913 (2003).
15. Brenner, S. The genetics of Caenorhabditis elegans. *Genetics* **77,** 71–94 (1974).
16. Nance, J., Munro, E. M., and Priess, J. R. *C. elegans* PAR-3 and PAR-6 are required for apicobasal asymmetries associated with cell adhesion and gastrulation. *Development* **130,** 5339–5350 (2003).
17. Rappaport, R. Cell division: direct measurement of maximum tension exerted by furrow of echinoderm eggs. *Science* **156,** 1241–1243 (1967).
18. Kruse, K., Joanny, J. F., Julicher, F., Prost, J., and Sekimoto, K. Asters, vortices, and rotating spirals in active gels of polar filaments. *Phys Rev Lett* **92,** 078101 (2004).
19. Kruse, K., Joanny, J. F., Julicher, F., Prost, J., and Sekimoto, K. Generic theory of active polar gels: a paradigm for cytoskeletal dynamics. *Eur Phys J E* **16,** 5–16 (2005).
20. Kruse, K., Zumdieck, A., and Julicher, F. Continuum theory of contractile fibers. *Eruophys Lett* **64,** 716–722 (2003).

21. Maupin, P. and Pollard, T. D. Arrangement of actin filaments and myosin-like filaments in the contractile ring and of actin-like filaments in the mitotic spindle of dividing HeLa cells. *J Ultrastruct Mol Struct Res* **94,** 92–103 (1986).
22. Pantaloni, D., Le Clainche, C., and Carlier, M. F. Mechanism of actin-based motility. *Science* **292,** 1502–1506 (2001).
23. Schroeder, T. E. The contractile ring and furrowing in dividing cells. *Ann N Y Acad Sci* **582,** 78–87 (1990).
24. Matzke, R., Jacobson, K., and Radmacher, M. Direct, high-resolution measurement of furrow stiffening during division of adherent cells. *Nat Cell Biol* **3,** 607–610 (2001).
25. Park, S., Koch, D., Cardenas, R., Kas, J., and Shih, C. K. Cell motility and local viscoelasticity of fibroblasts. *Biophys J* **89,** 4330–4342 (2005).
26. Kanbe, T., Kobayashi, I., and Tanaka, K. Dynamics of cytoplasmic organelles in the cell cycle of the fission yeast *Schizosaccharomyces pombe*: three-dimensional reconstruction from serial sections. *J Cell Sci* **94**(Pt 4), 647–656 (1989).
27. Huxley, A. F. and Niedergerke, R. Structural changes in muscle during contraction; interference microscopy of living muscle fibres. *Nature* **173,** 971–973 (1954).
28. Kruse, K. and Sekimoto, K. Growth of fingerlike protrusions driven by molecular motors. *Phys Rev E* **66,** 031904 (2002).
29. Dogterom, M. and Yurke, B. Measurement of the force-velocity relation for growing microtubules. *Science* **278,** 856–860 (1997).
30. Mogilner, A. and Oster, G. Force generation by actin polymerization II: the elastic ratchet and tethered filaments. *Biophys J* **84,** 1591–1605 (2003).
31. Dickinson, R. B., Caro, L., and Purich, D. L. Force generation by cytoskeletal filament end-tracking proteins. *Biophys J* **87,** 2838–2854 (2004).
32. Gerisch, G. and Weber, I. Cytokinesis without myosin II. *Curr Opin Cell Biol* **12,** 126–132 (2000).

6 Curved Activators and Cell-Membrane Waves

Barak Peleg, Andrea Disanza, Giorgio Scita, and Nir Gov

CONTENTS

6.1 Introduction ... 109
6.2 Materials and Methods ... 110
6.2.1 Drug Treatment and Staining ... 110
6.2.2 Cell Culture and Reagents ... 112
6.2.3 Immunofluorescence Microscopy and CDRs Counting ... 112
6.2.4 Time Lapse of CDRs ... 112
6.2.5 Model Details ... 112
6.2.6 Linear Stability Analysis ... 114
6.2.7 Non-Linear Simulations ... 117
6.2.8 Strong Concave Activator Approximation ... 118
6.3 Discussion ... 118
6.4 Results ... 127
6.4.1 Experimental Results ... 127
6.4.2 Theoretical Results ... 127
Keywords ... 130
References ... 130

6.1 INTRODUCTION

Cells exhibit propagating membrane waves which involve the actin cytoskeleton. One type of such membranal waves are Circular Dorsal Ruffles (CDR) which are related to endocytosis and receptor internalization. Experimentally, CDRs have been associated with membrane bound activators of actin polymerization of concave shape. Experimental evidence is presented for the localization of convex membrane proteins in these structures, and their insensitivity to inhibition of myosin II contractility in immortalized mouse embryo fibroblasts cell cultures. These observations lead us to propose a theoretical model which explains the formation of these waves due to the interplay between complexes that contain activators of actin polymerization and membrane-bound curved proteins of both types of curvature (concave and convex). Our model predicts that the activity of both types of curved proteins is essential for sustaining propagating waves, which are abolished when one type of curved activator is removed. Within this model waves are initiated when the level of actin polymerization induced by the curved activators is higher than some threshold value, which allows the cell to control CDR formation. It is demonstrated that the model can explain many features of CDRs, and give several testable predictions. This work demonstrates the importance of curved membrane proteins in organizing the actin cytoskeleton and cell shape.

Living cells have the ability to produce propagating waves on their membranes, which are traveling membrane undulations involving an accumulation of the actin cytoskeleton, that persist over microns and during minutes. Such membrane waves have been observed in a variety of cells, during cell spreading [1–3] and in response to excitation by soluble factors [4]. These waves are believed to play a role in cellular motility, probing of the surrounding matrix, endocytosis and internalization of membrane receptors [4]. In the damped liquid environment of the cell, these propagating waves are maintained by the constant supply of active forces from the cytoskeleton. The main type of active force at the membrane is the protrusive force due to the polymerization of actin filaments near the membrane.

The mechanisms responsible for these different waves are not well understood at present. Several theoretical models have been suggested to explain the propagation of actin waves on the membrane of cells [5, 6]. One kind of mechanism that was shown to drive membrane-cytoskeleton waves involves the recruitment to the membrane of actin polymerization by curved membrane proteins (activators). The coupling between the membrane shape and the protrusive force of actin polymerization was shown to produce damped waves when only concave activators are present [7]. In contrast, a model that was able to produce non-decaying waves relied on the addition of contractile forces produced by myosin II motors, in conjunction with only convex actin activators [8]. This model was shown to fit recent experiments [9], where myosin inhibition abolished the observed waves. Conversely, other types of membrane ruffles are insensitive to inhibition of actomyosin contractility or to the genetic removal of myosin II (Supporting movies 7 and 8 of [10]). In order to account for such waves that do not require myosin-driven contractility, it is explored in this chapter, whether only using the protrusive forces of actin polymerization can give rise to non-decaying membrane-cytoskeleton waves. A new mechanism is identified for such waves, which is based on the interplay between curved membrane proteins of both convex and concave shapes, and give a specific biological example where it may apply.

6.2 MATERIALS AND METHODS

6.2.1 Drug Treatment and Staining

In order to test whether CDR induced by PDGF stimulation are dependent on an intact actomyosin contractile system, mouse embryo fibroblasts (MEF) were serum-starved and pre-treated with vehicle or Y-27632 (10 μM, 30′), a specific inhibitor of ROCK kinase, that regulates myosin light chain kinase and MLC-based contractility [11], or Blebbistatin (50 μM, 30′), a small molecule inhibitor showing high affinity and selectivity toward myosin II [12] (Figure 1a,b). Cells were subsequently treated with 10 ng/ml of PDGF for 7 min, which potently and synchronously induces CDR formation [13] in MEFs. Cells were then fixed and stained with rhodamine-phalloidin to detect F-actin and visualize CDR. The percentage of MEFs exhibiting CDRs were counted. Data is expressed as mean + SD (Figure 1b). To detect the localization of IRSp53 in CDRs, cells were fixed and stained with anti-IRSp53 antibody (green) and rhodamine-phalloidin to detect F-actin (red) (Figure 1c).

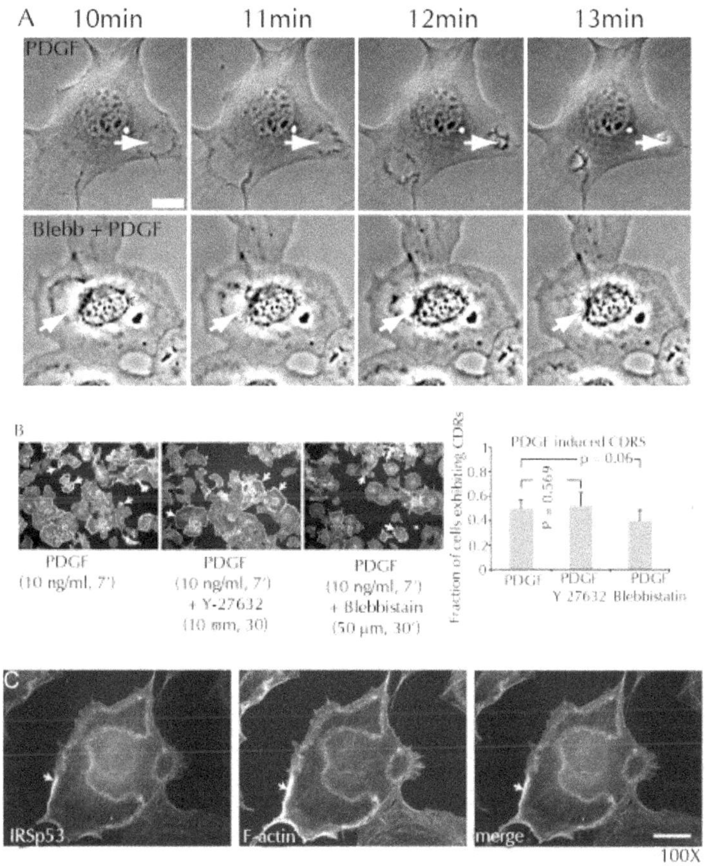

FIGURE 1 Experimental results. Experiments done in MEF cells which are stimulated by PDGF. (A) Time-lapse of CDRs dynamics. Still images of MEF cells serum-starved and pre-treated with vehicle (upper panels) or Blebbistatin (lower panels) and subsequently treated with PDGF to induce CDRs formation. CDR dynamics were recorded by time-lapse video microscopy (see also Movie S1 and Methods section). Bar, 20 μm. (B) The fraction of cells exhibiting CDRs is unaffected by treatment with two different myosin II inhibitors. P-values show no statistical significance. (C) IRSp53 is localized at CDRs. IRSp53 marked in green and actin in red. Bar 10 μm. Arrows denotes CDRs.

6.2.2 Cell Culture and Reagents

Mouse embryo fibroblasts (MEFs) used in the experiments were derived as described in [14] from Eps8 null mice. MEFs were cultured in DMEM-Glutamax-1 medium supplemented with 10% FBS, 1×Pen-Strep. IRSp53 knockout cells were spontaneously immortalized cells from IRSp53 knockout mouse embryos infected either with pBABE-puro or pBABE-puro-IRSp53 [15]. MEFs were cultured in DMEM-Glutamax-1 medium supplemented with 20% FBS, 1×Pen-Strep, and 1μm/ml puromycin. The monoclonal anti-IRSp53 was generated against the full-length His-tagged purified protein [16]. PDGF was from Immunological Science (Rome, Italy), Blebbistatin from Sigma-Aldrich (St. Louis, MO, USA), Y-27632 from Tocris Bioscience (Ellisville, MO, USA).

6.2.3 Immunofluorescence Microscopy and CDRs Counting

Cells seeded on gelatin were serum starved for two hours and then treated with PDGF for 7 min. Cells were then processed for indirect immunofluorescence microscopy. Briefly, cells were fixed in 4% paraformaldehyde for 10 min, permeabilized in 0.1% Triton X-100 and 0.2% BSA for 10 min, and then incubated with the primary antibody for 45 min, followed by incubation with the secondary antibody for 30 min. F-actin was detected by staining with rhodamine-phalloidin (Sigma-Aldrich, St. Louis, MO, USA) at a concentration of 6.7 U ml^{-1}. The number of cells exhibiting CDRs upon PDGF treatment was counted. At least 500 cells in each experiment performed in triplicate were analyzed (mean s.e.m.).

6.2.4 Time Lapse of CDRs

MEFs cells seeded on gelatin were serum-starved for two hours and then pre-treated with vehicle or Blebbistatin. Cells were treated with PDGF and subjected to time-lapse video microscopy at 37°C, 5% CO_2 using an Olympus IX81 microscope (40X objective) connected to a Photometrics cascade 1K camera. Images were taken every 5 s for 20 min. Reduction of the area of each CDR was monitored over time using Image-J software and from the relation between the area and time the reduction in the average radius could be extracted, by assuming a circular shape. Then the change in the CDR radius was used at the beginning of the shrinking, to calculate the velocity.

6.2.5 Model Details

The membrane is characterized by height undulations $h(\vec{r})$ (Monge representation in the limit of small undulations), while the area coverage fractions of the convex and concave activators are $\phi-(\vec{r})$ and $\phi+(\vec{r})$, with spontaneous curvature $H-<0$ and $H+>0$. The dynamics are governed by the Helfrich Hamiltonian [17] where the bending energy is proportional to the mismatch between the mean membrane curvature (∇^{2h}) and the spontaneous curvature of the curved activators (up to quadratic order)

$$\mathcal{H}=\int_S \frac{\kappa}{2}\left(\nabla^2 h-H_-\phi_- - H_+\phi_+\right)^2+\frac{\sigma_{\mathrm{eff}}}{2}(\nabla h)^2 d^2 r. \tag{1}$$

where κ is the membrane's bending modulus and σ_{eff} is an effective surface tension which includes contributions due to the spontaneous curvature and entropy of the activators (details in Text S1).

It is assumed that the pushing force of actin polymerization is linearly proportional to the activators' density

$$f_{\text{actin}}(\vec{r}) = A_{+}\left(\phi_{+}(\vec{r}) - \bar{\phi}_{+}\right) + A_{-}\left(\phi_{-}(\vec{r}) - \bar{\phi}_{-}\right) \tag{2}$$

where $A_{\pm}$ is a proportionality constant that gives a measure of the activity of the actin polymerization induced by the respective activator and $\bar{\phi}$ is the average concentration. It will be assumed in this work that the values of $A_{\pm}$ are uniform throughout the domain and constant in time. The $-A_{\pm}\bar{\phi}_{\pm}$ terms in Eq .2 are equivalent to a uniform displacement of the entire membrane (Galilean transformation) which does not change the shape evolution. Similar results were obtained when the analysis was carried out using an osmotic pressure restoring force (see Text S1).

The elastic forces acting on the membrane are derived variationally from the free energy, which is the energy (Eq. 1) plus the entropy of the activators. Together with the forces due to actin polymerization (Eq. 2) is obtained.

$$\frac{\partial h}{\partial t} = \frac{d}{4\eta}\left(-\frac{\delta \mathcal{F}}{\delta h} + f_{\text{actin}}\right) \tag{3}$$

Assuming local hydrodynamic interactions, where η is the viscosity of the fluid surrounding the membrane and d is the typical extent of the hydrodynamic interactions [8, 18], which represents the effective distance of fluid flow between the membrane and the cytoskeleton elements [19]. This approximation of local hydrodynamic interactions is more relevant for a membrane near a dense network of actin filaments, which is the situation for membranes that are deformed by the cortical actin cytoskeleton [20]. Note that Eq. 3 describes how the membrane shape is locally dependent on the activators' distribution which promote the actin protrusive force, leading to an increase in h (feedback scheme Figure 2c).

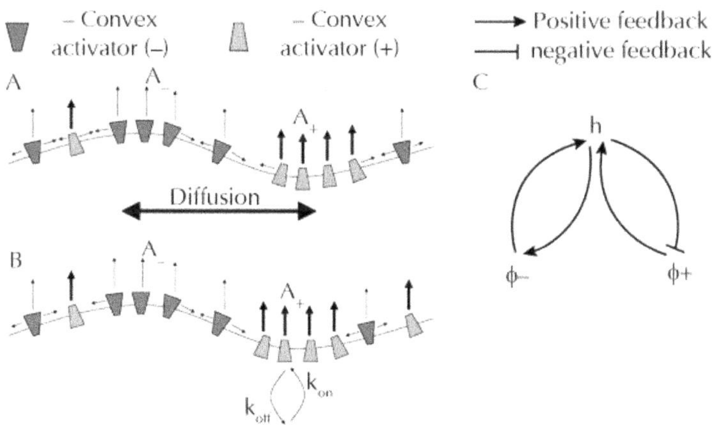

Figure 2. Schematic description of the model. (a) The activator diffuses in the membrane. (b) The activator adsorbs to the membrane from an infinite reservoir. (c) Feedback diagram describing the main interactions in our model, where positive and negative feedback loops combine to produce oscillations.

Two distinct cases were consider for the dynamics of the activators, either diffusive in the membrane or adsorptive from the cytoplasm. For the case of diffusive dynamics the total amount of activators is conserved, so the equation of motion derived from the free energy (details in Text S1) is given by

$$\frac{\partial \phi_i}{\partial t} = \frac{D_i}{n_i^s T} \vec{\nabla} \cdot \left[\phi_i \vec{\nabla} \left(\frac{\delta \mathcal{F}}{\delta \phi_i} \right) \right] \tag{4}$$

where $i = +, -$, D_i is the diffusion coefficient of the curved activator, n_i^s is the saturation concentration i.e. the maximal concentration at which these complexes cover the whole cell membrane and T is the temperature. Note that the current of activators in response to the local membrane curvature, is proportional to: $H_i \phi_i \vec{\nabla}(\nabla^2 h)$. This term in Eq. 4 describes how the diffusive activators' distribution depends on the local membrane shape (curvature), since this current of activators carries them towards regions where the membrane curvature matches their spontaneous shape (feedback scheme Figure 2c).

For the case of adsorptive dynamics, the rate constants of the binding/unbinding process are governed by a Boltzmann factor of the mismatch in the bending energy between the local membrane curvature and the activator's spontaneous curvature

$$\frac{k_{\text{on}}^i}{k_{\text{off}}^i} = \exp\left[\mu - \frac{\kappa}{n_i^s T} (\nabla^2 h - H_i)^2 \right] \tag{5}$$

where μ is the chemical potential describing the affinity for adsorption on a membrane of matching curvature, and the equation of motion for φ_{i} is of first-order kinetics in the form

$$\frac{\partial \phi_i}{\partial t} = k_{\text{on}}^i - k_{\text{off}}^i \phi_i \tag{6}$$

where it is assumed that the cytoplasmic concentration of curved activators is approximately constant and uniform due to the fast diffusion of proteins in the cytoplasm, compared to the typical oscillation time of the waves. For small undulations of the membrane, the equation is linear in the curvature $\nabla^2 h$. Eq. 6 describes how the adsorptive activators' distribution depends on the local membrane shape (curvature), since they adsorb in regions where the membrane curvature matches their spontaneous shape (feedback scheme Figure 2c).

6.2.6 Linear Stability Analysis

For the linear stability analysis the equations of motion are linearized, for all types of dynamics (Eqs. 3, 4, 6). We expand in small deflections around the uniform steady-state, where the membrane is flat and the uniform concentrations are $\bar{\phi}_{\pm}$. The domain of wave instability is bounded by the red and brown lines in Figure 3 (calculated in Text S1). In this region there are oscillatory unstable modes where: Re$\{\omega_i\} \neq 0$ and Re $\{\omega_i\} > 0$. The amplitude of these modes grow exponentially from small initial perturbations, and oscillate or propagate on the membrane surface. The system is stable below the red line, such that initial perturbations decay exponentially: $\mathrm{Re}\{\omega_i\} < 0$.

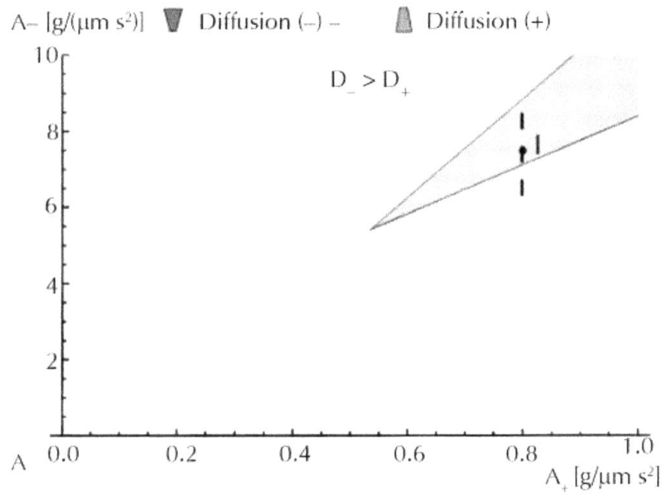

A

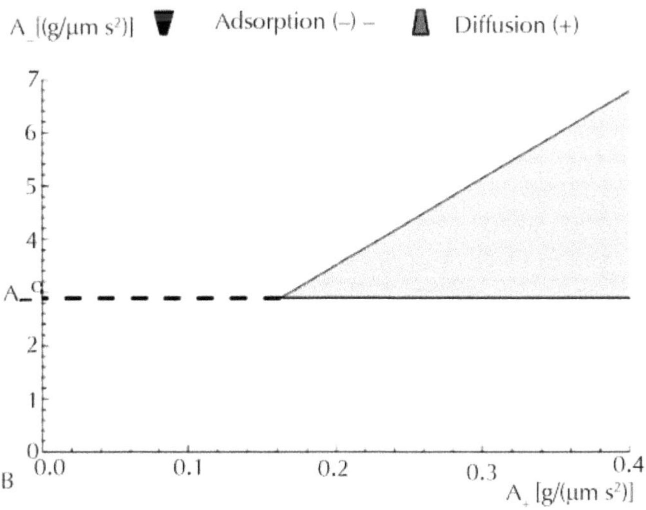

B

FIGURE 3 *(Continued)*

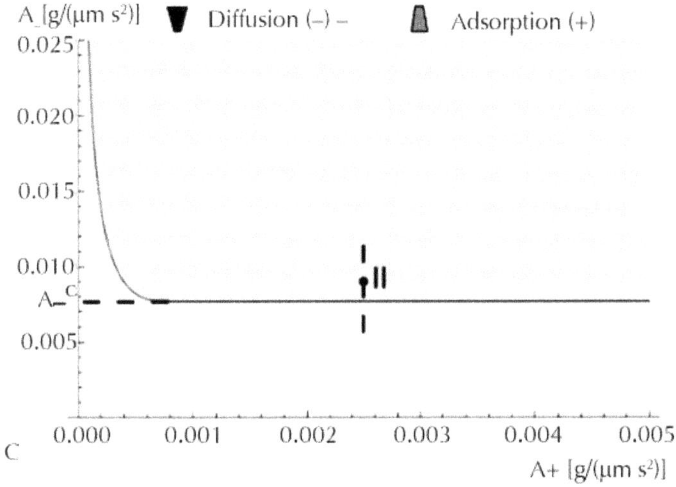

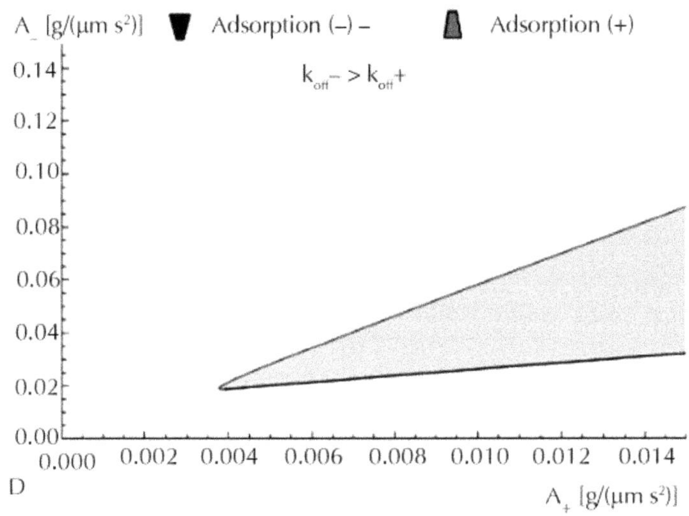

FIGURE 3 Wave instability phase diagram in the A_{-}–A_{+} plane. Regions marked in pink denote the unstable waves. (a) the diffusion(−)–diffusion(+) model, when $D_{-}>D_{+}$. (b) the adsorption(−)–diffusion(+) model. (c) the diffusion(−)–adsorption(+) model. (d) the adsorption(−)–adsorption(+) model when $k_{off}^{-} > k_{off}^{+}$. In (a) and (c) the dashed line marks the values along which the bifurcation graph (Figure 4) was plotted. In (b) and (c) the threshold value of A_{-} is denoted by A_{-}^{c}.

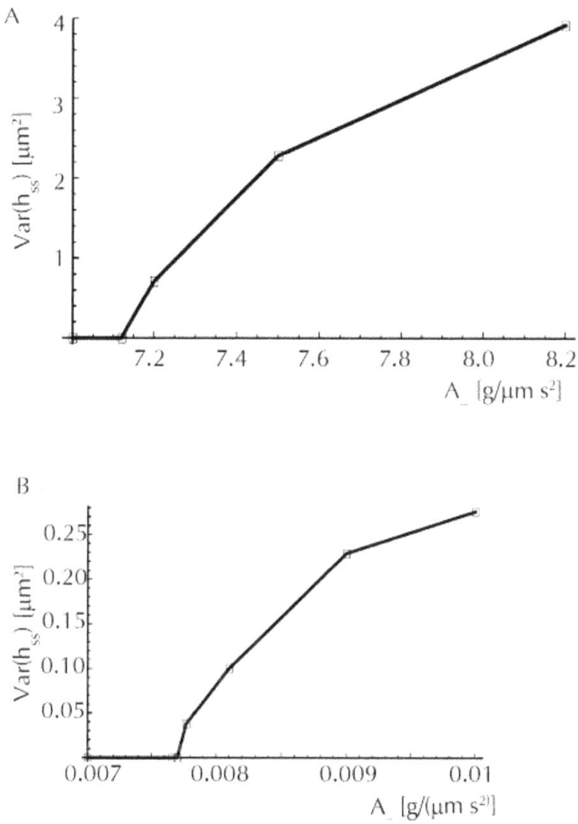

FIGURE 4 Bifurcation analysis. The mean square amplitude of the membrane height displacement in the two systems: (a) diffusion(−)–diffusion(+), (b) diffusion(−)–adsorption(+), along the vertical dashed lines in Figure 3a,c respectively. The amplitude of the steady-state waves continuously vanishes as we approach the wave instability transition line from above (supercritical bifurcation).

6.2.7 Non-Linear Simulations

The one-dimensional simulations are done using a finite-difference scheme for the full nonlinear model with translational symmetry, using Matlab. Periodic boundary conditions were used, and the initial perturbation in the membrane shape was Gaussian (uniform initial distributions of the activators). The exponential growth in the amplitude of the membrane wave is arrested in the real cell due to the finite membrane area, which is described by adding a non-linear tension term [21], in the form: $\sigma \propto \exp \beta(L/L_0 - 1)$, where L is the total membrane length, L_0 is the initial length and β is the non-linear coefficient. A value of β is used which limited the amplitude of the waves to be of order $1\mu m$, as is estimated for CDRs.

6.2.8 Strong Concave Activator Approximation

In the diffusive(−)–adsorptive(+) model, for strong concave activator levels ($A_+ >> 1$) a deeper understanding of the source of the wave velocity can be gained. In this limit Eqs. 3, 6 can be simplified, neglecting the effect of the forces due to the convex activator, and get

$$\frac{\partial h}{\partial t} \approx \frac{d}{4\eta} A_+ \phi_+ \tag{7}$$

$$\frac{\partial \phi_+}{\partial t} \approx k_{\mathrm{off}}^{+} \bar{\phi}_+ \left(1 + \frac{2H_+ \kappa}{n_+^s T} \nabla^2 h\right) \tag{8}$$

From these equations a wave equation can be derived of the form

$$\frac{\partial^2 h}{\partial t^2} = \frac{d}{4\eta} A_+ k_{\mathrm{off}}^{+} \bar{\phi}_+ \left(1 + \frac{2H_+ \kappa}{n_+^s T} \nabla^2 h\right) \tag{9}$$

with the wave velocity given in Eq. 10. In this limit the dispersion relation is acoustic-like, it is almost linear in q.

6.3 DISCUSSION

Experimental evidence given here demonstrates that CDRs contain curved membrane proteins of both curvatures which are furthermore known to be involved in the recruitment of actin polymerization to the membrane. In addition, myosin II contractility was shown not to be an essential component of such waves, and its inhibition does not change the wave velocity. Our theoretical model demonstrates that indeed actin protrusive forces induced by the interplay of these two types of membrane-bound curved activators is sufficient to give rise to propagating membrane waves (Figure 2c). Therefore this result suggests that this could be the dominant mechanism for CDRs.

The following can be made more quantitative comparisons between the waves that our model gives and the observed CDRs.

1. For the cases where the concave activator is adsorptive, the waves in our model have a typical wavelength of order of a few microns (for "rule of thumb" parameter values, Table 1), which is similar to the width of observed CDRs [22, 23].
2. The experimentally observed wave velocity is in reasonable agreement with the the range of velocities that are observed in the model.
3. The concave and convex activators are displaced within the propagating CDR, such that the convex activator is localized at the membrane protrusion, while the concave activators are localized where the membrane is depressed (Figure 5b,f). This may explain the observation that Tuba trails the actin front in CDRs [23].

TABLE 1 List of parameters used in the calculations.

Parameter	Units	Value[a]	Parameter	Units	Value[a]
T	K	300	$D_-{}^{b}$	$\mu m^2/s$	1.3, 1
$H_+ = H$	μm^{-1}	10	$D_+{}^{c}$	$\mu m^2/s$	1
$H_- = -\alpha H$	μm^{-1}	-1	κ	$g\ \mu m^2/s^2$	$5\ k_B T$
$\alpha = \|H_-\|/H_+$	a.u.	0.1	$\mu^{d,e}$	a.u.	-1
d	μm	0.1	$k_{off}^{-}{}^{d}$	s^{-1}	0.02, 0.01
$\eta = 100\eta_{water}$	$g/(\mu ms)$	10^{-4}	$k_{off}^{+}{}^{e}$	s^{-1}	0.01
$\bar{\phi}_-{}^{b}$	a.u.	0.5, 0.8	$n^s_- = n^s_+$	μm^{-2}	500
$\bar{\phi}_+{}^{c}$	a.u.	0.5, 0.8	β	a.u.	10

[a]Dynamic constants were estimated from [32] and spontaneous curvatures from [16,21]. Other values are of typical magnitude for cells.
[b]First number corresponds to diffusion(−)-diffusion(+) model and the second number corresponds to the diffusion(−)-adsorption(+) model.
[c]First number corresponds to diffusion(−)-diffusion(+) model and the second number corresponds to the adsorption(−)-diffusion(+) model.
[d]Relevant for adsorption(−)-diffusion(+) model.
[e]Relevant for diffusion(−)-adsorption(+) model.
doi:10.1371/journal.pone.0018635.t001

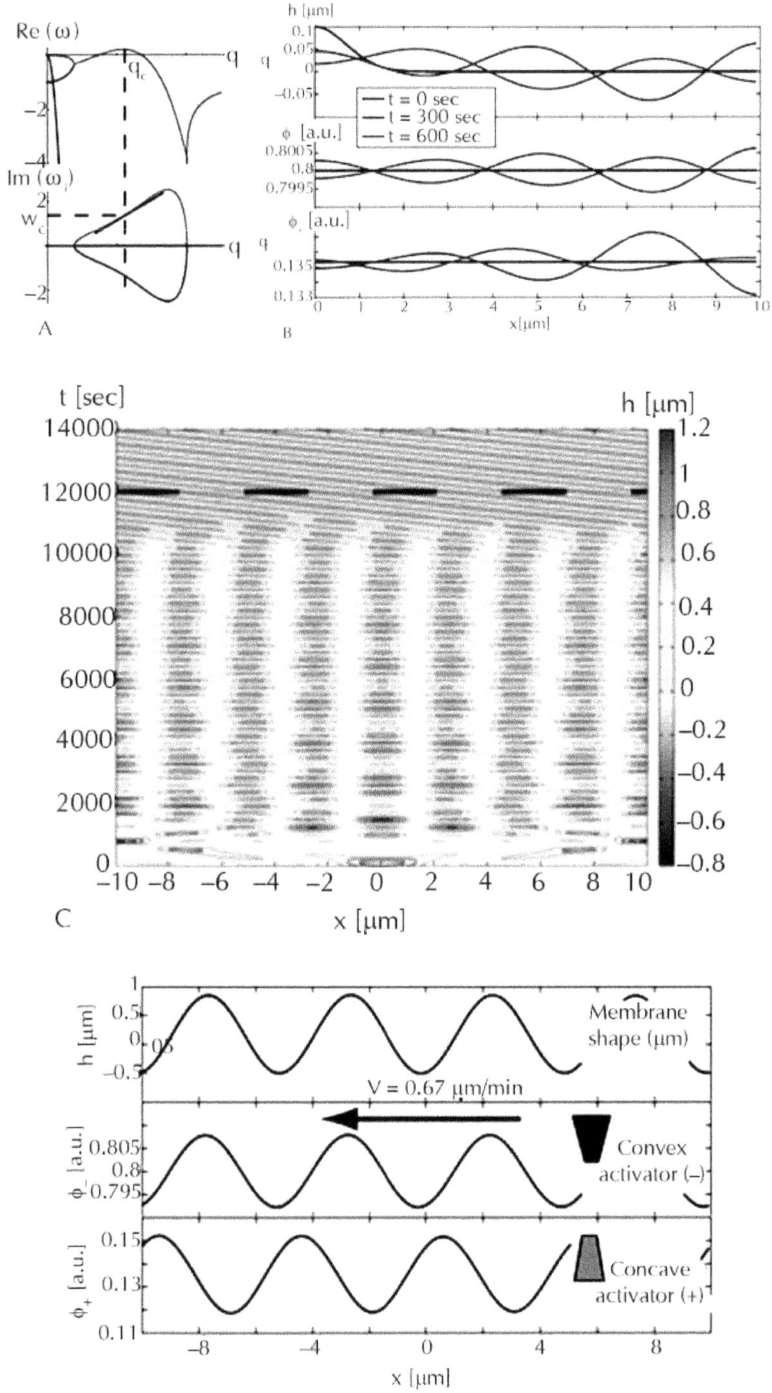

FIGURE 5 *(Continued)*

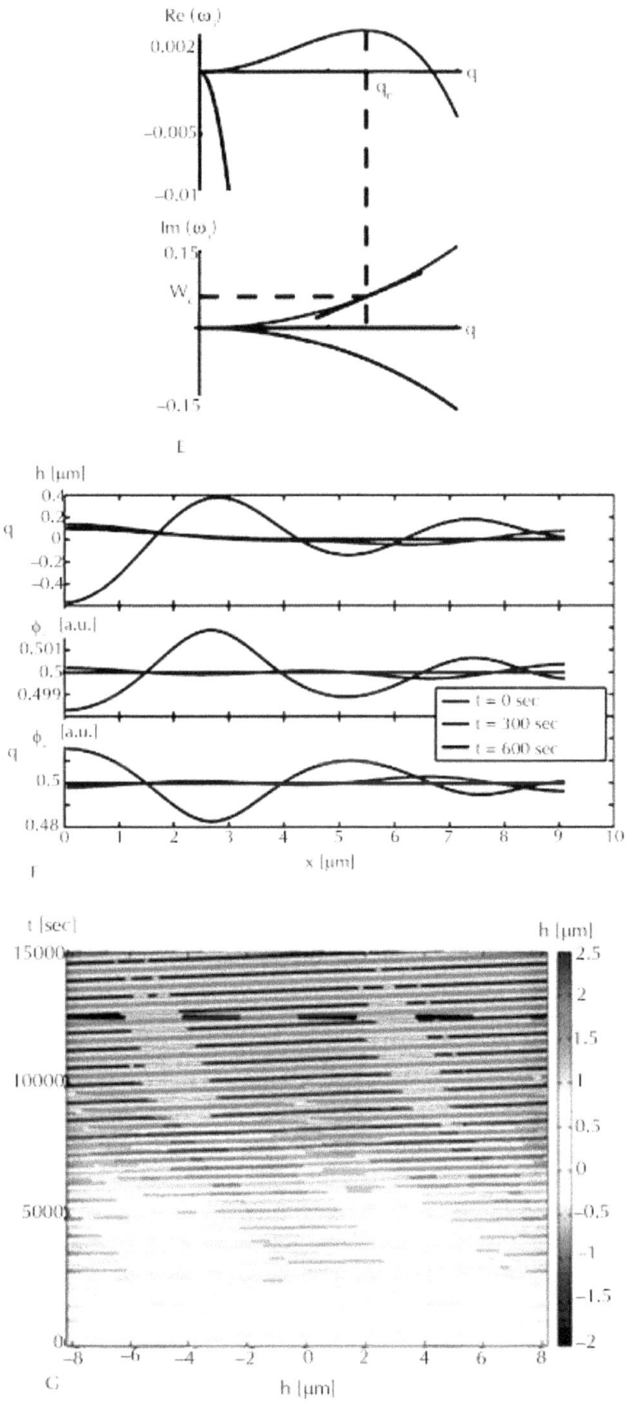

FIGURE 5 *(Continued)*

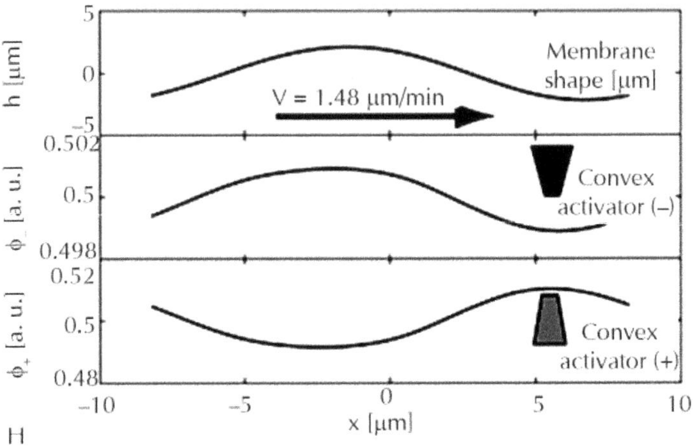

FIGURE 5 Linear stability and simulation results. (a–d) Results of the diffusion(−)–adsorption(+) system. (a) Dispersion relation of point marked II in Figure 3c. Vertical dashed line mark q_c and horizontal dashed line marks ω_c. The slope of the imaginary part of the dispersion relation at q_c gives us an estimate of the group velocity of the waves V. (b) Simulation for short times. One can see that the convex activators are in-phase with the membrane while the convex activators are in anti-phase. Due to symmetry only half of the domain is shown. (c) Kymograph depicting the membrane height displacement as a function of space and time. (d) Steady state wave at time t = 12,500 s (marked by the dashed line in (c)). Arrow shows direction of propagation. (e–h) Results of the diffusion(−)–diffusion(+) system. (e) Dispersion relation of point marked I in Figure 3a. Vertical dashed line marks q_c and horizontal dashed line marks ω_c. (f) Simulation for early times (as in (b)). (g) Kymograph depicting the membrane height displacement as a function of space and time. (h) Steady state wave at time t = 12,000 s (marked by the dashed line (g)). Arrow shows direction of propagation. The simulations are shown in Movies S2 and S3 respectively.

These comparisons support the validity of our model for CDR, and may further indicate that the concave complex (e.g. containing Tuba) is more slowly diffusing in the membrane compared to the convex complex (e.g. containing IRSp53).

Regarding the velocity of the waves in our model, Eq. 10 shows that it depends on both the passive parameters of the system (such as the membrane elasticity and fluid viscosity) and on the average concentration and activity of the concave activators ($\bar{\phi}_+, A_+$). This expression highlights that the wave phenomenon that is described is a result from an interplay between the active forces due to actin polymerization and the passive reaction of the system. Note that the approximate expression that is derived for the wave velocity (Eq. 10) is reminiscent of the expression that appears for myosin-II driven membrane waves (Eq. 4 in [8]).

Our model gives the following insight about the process of CDR excitation in cells. Before the cell is excited its internal parameters correspond to a point in the stable regime of the phase diagram (below the red line in Figure 3). When it is excited the stimulation changes the internal parameters, for example the activity of the actin activators ($A_{\pm}$), and above some threshold values the system crosses into the unstable-wave region. An alternative possibility could have been that the cell can be close enough to the transition line (in the stable regime), such that a large perturbation switches it to the propagating wave state. This route does not exist within our non-linear model, as illustrated in Figure 4. This means that the difference between a quiescent cell and an excited cell with CDRs is a real change in the internal state of the cytoskeleton activity, and not simply a large perturbation of the membrane-cytoskeleton organization.

Let us discuss some assumptions that have been used in this model. It is assumed that the actin polymerization induced by the curved activators ($A_{\pm}$) is spatially uniform. However, there are mechanisms in the cell that can make this parameter vary in space since it may depend on the local membrane curvature [24] and signaling pathways [25]. Our model demonstrates that even without this added level of complexity propagating waves can form. Furthermore, our simulations were done in a regime where the amplitude of the concentration undulations of the activators in the waves are small (Figure 5d,h), and as a result the waves are purely periodic in space. In comparison, the observed CDRs are solitary (Figure 1) and the actin activators are highly localized in the CDR. Nevertheless, the conditions that allow the system to support waves are independent of the amplitude of the wave (Figure 4), so our conclusions remain unaffected. As soon as the membrane tension is reduced and the membrane amplitude is allowed to form stronger gradients, a complete depletion of activators is obtained from certain regions of the membrane, and this indicates that the system has then the tendency to form isolated structures, similar to the solitary waves observed experimentally. A simulation of a solitary propagating structure, which shows that such structures indeed tend to form in our model, is shown in Movie S5. This regime remains to be explored in future studies.

The different versions of our model (Figure 3) give different behavior for the propagating waves, as can be seen in the final wavelengths in Figure 5. Future experiments may allow to distinguish between the different versions of our model. One example for such a discriminating observation between the models is shown in Figure 6, where the calculated dependence of the wave group velocity is plotted on the actin polymerization activity. This actin activity may be modified experimentally by using a variety of actin inhibitors or promoters, which would therefore change both A_- and A_+. The plotted trajectory is schematic, as it assumes a simple linear relation between the response of both types of activators to the drug.

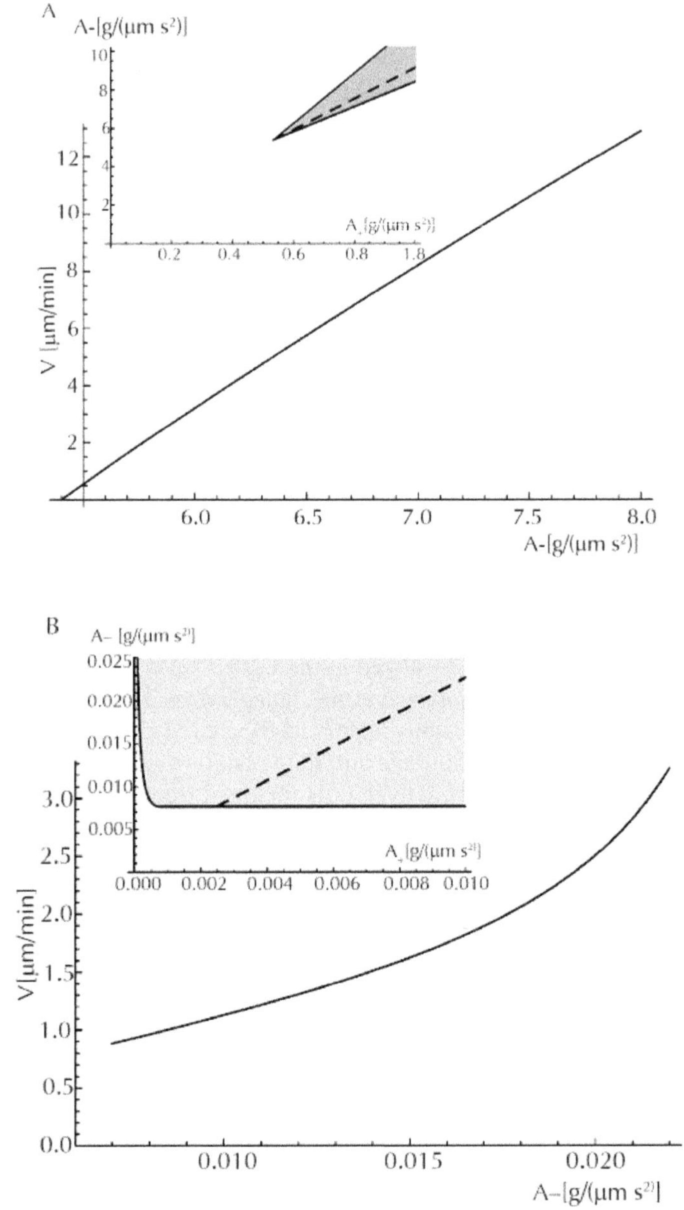

FIGURE 6 Group velocity dependence on A_-. The group velocity dependence along the dashed lines in the insets: (a) the diffusion(−)–diffusion(+) model, when $D_- > D_+$. (b) the adsorption(−)–diffusion(+) model. This trajectory represents the effects of addition of actin polymerization inhibitors or promoters. In both cases it is found that the wave velocity increases with the actin activity, but in a very different manner. This prediction can serve to differentiate between the different types of activator dynamics described by our model.

This model can be used to make the following list of observable predictions: (i) functional or genetic interference with one type of curved proteins (assuming non-redundant roles among proteins of the same type of curvature, see Text S1 section 2) should inhibit CDR formation, (ii) the two types of curved activator complexes are spatially displaced within the CDR, following the undulation in the membrane shape (Figure 5b,f), (iii) the phase diagrams shown in Figure 3 may be explored systematically by controlling the rate of actin polymerization in the cell (note that drugs such as Latrunculin A would change both A_- and A_+, Figure 6), (iv) the expression levels of the two types of activators may be regulated artificially and would change the behavior of the cell (shown in Figure 7a), (v) the CDR velocity should increase roughly as a square-root of the activity of the concave activator, A_+ (Eq. 10, Figure 7b), and (vi) change of the membrane tension will change the velocity of the CDR and the threshold value of A_- for wave instability (Figs. 7c,d respectively).

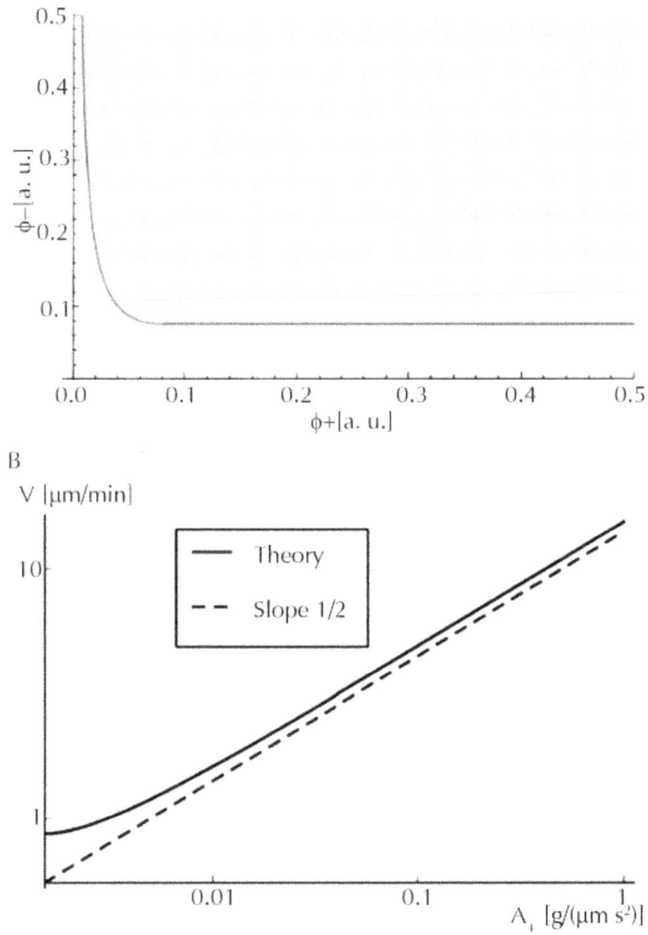

FIGURE 7 *(Continued)*

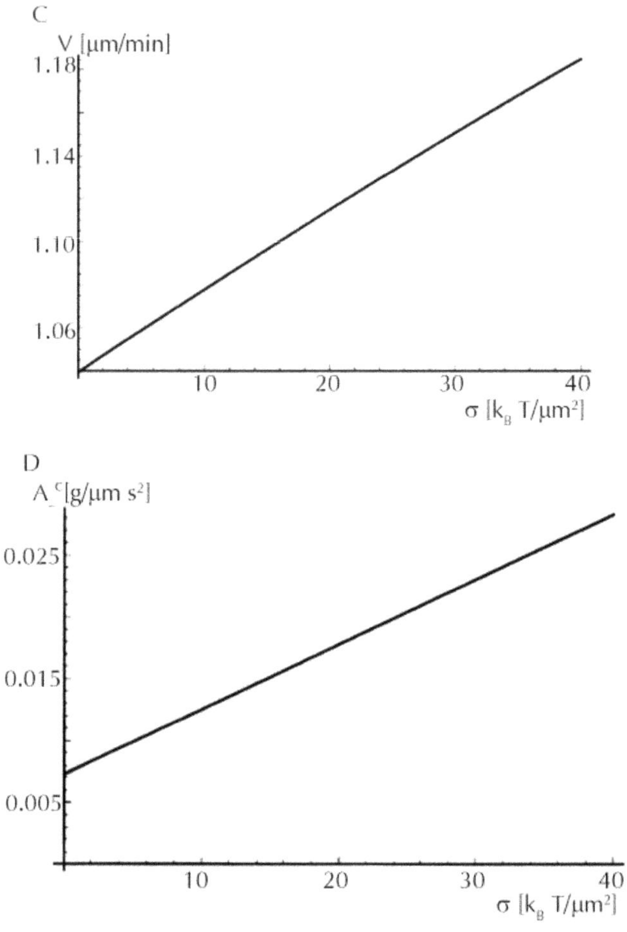

FIGURE 7 Predictions for the diffusion (−)–adsorption (+) model. (a) Wave instability phase diagram in the φ_{-}–φ_{+} plane. It is very similar to the phase diagram in the A_{-}–A_{+} plane (Figure 3c). (b) Log-log plot of the dependence of the group velocity at q_c (Figure 5a) on the parameter A_{+}, along the wave instability transition line in Figure 3c. The dashed line gives the approximate expression for the velocity, given in Eq. 10. (c) The dependence of the group velocity at q_c (for $A_{-} = 0.0076 g/\mu ms^2$, $A_{+} = 0.0035\ g/(\mu ms^2)$), on the membrane tension. (d) The dependence of the threshold value A_{-}^{c} (Figure 3c) on the membrane tension.

In Figure 7b the accuracy of the approximate expression for the wave group velocity given in Eq. 10 can be judged, as a function of A_{+}, by comparing to the group velocity at q_c.

A physical model is presented that demonstrates how actin protrusive forces induced by the interplay of membrane-bound curved activators of both convex and concave curvatures, can give rise to propagating membrane waves. This is a new mechanism for membrane-cytoskeleton waves, and may be the dominant driving force for CDRs. Our model explains many of the observed features of CDRs and provides testable experimental predictions. The theoretical model, together with the experimental observations, demonstrate the essential role played by curved membrane proteins that recruit actin polymerization as organizers of the cortical actin cytoskeleton. Unlike other cellular structures that have been shown to contain such proteins [26, 27], it is demonstrated that proteins of both curvatures are necessary to drive propagating waves.

6.4 RESULTS

6.4.1 Experimental Results

In this chapter, it is focused on the phenomenon of Circular Dorsal Ruffles (CDR), which form on the apical surface of cells as circular actin rings that eventually enclose, generating an endocytic vesicle [4] (Figure 1). These CDRs are involved in internalization of the membrane and its receptors, and are induced by ligand stimulation of membrane receptors, mainly of the tyrosine kinase family. These dynamic structures are driven by actin polymerization, which is initiated by membrane bound activators, such as N-WASP and WAVE complex [4, 22]. CDRs are formed in response to excitation of the cell by growth factor.

In order to test whether CDRs are dependent on actomyosin contractility, as suggested in [8], mouse embryo fibrobalsts were treated with two types of myosin II inhibitors (Y-27632 and Blebbistatin), and showed that CDRs are largely independent of actomyosin contractility (Figure 1a,b). The observed velocities for CDRs in normal and blebbistatin-treated cells are 2.3 ± 0.4 and 1.6 ± 0.6 μm/s respectively. This difference in velocities is not statistically significant (see Movie S1).

There has been evidence that the actin activator N-WASP is recruited to CDRs by a curved membrane protein called Tuba [23]. Tuba is a protein that contains the Bin/Amphiphysin/Rvs (BAR) domain [28], which is known to bend membranes in a concave shape [29]. In addition, a new experimental observations are presented that indicate the localization in CDRs of IRSp53 protein (Figure 1c), which contains the Missing–in–metastasis (MIM) domain, and induces convex membrane shape [30]. This protein was also shown to have the ability to recruit actin activating proteins [31].

6.4.2 Theoretical Results

Motivated by these observations, a model for CDRs is proposed here, which is based on the interplay between two types of protein complexes that contain an activator of actin polymerization and a curved membrane protein; one type is convex while the other is concave in shape (Figure 2). For example, one such concave complex may contain Tuba and N-WASP [23], and a convex complex may contain IRSp53 and WAVE [31]. Note that the minimal model is explored here that contains just one type of activator of each type of curvature (concave and convex), while in the real cell many different proteins of both curvatures coexist and may play a role in CDR formation, as it is indicated in Text S1.

In this model the following three components are included (Figure 2): the flexible cell membrane, and the concentration fields of the membrane-bound activators of the two types of curvatures. The membrane has the usual bending and stretching elasticity, and is assumed to be flat when there are no activators present. The activators induce a spontaneous curvature on the membrane, proportional to their local concentration.

The membrane is further pushed by actin polymerization, which is proportional to the local concentration of the activators. In turn, the dynamics of the activators is influenced by the membrane shape, causing the activators to aggregate where the local membrane shape more closely matches their spontaneous curvature. In the cell the activators both diffuse in the membrane and adsorb from the cytoplasm. In order to analyze the influence of the two processes separately and to keep the analysis simple, it will be assumed that each activator can be either diffusive or adsorptive but not both (Figure 2a,b). All possible sets of different types of dynamics are analyzed. This is a mean-field, continuum model, whereby the small-scale shape of the membrane is not described due to the individual activators, but treat only the averaged (coarse-grained) membrane shape.

The feedback mechanisms (Figure 2c) that operate in our model, couple the distribution of the curved activators on the membrane to the membrane shape (curvature). The activators tend to localize where the membrane has a curvature that matches their spontaneous shape, while they in turn modify the membrane shape due to the forces that they apply; one force is simply due to their shape which tends to curve the membrane, and the other, active force is due to the recruitment of actin polymerization, and is purely protrusive. The convex activators alone can give rise to a positive feedback with the local membrane deformation, whereby they tend to form membrane protrusions in which they are highly localized [7, 18], but do not propagate laterally. The concave activators alone give rise to a negative feedback with the membrane deformation, resulting in damped oscillations [7]. Combining the two types of activators can give rise to unstable waves, whereby the convex activators initiate a protrusion, which is then modified by the aggregation of concave activators that tend to inhibit the local instability, but end up only shifting it laterally in space. This is how the propagating waves arise in our model from the interplay between the positive and negative feedbacks of the two curved activators and the membrane shape.

The membrane is characterized by height undulations $h(\vec{r})$, while the area coverage fractions of the convex and concave activators are denoted by $\phi(\vec{r})$ and $\phi+(\vec{r})$. The proportionality factors relating the local concentration of activators to the protrusive actin force that they induce, are denoted by $A_{\pm}$ respectively. The activator dynamics will be denoted by the dynamics of the convex followed by the dynamics of concave activator, e.g. diffusion(−)–adsorption(+). We are looking for the regimes of parameters where the system supports undamped propagating waves. Linear stability analysis is used to map the regimes of parameters where the system becomes unstable, and complement this analysis with simulations that include the non-linearity due to conservation of the diffusive activators (Eq. 4). It is found that indeed the model which is described has regimes in which unstable waves arise, even in the limit of small perturbations (linear analysis).

The linear stability of the system is analyzed as a function of the activity levels of the two activators, i.e. in the A_{-}–A_{+} plane, in Figure 3 (parameters used in these calculations are given in Table 1). The system is chosen to analyze in terms of these parameters because cells can regulate the activity of the actin cytoskeleton through a variety of signaling pathways [4], and these are also experimentally accessible. In Figure 3 it shows only the regions of wave instability, and a more detailed analysis of these phase diagrams is given in Text S1. The following general conclusions can be drawn from the phase diagrams in Figure 3

1. When the dynamics of both activators is of the same type (both adsorptive or diffusive - a, d), it is seen that for unstable waves to arise the convex activator (φ_-) needs to have faster dynamics than the concave activator (φ_+). The convex activator is the one responsible for the instability in our system, as it has a positive-feedback with the membrane shape (Figure 2), and it therefore needs to respond faster to the membrane deformations, as compared to the concave activators which have a negative feedback with the membrane shape.
2. In all the cases it is found that unstable waves occur above some minimal value of both A_- and A_+. Note that for all the cases except the diffusion(−)–adsorption(+), the unstable waves disappear for A_- above some critical value (a,b,d).
3. When the activators have different types of dynamics (b, c) the transition from damped waves to unstable waves is given approximately by a constant threshold value of A_-, denoted by A^c (red line). In both cases this critical value increases with increasing membrane tension. Only for case (c), it is found that above a critical value of the membrane tension, unstable waves appear even for vanishing A_+.

Now it is explored in more details the cases of diffusive(−)–adsorptive(+) (a) and diffusive(−)–diffusive(+) (c) dynamics. In Figure 5, the dynamics of the waves are given for parameter values that support unstable waves (points marked II and I in Figure 3a, c respectively). The dispersion relation and the time evolution simulation of the waves are plotted both for short times and at the final steady-state, from an initial small perturbation. In the dispersion relations the modes that support unstable waves are characterized by having a non-vanishing imaginary part, and a positive real part.

From the dispersion relation for the diffusive(−)–adsorptive(+) case (Figure 5a) it is found that the unstable waves exist for a limited range of wavelengths, around q_c>0. It is shown in Figure 5b the result of a simulation for short times, where it is found that the most dominant wavelength that propagates away from the initial perturbation is indeed $\lambda_c = 2\pi/q_c$, which has the largest positive real part in the dispersion relation and is therefore the most unstable mode (Figure 5a). An approximate expression for q_c is given in Text S1. It is found from this expression that the wavelength λ_c depends more strongly on the activity of the convex activator, as $\lambda_c \propto A_-^{-1/2}$. It depends very weakly on the activity of the concave activator A_+.

A simulation for the long time evolution of the waves is shown in Figure 5c. It is found that the initial perturbation induces counter-propagating waves and therefore a standing-wave pattern fills the domain, at the most unstable wavelength λ_c, with an oscillation period which is close to that predicted by the linear dispersion relation (ω_c in Figure 5a). Eventually, numerical noise breaks the symmetry of the counter-propagating waves, and a single traveling wave persists at wavelength λ_c (Figure 5d). The time it takes the system to break the symmetry is determined by noise, which is not included explicitly in these calculations. The velocity of this wave is V~0.7μm/min, which is smaller by about 30% compared to the group velocity predicted by the slope of the dispersion relation at q_c (Figure 5a). A good approximation for the wave velocity is given by

$$V_{\text{approx}} \approx \sqrt{\frac{dA_+\phi_+\kappa H_+k^+_{\text{off}}}{2\eta n^s_+ T}} \tag{10}$$

See Materials and Methods section for the definition of the different parameters and the derivation of this expression. As is shown in Eq. 10, the velocity increases with the strength of the active forces (A_+), and the rate of activator turnover (k^+_{off}), as well as with the membrane bending modulus (κ). The velocity decreases for increasing fluid viscosity (η). From this approximation it is understood that the velocity depends very weakly on the activity of the convex activators (A_-). The accuracy of this approximate expression is discussed below.

In Figure 5e–h the analysis of the diffusive(−)–diffusive(+) system is plotted. The main difference in this system is that the unstable waves extend to infinite wavelengths (Figure 5e). At short times (Figure 5f) the most unstable wavelength (λ_c) dominates, but non-linear interactions eventually cause the largest wavelength possible in the domain to persist (Figure 5g,h). The velocity of this wave is V~1.48 μm/min, which is smaller by about 40% compared to the group velocity predicted by the slope of the dispersion relation at the wavelength of steady-state wave.

In both cases it is found that in the propagating waves the convex activator (φ_-) is in-phase with the membrane displacement, while the concave activator (φ_+) is almost in anti-phase (Figure 5d,h).

In Figure 4 the mean-square amplitude of the steady-state membrane waves is plotted as a function of the activity of the convex activators, moving along the vertical dashed lines in Figure 3a,c. It is found that the amplitude of the steady-state waves continuously vanishes as we approach the wave instability transition line (red lines in Figure 3a,c) from above (supercritical bifurcation).

KEYWORDS

- **Actin Cytoskeleton**
- **Blebbistatin**
- **Membrane Waves**
- **Rhodamine-Phalloidin**
- **Tuba**

REFERENCES

1. Döbereiner, H. G., Dubin-Thaler, B. J., Hofman, J. M., Xenias, H. S., Sims, T. N. et al. Lateral membrane waves constitute a universal dynamic pattern of motile cells. *Phys Rev Lett* **97,** 10–13 (2006).
2. Machacek, M. and Danuser, G. Morphodynamic profiling of protrusion phenotypes. *Biophys J* **90,** 1439–1452 (2006).

3. Coelho Neto, J., Agero, U., Oliveira, D. C. P., Gazzinelli, R. T., and Mesquita, O. N. Real-time measurements of membrane surface dynamics on macrophages and the phagocytosis of leishmania parasites. *Exp Cell Research* **303,** 207–217 (2005).
4. Buccione, R., Orth, J. D., and Mcniven, M. A. Foot and mouth: podosomes, invadopodia and circular dorsal ruffles. *Nat Rev Mol Cell Biol* **5,** 647–657 (2004).
5. Carlsson, A. E. Dendritic actin filament nucleation causes traveling waves and patches. *Phys Rev Lett* **104,** 4–7 (2010).
6. Weiner, O. D., Marganski, W. A., Wu, L. F., Altschuler, S. J., and Kirschner, M. W. An actin-based wave generator organizes cell motility. *PLoS Biol* **5,** e221 (2007).
7. Gov, N. S. and Gopinathan, A. Dynamics of membranes driven by actin polymerization. *Biophys J* **90,** 454–469 (2006).
8. Shlomovitz, R. and Gov, N. S. Membrane waves driven by actin and myosin. *Phys Rev Lett* **98,** 168103 (2007).
9. Chen, C. H., Tsai, F. C., Wang, C. C., and Lee, C. H. Three-dimensional characterization of active membrane waves on living cells. *Phys Rev Lett* **103,** 1–4 (2009).
10. Zhang, W. and Robinson, D. N. Balance of actively generated contractile and resistive forces controls cytokinesis dynamics. *Proc Natl Acad Sci USA* **102,** 7186–7191 (2005).
11. Hirose, M., Ishizaki, T., Watanabe, N., Uehata, M., Kranenburg, O. et al. Molecular dissection of the rho-associated protein kinase (p160rock)-regulated neurite remodeling in neuroblastoma n1e-115 cells. *J Cell Biol* **141,** 1625–1636 (1998).
12. Straight, A. F., Cheung, A., Limouze, J., Chen, I., Westwood, N. J. et al. Dissecting temporal and spatial control of cytokinesis with a myosin ii inhibitor. *Science* **299,** 1743–1747 (2003).
13. Lanzetti, L., Palamidessi, A., Areces, L., Scita, G., and Di Fiore, P. P Rab5 is a signalling gtpase involved in actin remodelling by receptor tyrosine kinases. *Nature* **429,** 309–314 (2004).
14. Scita, G., Nordstrom, J., Carbone, R., Tenca, P., Giardina, G. et al. Eps8 and e3b1 transduce signals from ras to rac. *Nature* **401,** 290–293 (1999).
15. Weiss, S. M., Ladwein, M., Schmidt, D., Ehinger, J., Lommel, S. et al. Irsp53 links the entero-hemorrhagic *e. coli* effectors tir and espfu for actin pedestal formation. *Cell Host Microbe* **5,** 244–258 (2009).
16. Disanza, A., Mantoani, S., Hertzog, M., Gerboth, S., Frittoli, E. et al. Regulation of cell shape by cdc42 is mediated by the synergic actin-bundling activity of the eps8-irsp53 complex. *Nat Cell Biol* **8,** 1337–1347 (2006).
17. Helfrich, W. *Z Naturforsch* **C 28,** 693–703 (1973).
18. Veksler, A. and Gov, N. S. Phase transitions of the coupled membrane-cytoskeleton modify cellular shape. *Biophys J* **93,** 3798–3810 (2007).
19. Gov, N., Zilman, A., and Safran, S. Hydrodynamics of confined membranes. *Phys Rev* E **70,** 011104 (2004).
20. Giannone, G., Dubin-Thaler, B. J., Rossier, O., Cai, Y., Chaga, O. et al. Lamellipodial actin mechanically links myosin activity with adhesion-site formation. *Cell* **128,** 561–575 (2007).
21. Sens, P. and Safran, S. A. Pore formation and area exchange in tense membranes. *Europhys Lett* **43,** 95–100 (1998).
22. Legg, J., Bompard, G., Dawson, J., Morris, H., Andrew, N. et al. N-wasp involvement in dorsal ruffle formation in mouse embryonic fibroblasts. *Mol Biol Cell* **18,** 678 (2007).
23. Kovacs, E. M., Makar, R. S., and Gertler, F. B. Tuba stimulates intracellular n-wasp-dependent actin assembly. *J Cell Sci* **119,** 2715–2726 (2006).
24. Takano, K., Toyooka, K., and Suetsugu, S. Efc/f-bar proteins and the n-wasp-wip complex induce membrane curvature-dependent actin polymerization. *EMBO J* **27,** 2817–2828 (2008).
25. Takenawa, T. and Suetsugu, S. The wasp-wave protein network: connecting the membrane to the cytoskeleton. *Nat Rev Mol Cell Biol* **8,** 37–48 (2007).
26. Suetsugu, S. The proposed functions of membrane curvatures mediated by the bar domain superfamily proteins. *J Biochem* **148,** 1–12 (2010).
27. Mattila, P. K. and Lappalainen, P. Filopodia: molecular architecture and cellular functions. *Nat Rev Mol Cell Biol* **9,** 446–454 (2008).

28. Salazar, M. A., Kwiatkowski, A. V., Pellegrini, L., Cestra, G., Butler, M. H., et al. Tuba, a novel protein containing bin/amphiphysin/rvs and dbl homology domains, links dynamin to regulation of the actin cytoskeleton. *J Biol Chem* **278,** 49031–49043 (2003).
29. Peter, B. J., Kent, H. M., Mills, I. G., Vallis, Y., Butler, P. J. G. et al. Bar domains as sensors of membrane curvature: the amphiphysin bar structure. *Science* **303,** 495–499 (2004).
30. Mattila, P. K., Pykäläinen, A., Saarikangas, J., Paavilainen, V. O., Vihinen, H. et al. Missing-inmetastasis and irsp53 deform pi(4,5)p2-rich membranes by an inverse bar domain-like mechanism. *J Cell Biol* **176,** 953–964 (2007).
31. Scita, G., Confalonieri, S., Lappalainen, P., and Suetsugu, S. Irsp53: crossing the road of membrane and actin dynamics in the formation of membrane protrusions. *Trends Cell Biol* **18,** 52–60 (2008).
32. Ambroggio, E., Sorre, B., Bassereau, P., Goud, B., Manneville, J. B. et al. Arfgap1 generates an arf1 gradient on continuous lipid membranes displaying flat and curved regions. *EMBO J* **29,** 292–303 (2010).

7 Actin Waves and Phagocytic Cup Structures

Günther Gerisch

CONTENTS

7.1 Introduction 133
7.2 Materials and Methods 134
7.3 Discussion.. 135
7.3.1 Relation of Wave Propagation in Dictyostelium to Other Actin-Based Cell Functions.. 135
7.3.2 Actin Waves are Comparable to Trigger Waves in Bistable Systems.... 137
7.4 Results 137
7.4.1 Structure and Dynamics of Self-Organizing Actin Waves 137
7.4.2 Actin Waves are Confined by Different States of Actin Organization and by Membrane Areas of Different Phosphoinositide Composition............ 139
7.4.3 Protein and PIP3 Patterns in Phagocytic Cups Correspond to the Patterns in Wave-Forming Cells.. 142
Keywords 143
Acknowledgments.. 143
References.. 143

7.1 INTRODUCTION

This chapter deals with actin waves that are spontaneously generated on the planar, substrate-attached surface of *Dictyostelium* cells. These waves have the following characteristics. (1) They are circular structures of varying shape, capable of changing the direction of propagation. (2) The waves propagate by treadmilling with a recovery of actin incorporation after photobleaching of less than 10 s. (3) The waves are associated with actin-binding proteins in an ordered 3-dimensional organization: with myosin-IB at the front and close to the membrane, the Arp2/3 complex throughout the wave, and coronin at the cytoplasmic face and back of the wave. Coronin is a marker of disassembling actin structures. (4) The waves separate two areas of the cell cortex that differ in actin structure and phosphoinositide composition of the membrane. The waves arise at the border of membrane areas rich in phosphatidylinositol (3,4,5) trisphosphate (PIP3). The inhibition of PIP3 synthesis reversibly inhibits wave formation. (5) The actin wave and PIP3 patterns resemble 2-dimensional projections of phagocytic cups, suggesting that they are involved in the scanning of surfaces for particles to be taken up.

This chapter is an overview of work supported by the *Deutsche Forschungsgemeinschaft* in the Priority Program "Optical analysis of the structure and dynamics of supramolecular biological complexes". Subject of this chapter has been the self-organization of actin into propagating waves. It reviews published results on the dynamics and molecular composition of these waves and will elaborate on some connotations that are not detailed elsewhere.

The starting question that addressed was: how are actin structures organized *de novo* in cells that have been depleted of polymerized actin? To inhibit actin polymerization, latrunculinA (LatA) has been used. In living cells, this scavenger of actin monomers results in the depolymerization of actin filaments at a rate determined by their turnover. The disassembly and re-assembly of actin structures has been monitored in *Dictyostelium* cells using a GFP- tagged construct (LimEΔ) that proved to be an appropriate label for recording the dynamics of filamentous actin structures in these cells [1]. The actin structures rapidly turn over, resulting in microscopically detectable breakdown of the actin network in the cell cortex within less than 20 s after LatA addition [2].

The removal of LatA enables the cells to regain normal actin organization and cell motility within less than an hour. Of interest here is the intermediate state of excessive wave formation before normal cell motility will recover. This state is preceded by the formation of mobile actin patches [3]. The burst of patches at the onset of actin polymerization is mostly due to the fact that clathrin-dependent endocytosis requires actin to polymerize in order for vesicles to be budded off. Therefore, many clathrin-coated pits are arrested at the plasma membrane as long as LatA is present. Accordingly, the first actin structures seen within about 5 min after removal of the drug are small, clathrin-induced patches. Later on, actin waves are generated in a spatial relationship to the synthesis of phosphatidylinositol-(3,4,5) trisphosphate (PIP3) in the plasma membrane. The time window of excess formation of actin waves has been used for the experimental analysis of their structure and mode of propagation [4].

7.2 MATERIALS AND METHODS

Cells of *Dictyostelium discoideum* strain AX2-214 were transfected with expression vectors encoding GFP- or mRFP-fusion proteins [4] and subjected to imaging at 23 ± 2°C in 17 mM Na/K-phosphate buffer, pH 6.0 (PB) according to [3]. For the recovery of actin polymerization, cells were pretreated for about 15 min with 5 μM latrunculin A (Invitrogen). Subsequently, the drug was replaced with PB. To inhibit PI3-kinases, a stock solution of 30 μM LY-294002 (Sigma) in DMSO was diluted to 50 μM in PB and added to the cells during the wave-forming stage of recovery. For the imaging of clathrin-coated structures in relation to reversible inhibition of wave formation, a Zeiss LSM 510 equipped with a 63x/1.4 oil apochromate objective was used. TIRF microscopy was applied to actin structures according to [5]. Spinning-disc confocal microscopy was performed as detailed in [4].

For phagocytosis experiments, *D. discoideum* cells, double-labeled as for wave formation, were exposed to living *Saccharomyces cerevisiae* [6]. In the phagocytic cups formed around these large particles, the accumulation of GFP-tagged proteins relative to filamentous actin labeled with mRFP-LimEΔ was recorded using the Zeiss LSM 510 confocal microscope.

7.3 DISCUSSION

7.3.1 Relation of Wave Propagation in *Dictyostelium* to Other Actin-Based Cell Functions

The actin waves in *Dictyostelium* cells are macromolecular complexes with a defined 3-dimensional architecture. While propagating, the waves continuously turn over their constituents. The organization of these waves shows that self-sustained actin structures can be generated without a membrane acting as a scaffold and site of activating factors in front of these waves. The most relevant feature of the actin waves is their resemblance to functional actin structures that promote the extension of a phagocytic cup (Figures 1 and 2). Similar patterns of actin-associated proteins are assembled during leading edge protrusion, phagosome rocketing [7], cell spreading or clathrin-dependent membrane internalization [8]. In all these cases is actin associated with the Arp2/3 complex. A single-headed myosin, myosin-IB, is enriched at the front of the dynamic actin structures close to the underlying membrane, and coronin is always recruited to regions of actin disassembly. Characteristic of the actin waves is their localization at the boundary between PIP3 rich and depleted areas of the plasma membrane. This localization coincides with the boundary between two states of actin organization, one resembling the front region of a freely migrating or chemotaxing cell, the other corresponding to the tail region rich in filamentous myosin-II [5].

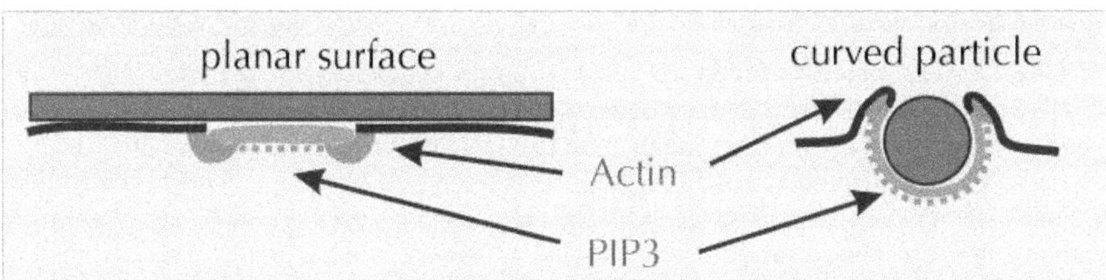

FIGURE 1 Comparison of actin-PIP3 patterns on a planar surface and in phagocytic cups. Left: autonomous actin waves (red) circumscribe a PIP3-rich area of the substrate-attached cell membrane (green). **Right**: similarly, an actin ring at the rim of a phagocytic cup surrounds the PIP3-rich membrane area invaginated in contact with a particle. Dotted red lines indicate accumulation of filamentous actin on top of PIP3-enriched membrane areas. This diagram is based on data reported in [9].

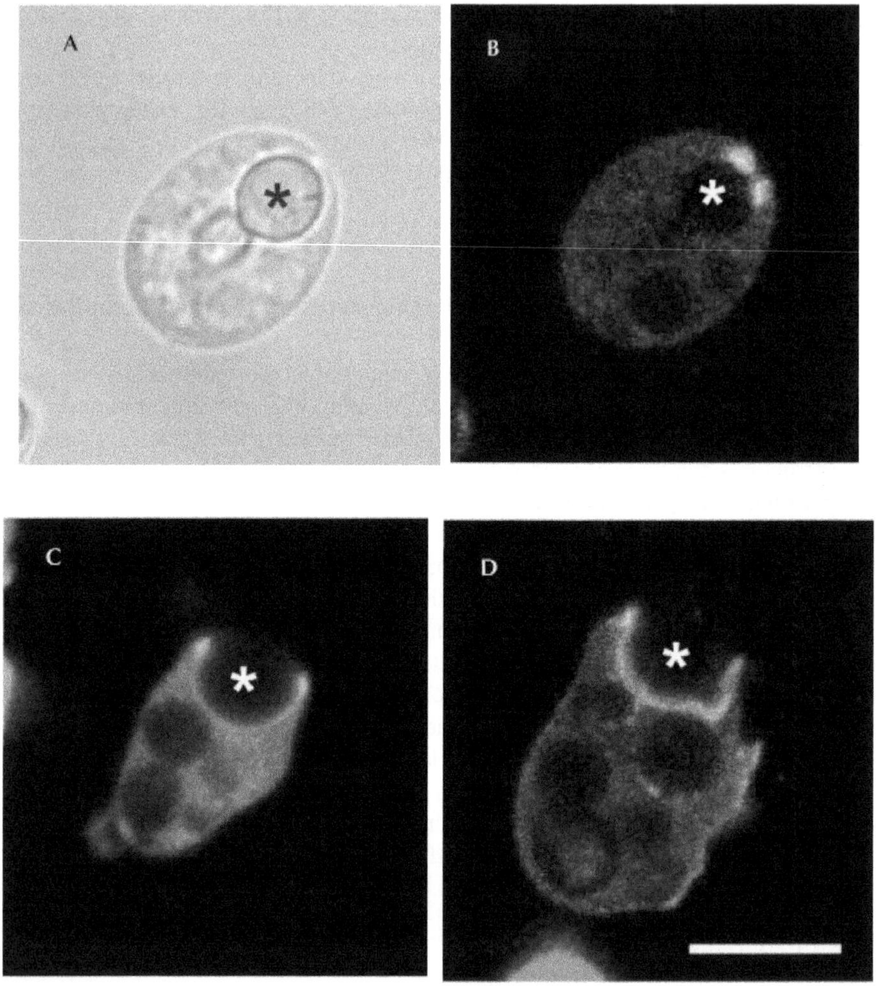

FIGURE 2 Arrangement of actin-associated proteins in phagocytic cups. During the uptake of yeast particles, cells were imaged to localize GFP-tagged myosin-IB, Arp2/3, or coronin (green) in combination with filamentous actin (red). Centers of the particles are indicated by asterisks. **A**, bright-field image of a late phagocytosis stage showing a particle being engulfed. **B**, double-fluorescence recording of the same cell showing GFP-myosin-IB enriched at the border of the cup close to the plasma membrane. **C**, uptake of a particle by a cell expressing GFP-Arp3, indicating coincident localization of the Arp2/3 complex and the labeled actin. **D**, the phagocytic cup in a cell expressing GFP-coronin shows coronin remote from the membrane at the interface between the actin layer and the cytoplasmic space. The progressing edge of the cup is free of coronin. Bar, 10 μm.

The actin waves studied in this chapter differ in their organization, dynamics and function from actin-based waves of Hem-1, a constituent of the Arp2/3-activating WAVE complex [10, 11]. Hem-1 waves propagate in neutrophils at intervals in one direction and are oriented by chemoattractant, whereas the actin waves in *Dictyostelium* are typically circular and related to the phagocytic activity of these cells.

7.3.2 Actin Waves are Comparable to Trigger Waves in Bistable Systems

The finding that the actin waves studied in *Dictyostelium* cells are confined at their front and back by two different states of actin organization is pertinent to the question of whether or not the actin waves are periodic structures formed in an excitable medium that, after a wave has passed, returns to its resting state ready to form a new wave. Different from these structures, the actin waves generated in *Dictyostelium* cells are confined at their front and back by two distinct states of actin organization. In this respect, the wave-forming actin cortex resembles those pattern-generating reaction-diffusion systems in which far from equilibrium bistability is established. Waves that separate the two phases in such systems are known as "trigger waves" [12]. A model specifying conditions under which bistability will arise in the actin cortex is provided by Carsten Beta in this issue [Beta C: A bistable model of actin dynamics. PMC Biophysics, submitted].

In summary, actin waves separate two large areas of the cell cortex that are distinguished by their actin structure and by the PIP3 content of the membrane. Actin structure and membrane composition of these areas are reciprocally interconverted when the actin waves propagate into one or the other direction. Since the wave patterns are formed in contact with a planar glass surface, they provide an opportunity to analyze phase transitions in the actin system under superior optical conditions using TIRF microscopy and to correlate these transitions with the regulation of phosphatidylinositides in the membrane.

7.4 RESULTS

7.4.1 Structure and Dynamics of Self-Organizing Actin Waves

The actin waves studied in this chapter are typically circular structures of varying shape (Figure 3). They change their shape by propagating at the substrate-attached surface of *Dictyostelium* cells with velocities of about 100 nm per second [4]. The waves propagate in a treadmilling mode, with net polymerization of actin at their front and net depolymerization at the back. Even when the waves keep their overall shape during propagation, their constituting proteins are exchanged: after photo bleaching, half-maximal fluorescence recovery of actin has been observed within 4 s, and of myosin-IB within 2 s [4]. The Arp2/3 complex is distributed throughout the wave, indicating that the entire wave structure is dominated by a dense fabric of branched actin filaments. Three-dimensional scanning of protein distributions using spinning disc confocal microscopy revealed distinct patterns for myosin-IB and coronin in cross-sections through a wave. Myosin-IB, a single-headed motor protein capable of binding to the plasma membrane, is enriched at the front of the wave and at the substrate-attached area of the membrane. Coronin occupies the sloping roof of the wave at its cytoplasmic face (Figure 4, left panel). Assuming that membrane-anchored myosin-IB clusters are sites of active actin polymerization whereas coronin is recruited to sites of depolymerization, two gradients of actin polymerization can be proposed: one gradient declining from the front to the back of a wave, the other from the substrate-attached membrane to the roof of the wave (Figure 4, right panel).

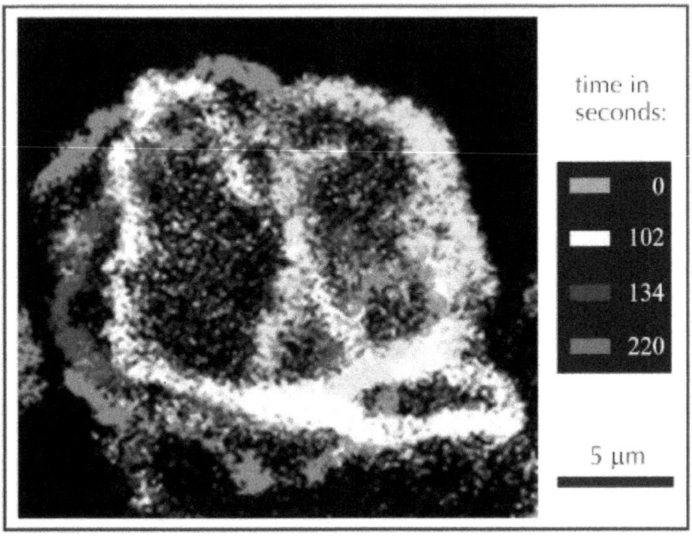

FIGURE 3 Shape dynamics of actin waves viewed on a substrate-attached cell surface. Wave images have been recorded from a single cell of *Dictyostelium discoideum* at the indicated times by spinning-disc confocal microscopy. During propagation of the wave, the cell showed negligible net movement. The images are color-coded and superimposed on top of each other. Shape changes within a perlod of less than 2 min are best recognized by comparing the green and white images. Waves can fuse or they may split into two as in the red image at the end of the recorded sequence. The images are taken from Figure 3 in [5].

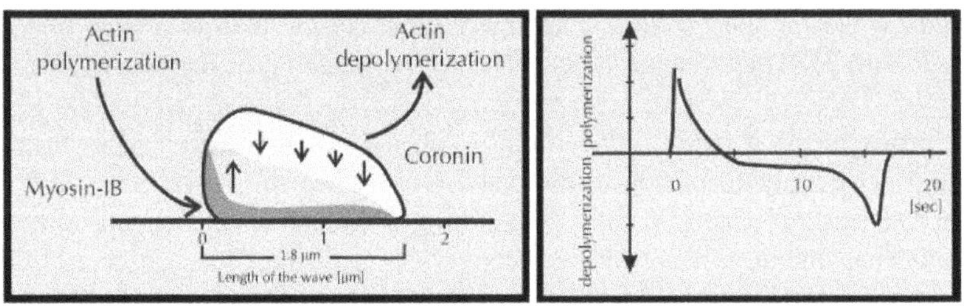

FIGURE 4 Spatial organization of actin waves and the temporal pattern of actin polymerization and depolymerization. Left panel: schematic cross-section through an actin wave showing the region at the front and close to the membrane where myosin-IB is enriched (green) and the sloping roof of the wave to which coronin is recruited (yellow). Vertical arrows indicate up and down regulation of actin polymerization. Data suggest two gradients of actin polymerization, one falling from the substrate-attached membrane to the top of the wave, the other from its front to the tail. **Right panel**: translation of the spatial profile into a temporal sequence of actin net polymerization and depolymerization. The profile illustrates the sequence of changes that occur when a wave passes over a point on the membrane. The data published in [4] suggest that an initial phase of high-rate actin polymerization turns into a longer phase of depolymerization.

7.4.2 Actin Waves are Confined by Different States of Actin Organization and by Membrane Areas of Different Phosphoinositide Composition

In order to probe for actin structures in the cortex of wave-forming cells, we used dual-color fluorescence labeling of proteins that associate with different arrangements of actin filaments [5]. The area circumscribed by a wave turned out to be enriched in the Arp2/3 complex, indicating a dominance of dense dendritic actin structures in this inner area. In contrast, cortexillin I, a protein that bundles actin filaments in anti-parallel direction, prevails in the external area, together with myosin-II, a conventional myosin that forms bipolar filaments (Figure 5). Since the waves can change the direction of propagation, the two areas reciprocally increase or decrease in size.

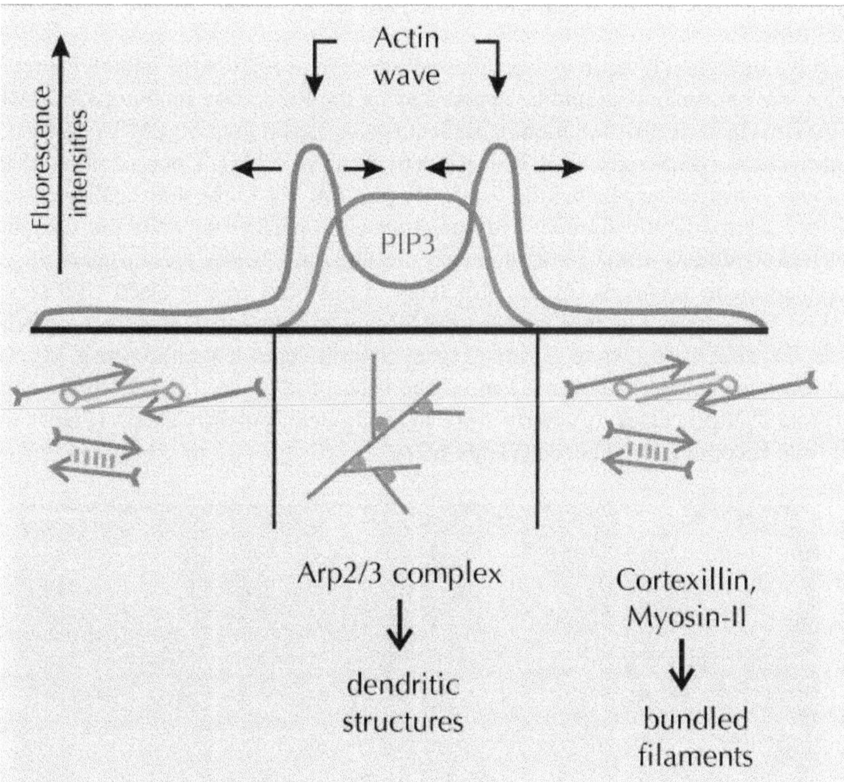

FIGURE 5 Spatial relationship of actin organization to the presence or absence of PIP3 in the underlying membrane. The diagram represents a line scan through an actin wave pattern on a planar glass surface. On top, the relationship of actin- and PIP3-labels is shown. The fluorescence intensity of actin (red) peaks at the position of the wave formed at the border of a PIP3-rich area of the cell membrane (green). Horizontal arrows point into the changing directions of wave propagation. On the bottom, actin structures are shown to differ between the PIP3-rich area circumscribed by the wave and the external area depleted of PIP3. The Arp2/3 complex known to nucleate branches of actin filaments dominates in the PIP3-rich area, while proteins associated with anti-parallel bundles of filaments are prevailing in the external area. This diagram summarizes data published in [5] and [9].

Labeling of phosphoinositides revealed that actin waves are formed at the boundary of a membrane area rich in PIP3 [9]. Enrichment of the Arp2/3 complex in this inner area suggests that PIP3 directs the assembly of actin into dendritic structures, while in the external area depleted of PIP3 a basal network of bundled actin filaments is dominating. The reciprocal expansion and shrinkage of these areas demonstrates a connection between PIP3 regulation and conversion of one state of actin organization into the other (Figure 5). A linkage of actin waves to PIP3 in the membrane has previously been found under other conditions [13]. This linkage is in line with the control of actin polymerization through a positive feedback loop involving PIP3 and Ras [14]. However, under our conditions the actin waves do not coincide with regions of strongest PIP3 accumulation but with zones where PIP3 forms the steepest gradients in the membrane. The control mechanisms underlying this complex behavior remain to be elucidated.

If the formation of actin waves relies on a pattern of PIP3 in the underlying membrane, wave formation should be suppressed by the PI3-kinase inhibitor LY-294002 (Figure 6). In fact, the formation of actin waves is sensitive to 50 μM LY-294002, a concentration which reduces PI3-kinase activity in vivo [15]. Upon addition of the inhibitor, waves disappear within 2 min (Figure 6A, B). Only short actin-rich protrusions are transiently formed, and numerous actin patches decorate the substrate-attached membrane of the LY-treated cells. To find out whether these patches are rudimentary actin waves or distinct structures, it has been co-expressed GFP-tagged clathrin light-chains together with the mRFP-tagged actin label, since actin patches of similar size are known to be involved in clathrin-dependent endocytosis [8]. The two labels coincided, indicating that, at the concentration used, LY-294002 neither prevents clathrin-coated structures from inducing actin polymerization (Figure 6C) nor from becoming internalized (Figure 6D).

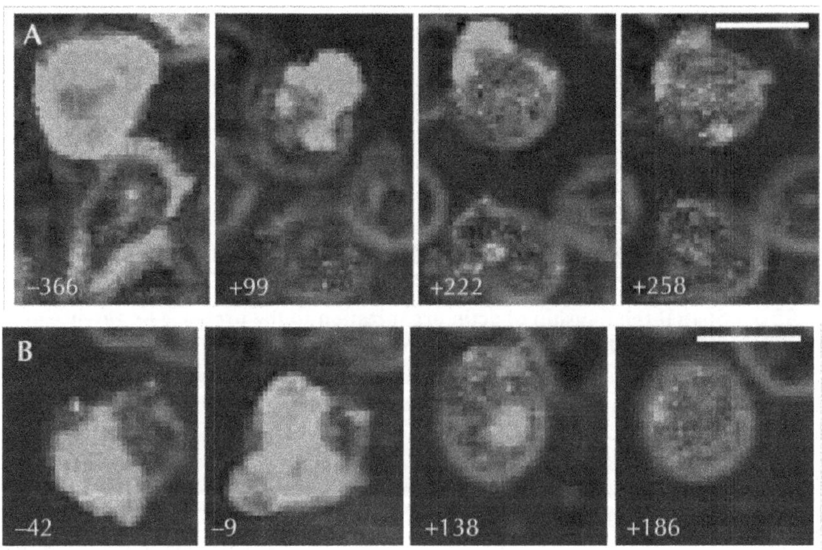

FIGURE 6 *(Continued)*

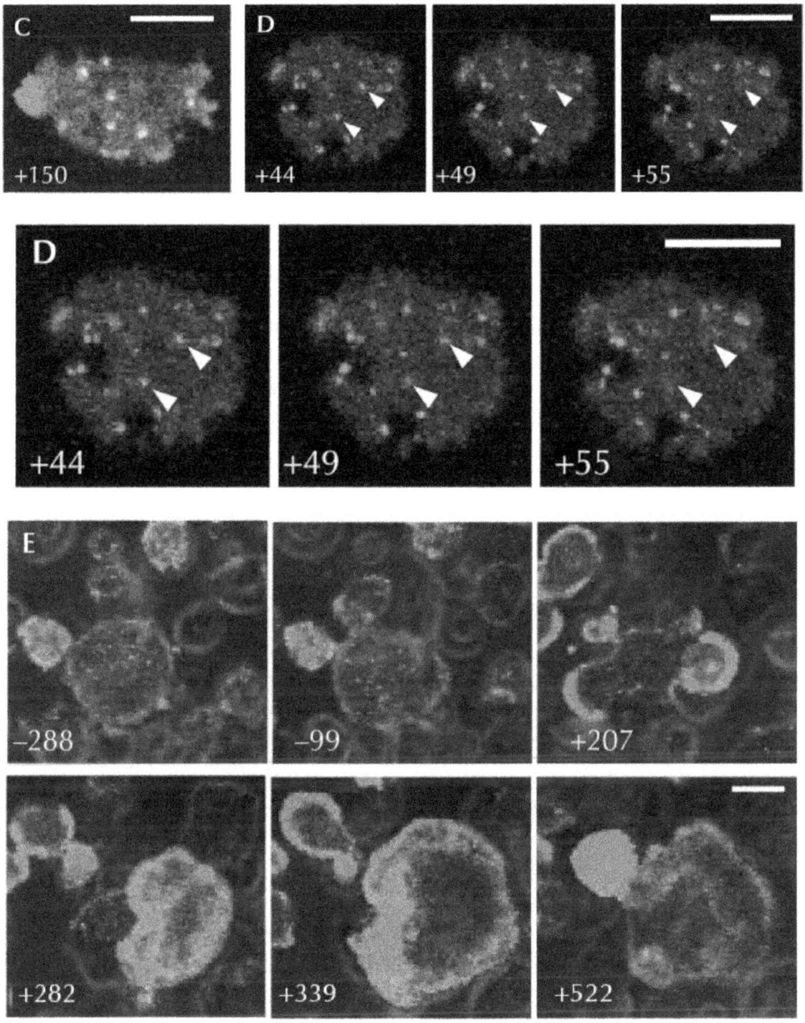

FIGURE 6 Reversible suppression of actin wave formation by the PI3-kinase inhibitor LY-294002. Cells recovering from LatA treatment were allowed to form waves before 50 μM LY-294002 were added. In (A to D) time is indicated in seconds before and after addition of the inhibitor (t = 0), in (E) before and after its removal. In (A, B, and E) the LimEΔ-GFP label for actin (green) is superimposed on phase-contrast images showing cell shape (red). In (C and D) cells are double-labeled with mRFP-LimEΔ for actin (red) and GFP-clathrin light chains for clathrin-coated structures (green). **A**, **B**, actin wave formation is suppressed by LY-294002 while small actin patches are persisting. **C**, co-localization of the actin and clathrin labels in six patches at the substrate-attached surface of an LY-treated cell. **D**, clathrin and actin dynamics in an LY-treated cell. Arrowheads point to clathrin-coated structures (44 s) that recruit actin thus turning from green to red (49 s) and subsequently disappear from the membrane (55 s). **E**, recovery of wave formation in a big cell after the removal of Ly-294002. Bars, 10 μm in (A, B, and E); 5 μm in (C and D).

The suppression of actin waves is readily reversible. Within 3 min after removal of the inhibitor, profuse formation of actin waves recovered (Figure 6E). In conclusion, the self-organization of actin waves is distinguished by its strong dependence on PIP3 from the activity of clathrin-induced actin patches, such that clathrin-dependent endocytosis can be separated from wave formation by the inhibition of PIP3 synthesis.

7.4.3 Protein and PIP3 Patterns in Phagocytic Cups Correspond to the Patterns in Wave-Forming Cells

Like other phagocytes, *Dictyostelium* cells interact with adhesive surfaces in two ways: on a planar surface they spread and migrate, on a convex surface a circular protrusion is induced to try and engulf a particle. By progression of its rim along the particle surface, this phagocytic cup envelops and eventually encloses the particle by separation of the phagosome membrane from the plasma membrane. Filamentous actin accumulates between the outer and inner leaflet of the cup membrane and is most strongly enriched at the rim of the cup, the site of its protrusion. The leaflet of the cup membrane that is in contact with the particle becomes rich in PIP3 [15, 16].

With regard to the biological function of the actin waves generated on a planar surface, it is appealing to consider them as devices to search for particles to be taken up. In fact, the state of excessive wave formation in cells recovering from actin depolymerization is characterized by a high propensity for taking up particles with a minimum of net movement of the cells on a planar surface [9]. If the particle-induced actin and PIP3 patterns in a phagocytic cup are projected onto a plane, they coincide with the spontaneously generated patterns in wave-forming cells (Figure 1). More complicated patterns are observed in cups formed around long rod-shaped particles. These patterns correspond to toroid-like actin-wave and PIP3 patterns that are formed in large cells on a planar surface [9].

To underscore the similar organization of spontaneously generated wave patterns and particle-induced phagocytic cups, the characteristic localization of three actin-associated proteins in cups has been compared with their localization in waves as reported in [4]. In the cups, myosin-IB is enriched close to the membrane at the very border of the protruding cup (Figure 2). The Arp2/3 complex is distributed throughout the actin layer of the cup. Coronin is again recruited to regions where actin is supposed to depolymerize: at the sides and the bottom of the cup, specifically at the boundary between actin layer and cytoplasmic space remote from the membrane (Figure 4).

For the optical analysis of the pattern dynamics, the planar arrangement of pattern elements in wave-forming cells has an advantage; phase transitions in the cell cortex can be recorded by TIRF microscopy at high temporal and spatial resolution. At each point of the substrate-attached surface of a wave-forming cell the actin structure undergoes transitions from one state to the other. The temporal sequences of these transitions are variable; they may be irregular or ordered into patterns with periods of about 5 min [5].

KEYWORDS

- **Actin**
- ***Dictyostelium* cells**
- **Phagocytic cup**
- **Phosphoinositide Composition**
- **Wave Propagation**

ACKNOWLEDGMENTS

The work reported here has been supported by SPP 1128 of the *Deutsche Forschungsgemeinschaft* and the *Max-Planck-Gesellschaft*. It was performed in collaboration with the Light Microscope Facility of the MPI-CBG in Dresden, The Oklahoma Medical Research Foundation, and the Nikon Imaging Center of the University of Heidelberg. I thank Carsten Beta, University of Potsdam for discussions on bistability, Mary Ecke and Jana Prassler for expert assistance, and Terry O'Halloran, University of Texas, Austin, for the GFP-clathrin light-chain vector.

REFERENCES

1. Schneider, N., Weber, I., Faix, J., Prassler, J., Müller-Taubenberger, A., Köhler, J., Burghardt, E., Gerisch, G., and Marriott, G. *Cell Motil Cytoskeleton* **56,** 130–139 (2003).
2. Diez, S., Gerisch, G., Anderson, K., Müller-Taubenberger, A., and Bretschneider, T. *Proc Natl Acad Sci USA* **102,** 7601–7606 (2005).
3. Gerisch, G., Bretschneider, T., Müller-Taubenberger, A., Simmeth, E., Ecke, M., Diez, S., and Anderson, K. *Biophys J* **87,** 3493–3503 (2004).
4. Bretschneider, T., Anderson, K., Ecke, M., Müller-Taubenberger, A., Schroth-Diez, B., Ishikawa-Ankerhold, H. C., and Gerisch, G. *Biophys J* **96,** 2888–2900 (2009).
5. Schroth-Diez, B., Gerwig, S., Ecke, M., Hegerl, R., Diez, S., and Gerisch, G. *HFSP J* **3,** 412–427 (2009).
6. Clarke, M., Maddera, L., Engel, U., and Gerisch, G. *PLoS ONE* **5**(e8585), 1–14 (2010).
7. Clarke, M., Müller-Taubenberger, A., Anderson, K. I., Engel, U., and Gerisch, G. *Mol Biol Cell* **17,** 4866–4875 (2006).
8. Heinrich, D., Youssef, S., Schroth-Diez, B., Engel, U., Aydin, D., Blümmel, J., Spatz, J. P., and Gerisch, G. *Cell Adh Migr* **2,** 58–68 (2008).
9. Gerisch, G., Ecke, M., Schroth-Diez, B., Gerwig, S., Engel, U., Maddera, L., and Clarke, M. *Cell Adh Migr* **3,** 373–382 (2009).
10. Weiner, O. D., Marganski, W. A., Wu, L. F., Altschuler, S. J., and Kirschner, M. W. *PLoS Biol* **5,** 2053–2063 (2007).
11. Millius, A., Dandekar, S. N., Houk, A. R., and Weiner, O. D. *Curr Biol* **19,** 253–259 (2009).
12. Mikhailov, A. S. *Foundation of Synergetics I*. Springer, Berlin, Heidelberg, New York, pp. 15–32 (1994).
13. Asano, Y., Nagasaki, A., and Uyeda, T. Q. P. *Cell Motil Cytoskeleton* **65,** 923–934 (2008).
14. Sasaki, A. T., Janetopoulos, C., Lee, S., Charest, P. G., Takeda, K., Sundheimer, L. W., Meili, R., Devreotes, P. N., and Firtel, R. A. *J Cell Biol* **178,** 85–191 (2007).
15. Dormann, D., Weijer, G., Dowler, S., and Weijer, C. J. *J Cell Sci* **117,** 6497–6509 (2004).
16. Loovers, H. M., Kortholt, A., de Groote, H., Whitty, L., Nussbaum, R. L., and van Haastert, P. J. M. *Traffic* **8,** 618–628 (2007).

8 Yeast and Scaling

Daniel Riveline

CONTENTS

8.1 Introduction 145
8.2 Methods 146
8.2.1 Two Laws for Fission Yeast Shape 146
8.3 Discussion 147
8.3.1 The Role of Molecular Mechanisms in this Framework 147
8.4 Results 147
8.4.1 Curvature at the Cell Ends: a Low Value for Membrane Surface Tension as the Motor for Recruiting the Growth Machinery 147
8.4.2 Estimate of the Pressure Difference with the Use of Cell Wall Tension 148
8.4.3 Length at Mitosis: The Septum Location 149
8.4.4 New-End Take Off (NETO) Length 150
8.4.5 The C-Shape 152
8.5 Conclusion 154
Keywords 154
Acknowledgments 154
References 155

8.1 INTRODUCTION

Lengths and shapes are approached in different ways in different fields: they serve as a read-out for classifying genes or proteins in cell biology whereas they result from scaling arguments in condensed matter physics. Here, a combined approach with examples illustrated for the fission yeast *Schizosaccharomyces pombe is* proposed.

Cells are regulated by highly connected signalling pathways [1]: activation and inhibition cascades are constantly changing the cell responses to its environment and to its own dynamics. In order to isolate independent signalling modules, there is a requirement to identify simple and reliable readouts. Levels of molecular activity such as proteins phosphorylation and dephosphorylation are efficient for this purpose. However microscopic cellular lengths and shapes have also been proven to be powerful readouts for classifying networks in cellular control. For example, genes deletions lead to classes of strains having different lengths [2] and modified shapes [3]. Genes leading to a similar phenotype are then grouped into a functional biological module.

Similar microscopic measurements are usually treated by *scaling* arguments in condensed matter physics. Key parameters of the system are extracted, and lengths or shapes formulae are derived using appropriate combinations of parameters. This approach has proven its efficiency for a variety of systems, ranging from whole organisms [4] to polymer physics [5] and wetting phenomena [6]. Since the selected parameters have to completely capture the matter properties of the system under study, these scaling laws reflect the physical relations bound to the problem. As a result, these laws provide satisfactory physical explanations for the measured lengths and shapes, beyond the fact that the derived formulae are constrained by the dimensional analysis of the parameters units. In addition, these scaling laws allow to predict changes in lengths and shapes caused by the variations of selected - and often unexpected - parameters.

It is proposed here to couple both genetic and mesoscopic approaches on a unicellular organism, the fission yeast *S. pombe*. The fission yeast cell is a rod of 15 µm length and 4 µm diameter with a rigid wall. Cells grow by elongation from the hemispherical ends and divide by medial fission. Wall tension and pressure difference between the inside and the outside of the cell are the main physical parameters used for explaining phenotypes [7]. The derived read-outs are here curvature at cell ends, cell radius, cytokinetic ring centering, lengths at "NETO" (New End Take-Off), C shape (ban mutants, see below). The relations are derived, and data illustrating the results are given; in addition, experiments are suggested for probing the laws in future works. The main contribution of this paper is to propose a quantitative framework to understand the microscopic read-outs, while suggesting new approaches for classifying genes.

8.2 METHODS

8.2.1 Two Laws for Fission Yeast Shape

According to the Pascal principle, the difference in pressure between the inside and the outside of the cell is constant

$$\Delta P = const \tag{1}$$

This property imposes a constant *global* pressure around the cell. The force associated with this pressure is perpendicular to the wall.

In contrast, the Young-Laplace equation imposes that *local* surface properties dictate local shapes:

$$\Delta P = \gamma_{local}\left(\frac{1}{R_1} + \frac{1}{R_2}\right) \tag{2}$$

where γ_{local} is the local surface tension, and R_1 and R_2 are the principal radii of curvature. This relation states that the pressure force perpendicular to the surface of the cell is balanced by the local elastic properties of the wall. As a result, this equation suggests that the cell shapes are directly set by the global pressure difference and wall local surface tensions.

8.3 DISCUSSION

8.3.1 The Role of Molecular Mechanisms in this Framework

Molecular mechanisms are usually presented for explaining the lengths and shapes of yeast cells [8]. They indeed play a key role in the signalling pathways leading to the read-out observed under the microscope. The same statement applies to the active cytoskeleton: for example, endocytosis at the growing ends via actin mediated transport by patches and filaments [9], the closure of the cytokinetic ring by acto-myosin motors in septum formation [10], or the restrictions of growing ends locations by microtubules [11]. All are involved in the creation of the wall tension. However they are required *intermediates* for assembling the wall and generating tension at the proper locations and phases in the cell cycle, and they do not determine or explain the measured shapes and lengths in a physical sense. The purpose of this work is to suggest coupled approaches where molecular mechanisms in signalling pathways will be characterized simultaneously with the corresponding mesoscopic measurements.

8.4 RESULTS

8.4.1 Curvature at the Cell Ends: a Low Value for Membrane Surface Tension as the Motor for Recruiting the Growth Machinery

The cell growth machinery assembles at one end of the cell after septation [12]. Key cytoplasmic proteins of this machinery leading to synthesis and local deposition of cell wall material are distributed around the hemispherical end (see for example [13–15]). This spatial organisation and the exclusion from the side of the cell long axis are surprising. It is not due to the microtubule cytoskeleton, since the same machinery operates in the absence of microtubules [16]. It is proposed that surface tension at *the membrane* may explain this preferred location for assembly: following Young-Laplace equation, the tension around the cap is twice lower than the tension along the side of the cell (see Figure 1); the growth machinery is thus preferentially inserted around this hemispherical cap.

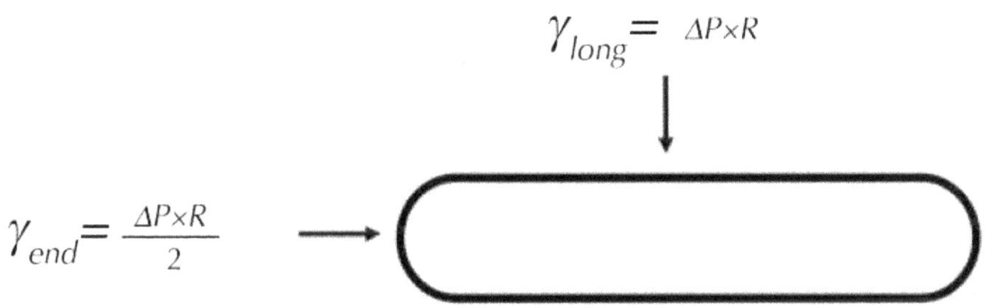

FIGURE 1 Scaling for lengths during cell extension; the tension γ_{local} is different between the hemispherical ends and the cylindrical longitudinal side (R is the cell radius; ΔP is the pressure difference).

An estimate can be given for the membrane tensions. Assuming that yeast membrane lipid composition is similar to mammalian cell membrane, 10^{-4} N/m can be used as tension value measured for fibroblasts (see [17, 18]). This value is applied to the tension at the hemispherical end of the yeast cell. Following the argument, the longitudinal tension is about 2.10^{-4} N/m. Note that this estimate is presented as a reasonable order of magnitude. Membrane tension measurements on fission yeast cells without a wall (*cytoplasts* [19]) will be required for confirming this value.

The following shapes mutants are consistent with this surface tension argument. Strains with T shapes have been documented in various conditions: they are obtained either by genetic modifications [16] or by removal of microtubules [20]. These strains exhibit a new growth zone in the side of the cell, with the same radius as regular growing ends: a hemispherical deformation appears which leads to further recruitment of the cell wall machinery; this step is followed by further growth. Additional growth zones appear along the sides of the cell with the same mechanism [20], that is local deformation of the cells, followed by elongation. It is proposed that the local reduced tension promotes the local recruitment of the machinery. Microtubules in wild type cells would restrict the remodelling of the wall exclusively at the ends of the cell; in these T-shaped cells, however, local wall remodelling on the side would trigger the local deformation due to the pushing force of the pressure.

In order to test this result, the following experiments could be performed: (i) decreasing the wall thickness locally by spraying a wall digesting enzyme (see [19]) close to the cell should promote a new local growing end (for the method of local spray, see for example [21]); the pressure will have promoted the local deformation of the cell, followed by the recruitment of the growth machinery; and (ii) forcing the cells into closed microfabricated patterns like in [22] with designed hemispherical ends should alter growth in both ways: a cell end with an imposed curvature smaller than wild type ends should promote growth, whereas an end with a larger curvature should block further cell elongation.

8.4.2 Estimate of the Pressure Difference with the Use of Cell Wall Tension

The outer layer of yeast is now considered, the cell *wall* and its associated surface tension. Note that this layer is close but distinct from the cell *membrane* mentioned in the previous paragraph. Taking the expressions of the radius on the long axis

$$\gamma_{long} = \Delta P R \tag{3}$$

(see Figure 1), two key features can be derived for fission yeast: (i) since the cell diameter is constant during cell growth, pressure difference remains *constant* during cell growth; (ii) This pressure difference can be estimated; surface tension is the product of the wall Young modulus E by the wall thickness w, so

$$\Delta P = \frac{E}{R} w \tag{4}$$

Based on whole cell measurements for E of 100 MPa [23, 24], and taking a wall thickness w of 200 nm [25], a pressure difference of about 10 MPa is obtained. Direct measurements similar to experiments on molds by Money et al [26] should allow to probe this estimate for fission yeast.

8.4.3 Length at Mitosis: The Septum Location

When cells reach mitosis, an acto-myosin ring is assembled around the central part of the cell [10]. The contraction of this ring associated with the local addition of cell wall leads to the formation of a septum and to the subsequent separation of sister cells. Strikingly this septum is located in the vicinity of the middle of the cell (see Figure 2). It is shown here that simple arguments can determine its location.

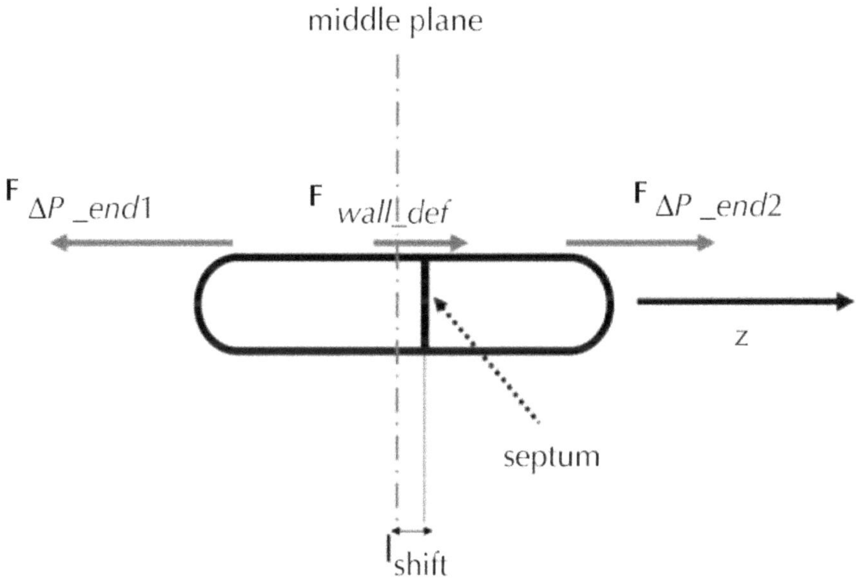

FIGURE 2 Scaling for the septum location at mitosis; the cytokinetic ring contraction leads to the septum formation; its location is shifted from the cell central plane by a distance lshift. Forces at the cell wall along the z-axis are represented.

It is proposed that the cell is under pressure while no wall is added at this stage of the cycle. The wall is then undergoing a longitudinal deformation: the pressure imposes traction forces at both ends; the wall is deformed along a distance l_{ext} (like a spring being pulled at both ends). l_{shift} is called as the distance between the middle of the cell and the location where forces are balanced.

The forces should be balanced using the Young-Laplace equation along the z axis (see Figure 2); that are at both ends:

$$F_{\Delta P\_end1} = \gamma_{end1} R_1 \text{ and } F_{\Delta P\_end2} = \gamma_{end2} R_2 \quad (5)$$

It is assumed that the wall elasticity is isotropic. The force associated with the deformation of the wall is given by:

$$F_{wall\_def} = \gamma_{long} l_{ext} \tag{6}$$

The z-location where forces are balanced is given by:

$$\gamma_{end2} R_2 + \gamma_{long} l_{shift} - \gamma_{end1} R_1 = 0 \tag{7}$$

So

$$l_{shift} = \frac{1}{\gamma_{long}} (\gamma_{end1} R_1 - \gamma_{end2} R_2) \tag{8}$$

Since

$$\gamma_{long} = \Delta P R \tag{9}$$

(see Eq. 3),

$$l_{shift} = \frac{1}{\Delta P R} (\gamma_{end1} R_1 - \gamma_{end2} R_2) \tag{10}$$

is obtained and have

$$\frac{R_2}{R} \cong \frac{R_1}{R} = 1;$$

It is concluded that the septum location is shifted from the center by the distance l_{shift} given by:

$$l_{shift} = \frac{\gamma_{end1} - \gamma_{end2}}{\Delta P} \tag{11}$$

Qualitatively, it suggests that septa are closer to ends with a larger radius, which is what is experimentally observed in cells with ends of different radii (see for example in [27]).

8.4.4 New-End Take Off (NETO) Length

After fission, cell growth is monopolar (see Figure 3a). Later in the cycle, above a threshold length, both ends assemble the growth machinery and elongate. This phenomenon was named New End Take-Off (NETO) because the new growing end is elongating only above this length [12]. It is proposed that NETO is due to a threshold deformation occurring at this new end wall, which reduces the curvature at new end;

following my hypothesis, the growth machinery is assembled at this new end, which promotes its elongation.

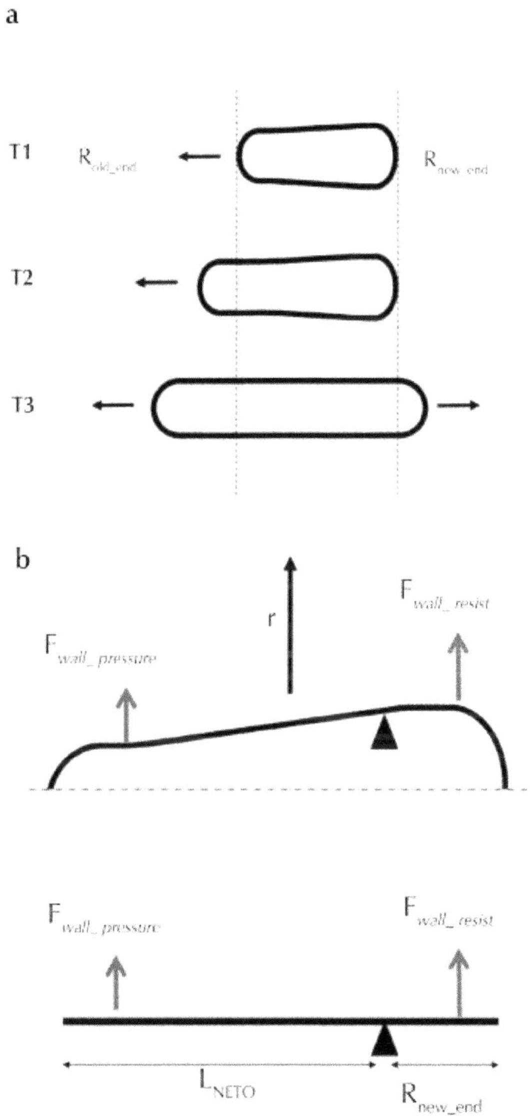

FIGURE 3 Scaling for NETO length: a/ Right after cytokinesis, only the "old end" elongates (T1 and T2); both radii of curvature Rold_end and Rnew_end are different; above the NETO length, both ends elongate (T3), with similar radii (dotted lines indicate ends locations for a cell attached on a substrate); b/ The equivalent mechanical model (top): the force applied at the old end wall with the lever arm L is opposed by the force of the resisting wall at the new end at a distance Rnew_end; this torque promotes the bending at the new end at NETO; 1-D representation is shown for simplicity (bottom).

Several features support this hypothesis: (i) following the Pascal principle, the pressure difference is the same in the cell; as a result, elongation should always occur at both ends; (ii) the old end radius of curvature is smaller than the new end radius of curvature before NETO (see [15, 28, 29]), while having an equal wall thickness (see for example electron microscopy images from Masako Osumi group [25]): the new end appears to be too rigid below a threshold cell length; in contrast, both radii are about the same after NETO.

It is proposed that the force at the old end triggers mechanically the bending at the new end after NETO, which reduces the radius of curvature at this end. As a consequence, tension is locally reduced, and the growth machinery is recruited locally, as suggested above: this "new end" elongates.

A simple model allows to extract the NETO length above which the radius of the new end decreases. The force associated with the pressure is perpendicular to the wall. As a result, two opposed torques appear along the longitudinal side of the cell at the wall (see Figure 3b). Specifically, two main forces along the radial axis are exerted on the wall at a distance L and Rnew_end respectively of a virtual pivot: the pushing force at the old end

$$F_{wall\_pressure} = \Delta P \pi R^2_{old\_end} \tag{12}$$

and the elastic force at the new end

$$F_{wall\_resist} = E \pi R^2_{new\_end} \tag{13}$$

At NETO, It suggests that the torques are equal:

$$F_{wall\_pressure} L_{NETO} \cong F_{wall\_resist} R_{new\_end} \tag{14}$$

By replacing both forces with their expressions (12) and (13), It can be written:

$$L_{NETO} \propto \frac{E}{\Delta P} \frac{R^3_{new\_end}}{R^2_{old\_end}} \tag{15}$$

Assuming Rnew_end = 2.2 μm and Rold_end = 2.0 μm, LNETO~20 μm is obtained.

Note that this model yields the proper order of magnitude for LNETO [12]. A thorough treatment of the model beyond the scope of this paper should allow the derivation of the prefactor for LNETO expression. This scaling law (15) could be probed in future experiments with thick mutants (see [30]): the length at NETO should increase with the cell radius.

8.4.5 The C-Shape

This approach can be used also to explain mutants shapes. For fission yeast, *ban* mutants with a curved shape (Figure 4) were isolated [13]. It is proposed that the cell wall buckles when a threshold pressure is imposed on the inner wall. The fission yeast is considered as a hollow cylinder of inner radius Rin of 1.8 μm and an outer radius Rout of 2.0 μm. The Euler formula gives the maximum axial load that a long, slender, ideal column can carry without buckling [31, 32]. It is set by

$$F_c = \pi^2 \left(\frac{EI}{L^2} \right) \tag{16}$$

with Fc critical force, E the Young modulus, L the length, and I the geometrical moment of inertia of cross section. This equation can be adapted directly by taking the threshold pressure given by

$$\Delta(\Delta Pc) = \frac{F_c}{A} \tag{17}$$

with A the surface of the cell wall under load $A = R_{in}L$. Above this threshold pressure, the cell buckles.

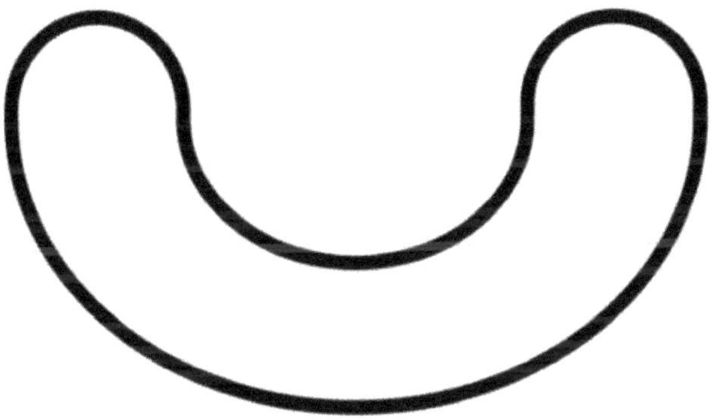

FIGURE 4 Scaling for a shape mutant: ban mutants exhibit a curved shape, suggesting a buckling phenomenon of the cell wall.

By replacing I by its expression [32], the increase can be estimated in pressure which triggers the cell buckling:

$$\Delta(\Delta Pc) = \frac{\pi^2}{8} \frac{E}{R_{in} L^3} \left(R_{out}^4 - R_{in}^4 \right) \tag{18}$$

Taking L = 10 μm, E = 100 MPa, Rout = 2.0 μm and Rin = 1.8 μm,

$$\Delta(\Delta Pc) \sim 0.3 MPa$$

is obtained

Note that this change in pressure is small compared to my estimate of ΔP = 10 MPa (see Eq. 4). It suggests a fine tuned connection between pressure differences and wall material addition during normal growth. In contrast, a delay in wall addition could cause the observed buckling of the ban mutants.

An experimental set-up similar to the study of microtubule buckling [31, 32] will allow to probe this prediction. By using two pipettes – a rigid one and a flexible one [33]-, a single yeast cell could be held and forced to buckle; by measuring the deflection of the flexible calibrated pipette, the estimate could be checked. In addition, varying the length of the cell undergoing buckling will permit to probe the relation (18): qualitatively, a longer cell will buckle for smaller applied forces.

8.5 CONCLUSION

Scaling arguments for typical read-outs used in fission yeast cellular studies are presented. Similar arguments should hold for other cell types with the appropriate modifications. For example, the wall tension for yeasts is due to its rigid wall whereas the tension for mammalian cells envelope is due to the cortical actin cytoskeleton [34] in addition, tugor pressure for yeast cells should be replaced by acto-myosin stress in mammalian cells [35, 36], by actin mediated forces in filopodia and lamellipodia [37], or by specific poroelasticity frameworks [38].

The rod shape of fission yeast is important for the arguments, but more than this specific shape, it is its *broken symmetry* which is essential for the reasoning. As a result, this treatment could be extrapolated to other cells. For budding yeast for example, the statement about the difference in surface tension could be used once the bud has emerged [39]. For mammalian cells, similar scaling arguments were tested on the actin cortex remodelling when a broken symmetry was generated [21].

In addition to lengths and shapes of this study, other microscopic read-outs for yeast would follow this logic. For example, it was recently shown that the volumes ratio of nucleus and cytoplasm was conserved in *S. pombe* [40] and in *S. cerevisiae* [41]. This conserved ratio may be derived using laws of chemical physics for dialysis. Altogether this scaling approach for cellular systems should allow to combine microscopic read-outs resulting from signalling networks together with quantitative matter properties.

KEYWORDS

- **Cytoskeleton**
- **Fission Yeast**
- **Molecular Mechanism**
- ***S. cerevisiae***
- ***S. pombe***

ACKNOWLEDGMENTS

I would like to thank Paul Nurse for insightful discussions and for his reading of the manuscript; Albert Libchaber for fruitful discussions; and Axel Buguin, Felice Kelly, Mark Lekarew, James Moseley, Frank Neumann, and the Nurse lab for stimulating discussions and comments.

REFERENCES

1. Nurse, P. Life, logic and information. *Nature* **454,** 424–426 (2008).
2. Nurse, P. Genetic control of cell size at cell division in yeast. *Nature* **256,** 547–551 (1975).
3. Hayles, J. and Nurse, P. A journey into space. *Nat Rev Mol Cell Biol* **2,** 647–656 (2001).
4. d'Arcy Thompson, W. *On growth and form*. Cambridge, Cambridge Univ. Press, (1917).
5. de Gennes, P. G. *Scaling concepts in polymer physics*. Ithaca, NJ, Cornell University Press, (1979).
6. de Gennes, P. G., Brochard Wyart, F., and Quéré, D. *Gouttes, bulles, perles et ondes*. Paris, Berlin, (2005).
7. Boudaoud, A. Growth of walled cells: from shells to vesicles. *Phys Rev Lett* **91,** 018104 (2003).
8. Moseley, J. B. and Nurse, P. Cdk1 and cell morphology: connections and directions. *Curr Opin Cell Biol* **21,** 82–88 (2009).
9. Galletta, B. J. and Cooper, J. A. Actin and endocytosis: mechanisms and phylogeny. *Curr Opin Cell Biol* **21,** 20–27 (2009).
10. Vavylonis, D., Wu, J. Q., Hao, S., O'Shaughnessy, B., and Pollard, T. D. Assembly mechanism of the contractile ring for cytokinesis by fission yeast. *Science* **319,** 97–100 (2008).
11. Chang, F. and Peter, M. Yeasts make their mark. *Nat Cell Biol* **5,** 294–299 (2003).
12. Mitchison, J. M. and Nurse, P. Growth in cell length in the fission yeast Schizosaccharomyces pombe. *J Cell Sci* **75,** 357–376 (1985).
13. Verde, F., Mata, J., and Nurse, P. Fission yeast cell morphogenesis: identification of new genes and analysis of their role during the cell cycle. *J Cell Biol* **131,** 1529–1538 (1995).
14. Martin, S. G., McDonald, W. H., Yates, J. R., and Chang, F. Tea4p links microtubule plus ends with the formin for3p in the establishment of cell polarity. *Dev Cell* **8,** 479–491 (2005).
15. Ge, W., Chew, T. G., Wachtler, V., Naqvi, S. N., and Balasubramanian, M. K. The novel fission yeast protein Pal1p interacts with Hip1-related Sla2p/End4p and is involved in cellular morphogenesis. *Mol Biol Cell* **16,** 4124–4138 (2005).
16. Sawin, K. E. and Nurse, P. Regulation of cell polarity by microtubules in fission yeast. *J Cell Biol* **142,** 457–471 (1998).
17. Raucher, D. and Sheetz, M. P. Characteristics of a membrane reservoir buffering membrane tension. *Biophys J* **77,** 1992–2002 (1999).
18. Sens, P. and Turner, M. S. Budded membrane microdomains as tension regulators. *Phys Rev E Stat Nonlin Soft Matter Phys* **73,** 031918 (2006).
19. Takagi, T., Ishijima, S. A., Ochi, H., and Osumi, M. Ultrastructure and behavior of actin cytoskeleton during cell wall formation in the fission yeast Schizosaccharomyces pombe. *J Electron Microsc (Tokyo)* **52,** 161–174 (2003).
20. Castagnetti, S., Novak, B., and Nurse, P. Microtubules offset growth site from the cell centre in fission yeast. *J Cell Sci* **120,** 2205–2213 (2007).
21. Paluch, E., Piel, M., Prost, J., Bornens, M., and Sykes, C. Cortical actomyosin breakage triggers shape oscillations in cells and cell fragments. *Biophys J* **89,** 724–733 (2005).
22. Terenna, C. R., Makushok, T., Velve-Casquillas, G., Baigl, D., Chen, Y. et al. Physical mechanisms redirecting cell polarity and cell shape in fission yeast. *Curr Biol* **18,** 1748–1753 (2008).
23. Sato, M., Kobori, H., Ishijima, S. A., Feng, Z. H., Hamada, K. et al. Schizosaccharomyces pombe is more sensitive to pressure stress than Saccharomyces cerevisiae. *Cell Struct Funct* **21,** 167–174 (1996).
24. Smith, A. E., Zhang, Z., Thomas, C. R., Moxham, K. E., and Middelberg, A. P. The mechanical properties of Saccharomyces cerevisiae. *Proc Natl Acad Sci USA* **97,** 9871–9874 (2000).
25. Konomi, M., Fujimoto, K., Toda, T., and Osumi, M. Characterization and behaviour of alpha-glucan synthase in Schizosaccharomyces pombe as revealed by electron microscopy. *Yeast* **20,** 427–438 (2003).
26. Money, N. P. and Harold, F. M. Extension growth of the water mold Achlya: interplay of turgor and wall strength. *Proc Natl Acad Sci U S A* **89,** 4245–4249 (1992).

27. Celton-Morizur, S., Racine, V., Sibarita, J. B., and Paoletti, A. Pom1 kinase links division plane position to cell polarity by regulating Mid1p cortical distribution. *J Cell Sci* **119,** 4710–4718 (2006).
28. Bahler, J. and Nurse, P. Fission yeast Pom1p kinase activity is cell cycle regulated and essential for cellular symmetry during growth and division. *Embo J* **20,** 1064–1073 (2001).
29. Snaith, H. A., Samejima, I., and Sawin, K. E. Multistep and multimode cortical anchoring of tea1p at cell tips in fission yeast. *Embo J* **24,** 3690–3699 (2005).
30. Das, M., Wiley, D. J., Medina, S., Vincent, H. A., Larrea, M. et al. Regulation of cell diameter, For3p localization, and cell symmetry by fission yeast Rho-GAP Rga4p. *Mol Biol Cell* **18,** 2090–2101 (2007).
31. Kurachi, M., Hoshi, M., and Tashiro, H. Buckling of a single microtubule by optical trapping forces: direct measurement of microtubule rigidity. *Cell Motil Cytoskeleton* **30,** 221–228 (1995).
32. Gittes, F., Mickey, B., Nettleton, J., and Howard, J. Flexural rigidity of microtubules and actin filaments measured from thermal fluctuations in shape. *J Cell Biol* **120,** 923–934 (1993).
33. Almagro, S., Dimitrov, S., Hirano, T., Vallade, M., and Riveline, D. Individual chromosomes as viscoelastic copolymers. *Europhys Lett* **63,** 908 (2003).
34. Pollard, T. D. and Borisy, G. G. Cellular motility driven by assembly and disassembly of actin filaments. *Cell* **112,** 453–465 (2003).
35. Balaban, N. Q., Schwarz, U. S., Riveline, D., Goichberg, P., Tzur, G. et al. Force and focal adhesion assembly: a close relationship studied using elastic micropatterned substrates. *Nat Cell Biol* **3,** 466–472 (2001).
36. Riveline, D., Zamir, E., Balaban, N. Q., Schwarz, U. S., Ishizaki, T. et al. Focal contacts as mechanosensors: externally applied local mechanical force induces growth of focal contacts by an mDia1-dependent and ROCK-independent mechanism. *J Cell Biol* **153,** 1175–1186 (2001).
37. Prost, J., Barbetta, C., and Joanny, J. F. Dynamical control of the shape and size of stereocilia and microvilli. *Biophys J* **93,** 1124–1133 (2007).
38. Charras, G. T., Yarrow, J. C., Horton, M. A., Mahadevan, L., and Mitchison, T. J. Non-equilibration of hydrostatic pressure in blebbing cells. *Nature* **435,** 365–369 (2005).
39. Nelson, W. J. Adaptation of core mechanisms to generate cell polarity. *Nature* **422,** 766–774 (2003).
40. Neumann, F. R. and Nurse, P. Nuclear size control in fission yeast. *J Cell Biol* **179,** 593–600 (2007).
41. Jorgensen, P., Edgington, N. P., Schneider, B. L., Rupes, I., Tyers, M. et al. The size of the nucleus increases as yeast cells grow. *Mol Biol Cell* **18,** 3523–3532 (2007).

9 Cellular Shapes from Protrusive and Adhesive Forces

Roie Shlomovitz, Kathrin Schloen, Theresia Stradal, and Nir S. Gov

CONTENTS

9.1 Introduction 157
9.2 Materials and Methods 163
9.2.1 Experimental Methods 163
9.2.2 Model Details 163
9.2.3 Equations of Motion 165
9.2.4 Linear Stability Analysis 168
9.2.5 Flat Geometry 168
9.2.6 Round Cell Geometry 169
9.2.7 Numerical Simulations 169
9.3 Discussion 170
9.3.1 Cellular Shapes 172
9.3.2 Evolution of Cellular Protrusions: Coalescence 179
9.3.3 Cell Motility 184
9.4 Results 185
9.4.1 Shape Evolution Driven by Adhesion (Aactin = 0,a>0) 186
9.4.2 Shape Evolution Driven by Actin Polymerization (Aactin = 0,a>0) 186
9.4.3 Dynamics of the Approach to the Steady-State Shape: Coalescence 187
9.4.4 Shape Evolution Driven by both Actin Polymerization and Adhesion . 187
9.5 Conclusions 188
Keywords 189
Acknowledgments 189
References 189

9.1 INTRODUCTION

The forces that arise from the actin cytoskeleton play a crucial role in determining the cell shape. These include protrusive forces due to actin polymerization and adhesion to the external matrix. A theoretical model for the cellular shapes resulting from the feedback between the membrane shape and the forces acting on the membrane, mediated by curvature-sensitive membrane complexes of a convex shape is presented. In previous theoretical chapters the regimes of linear instability were investigated where spontaneous formation of cellular protrusions is initiated. Here we calculate the evolution of a two dimensional cell contour beyond the linear regime and determine the final steady-state shapes arising within the model. We find that shapes driven by adhesion or by actin polymerization (lamellipodia) have very different morphologies, as observed in cells. Furthermore, we find that as the strength of the protrusive forces diminish, the system approaches a stabilization of a periodic pattern of protrusions. This result can provide an explanation for a number of puzzling experimental observations regarding cellular shape dependence on the properties of the extra-cellular matrix.

Cells have highly varied and dynamic shapes, which are determined by internal forces generated by the cytoskeleton. These forces include protrusive forces due to the formation of new internal fibers and forces produced due to attachment of the cell to an external substrate. A long standing challenge is to explain how the myriad components of the cytoskeleton self-organize to form the observed shapes of cells. A theoretical study of the shapes of cells is presented that are driven only by protrusive forces of two types; one is the force due to polymerization of actin filaments which acts as an internal pressure on the membrane, and the second is the force due to adhesion between the membrane and external substrate. The key property is that both forces are localized on the cell membrane by protein complexes that have convex spontaneous curvature. This leads to a positive feedback that destabilizes the uniform cell shape and induces the spontaneous formation of patterns. We compare the resulting patterns to observed cellular shapes and find good agreement, which allows us to explain some of the puzzling dependencies of cell shapes on the properties of the surrounding matrix.

The factors that determine the local and global shape of a cell, are numerous, including the internal state of the cell, with respect to the cell cycle and metabolism, and the properties of the extra-cellular matrix (ECM). Cells that are round while floating in solution, change their shapes dramatically when in contact with a solid substrate [1–5]. On a two dimensional surface some cells spread uniformly, while others form elongated extensions (filopodia), or form motile fan-shaped lamellipodia. Inside a three dimensional matrix, cells extend protrusions through their ability to penetrate between the matrix filaments, and by degrading the surrounding material [6–8]. These processes have been widely studied in recent years due to the interest in cell motility in normal and cancerous cells, and in relation to the observed dependence of stem-cell differentiation on the properties of the surrounding matrix. Providing a unified model for this large variety of cellular behaviors is difficult, and it is aimed here to explore the consequences of a relatively simple model, which describes some of the principle forces acting on the cell membrane.

There are several examples of puzzling cellular shape dependencies that have been observed in recent years; (i) Developing neuronal cells have been shown to produce more (less) numerous and shorter (longer) protrusions, when the cells had less (more) actin filament polymerization [9]. (ii) In [7] cells encapsulated in a three-dimensional matrix have been found to have more (less) numerous and shorter (longer) protrusions, when the surrounding gel was stiffer (softer) and therefore harder (easier) to degrade.

While the two examples given above studied the static shapes of cells, there are several studies which investigated the dynamics of cellular shape changes; (i) In [10] the polarization of adhering cells was followed in time, and it was observed that cells initially form numerous and short adhesion "spikes" along the cell perimeter, which later (up to 24hrs) reorganized into two large adhesion regions at the opposite poles of the final elongated cell shape. (ii) In [11] it was observed that cells on a flat substrate, can spontaneously change their shape and cytoskeleton organization between three prototypical forms, which are round (featureless), spiky or ruffled. The cells seemed to randomly switch between these three morphologies over the time course of the experiments.

In recent years, experiments have implicated a large family of curved membrane proteins, for example those containing Bin/Amphiphysin/Rvs (BAR) and IRSp53-Missing-In-Metastasis (IMD) domains, as responsible for sensing (and inducing) concave or convex curvature [12]. Such curved proteins bind preferentially to curved membranes [13–15] and the curved domains have also been shown to tubulate membranes [16]. Furthermore, these curved proteins are known to form membrane-bound protein complexes (membrane protein) that include actin-activating components, such as WASP and WAVE [17, 18], and have been found to localize and induce cellular protrusions (filopodia) [19–22], and at the leading edge of lamellipodia [23, 24].

In several previous theoretical studies it is described how such membrane complexes that have both convex curvature and promote actin polymerization, can induce the spontaneous initiation of membrane protrusions [25, 26]. The intrinsic curvature is an essential since it completes the positive feedback between the membrane shape and local density of membrane proteins; only due to the curvature sensitivity of the proteins do they flow towards the protruding curved parts of the membrane, thereby increasing the cytoskeletal forces acting there on the membrane and leading to the instability and spontaneous formation of protrusions.

Furthermore, it has been shown that the adhesion molecules that connect the cell membrane to the external substrate (such as integrins) aggregate at regions of high convex membrane curvature [5, 6, 27, 28], at the leading edge of motile cells and at cellular protrusions, such as microvilli [29, 30] and filopodia [31].

Thus, it has been previously proposed [26] to treat the adhesion molecules as part of the same convex membrane protein that is also responsible for the recruitment of actin polymerization to the membrane (this simplification is discussed further in the Model Details). The protrusive force in this model can therefore originate either from the reduction in the effective membrane tension due to the adhesion with the extracellular matrix or from the force of actin polymerization (Figure 1a). For simplicity, it is assumed that these are the two dominant forces that determine the cell shape; direct contractile forces applied to the membrane are neglected. Also, the role of microtubules (MT) in determining the cell shape is not described here.

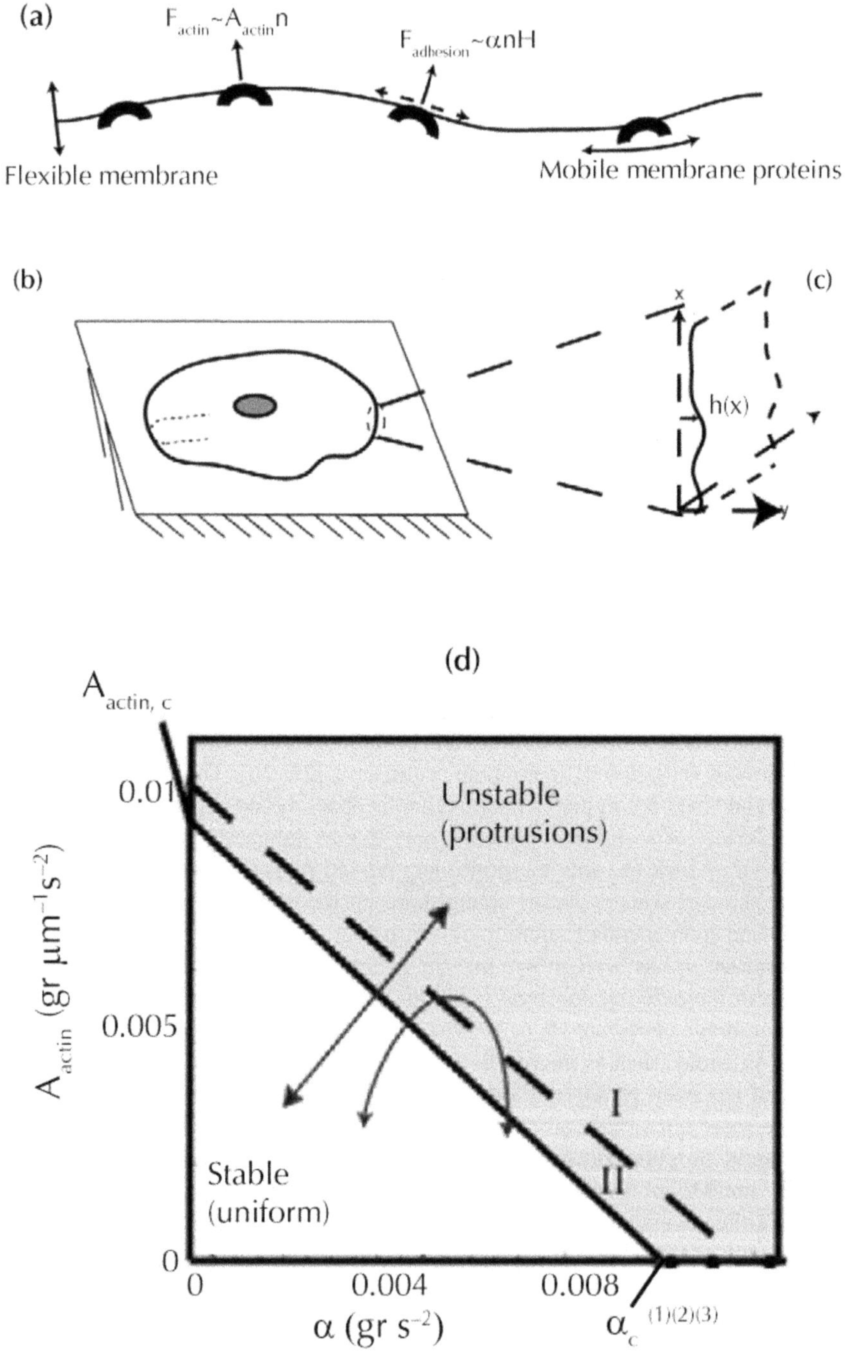
(a)
F actin ~A actin n
F adhesion ~αnH
Flexible membrane
Mobile membrane proteins
(b)
(c)
x
h(x)
y
(d)
A actin, c
0.01
0.005
0
A actin (gr μm⁻¹s⁻²)
Unstable
(protrusions)
Stable
(uniform)
I
II
0
0.004
0.008
α (gr s⁻²)
α c (1)(2)(3)

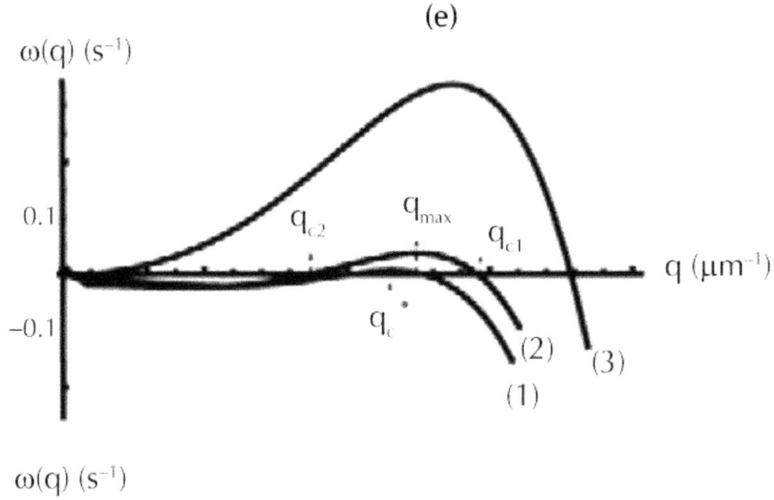

FIGURE 1 Model scheme and linear stability diagram. (a) Schematic description of the model ingredients: a one dimensional flexible membrane contour, with convex and mobile membrane proteins, which induce normal protrusive forces, due to actin (F_{actin}) and due to adhesion-driven tension reduction (dashed arrows, $F_{adhesion}$). Both forces are linearly proportion to the local membrane protein concentration n, which in this coarse-grained model is treated as a uniform field on length-scales larger than those of the individual proteins. The one-dimensional membrane contour geometries that is calculated (b) round geometry representing the outer contour of a spread cell on a flat substrate, and (c) the flat geometry which describes either a segment of the cell contour or a membrane with translational symmetry. The variable $h(x)$ gives the local height deformation of the membrane, relative to its uniform configuration. The curvature at the rim along the cell thickness is indicated by the thin dotted line in (b), and is not considered in two-dimensional analysis. (d) The phase diagram obtained from the linear stability analysis as a function of the actin protrusive force (A_{actin}) and adhesive strength (x). Two regimes are found: Stable (uniform) state (below the solid line), and unstable above (gray region, for A_{actin} or $\alpha > \alpha_c$). In the unstable regime, type *I* dispersion relation above the dashed line has been found, and type *II* dispersion relation between the solid and dashed lines. The correlation between the actin polymerization and the adhesion strengths is illustrated by two possible trajectories (lines with arrows) in this phase space. (e) The dispersion relations for three different values of α (numbered and marked by bold dots along the α axis in (c)). Negative values of the dispersion corresponds to stable modes, and positive values corresponds to unstable modes. (f) An example of the dispersion relation for the round cell; only the values at integer qR_0 play a role. The real part (solid line) is zero at $qR_0=1$ which is an asymmetric mode of translation, while it can be unstable for higher modes.

This model does not explicitly describe the dynamics of the cytoskeleton in the cell interior, only close to the membrane. Specifically, the role of contractility induced by myosin-II in the actin network, is not directly accounted for. However, it is possible to effectively take the role of this contractility into account in the present model through the following parameters; (i) The direct contraction of the membrane inwards due to myosin activity, may be included in the values of the effective tension parameter (σ) and the effective bulk modulus for the cell's projected area (K). Experiments indicate that both parameters are stiffer when myosin contractility is present, and are softer when it is absent [32–34]. (ii) In addition, myosin activity is critical for the formation and maturation of adhesion contacts [35]. Therefore the adhesion strength parameter (x) is very much dependent on the myosin activity, again qualitatively. (iii) Myosin activity can furthermore affect the turn-over of the actin and thus modify the rate of actin polymerization, making the parameter (A_{actin}) also myosin dependent [36]. While the exact relation between these parameters and the activity of myosin is not known, qualitatively the dependencies should be as described here.

The model presented here is meant to explore the dynamics of cellular shapes driven by the coupling of the cytoskeletal forces with the membrane through curved proteins that can recruit the cytoskeleton activity to the membrane. As a step in this direction, it is therefore important to first understand the behavior of this coupling and feedback, before adding to the model further layers of realism and complexity (which are currently absent). This is the basic philosophy of our approach. The treatment that is presented here is general and is not limited to a particular set of curved membrane proteins.

This model is a coarse-grained model, whereby the detailed of the molecular-scale level is not described. The minimal length-scale along the membrane that is relevant to this model is of order 100 nm. The model is written as a set of equations of motion for the continuum fields that describe the membrane shape and density of membrane proteins, including the actual forces acting on the membrane, and the details of the membrane elasticity.

Other coarse-grained models were recently proposed [37, 38]. These models take a much more detailed description of the actin gel that is pushing the membrane, and the dynamics within the gel away from the membrane itself. Another recent model [39] treats the shape evolution of the cell in terms of a spreading layer of fluid, and relates the instabilities that initiate protrusions to the behavior of such a fluid. These models do not contain however the key component that our model was set up to explore, which is the role of curved membrane proteins that recruit the cytoskeleton forces to the membrane. Other types of models, such as [40], deal with an even more coarse-grained view of the cell. In such models the actual forces acting on the membrane and the membrane elasticity are not explicitly calculated. They are replaced by a kinematic model for the shape evolution, taking into account the biochemical signals that act locally on the cell membrane. These signals represent external and internal pathways that eventually control the cytoskeleton and lead to membrane motion. The huge complexity of the cytoskeleton and membrane dynamics makes it highly beneficial to explore many simplified models, each exploring the consequences of a small set of mechanisms and at different length and time-scales. From the study of these various models we will gain a deeper understanding regarding the many entangled mechanisms that interact within the real cell. It may well be the case that different mechanisms are dominant under different conditions, and therefore control cell morphology under these circumstances.

The model which is presented here was previously analyzed in the linear limit of small deviations from a uniform flat state [25, 26], which therefore only gave information about the initiation stage of membrane protrusions. In the present chapter the dynamics of the protrusions beyond the linear limit, and for closed shapes are explored. Despite the simplicity of this model, which does not describe all the forces that can arise within cells, it may provide a general understanding of the shapes driven by the coupling of the cytoskeleton to the membrane, and shed light on the above mentioned puzzling experimental observations.

9.2 MATERIALS AND METHODS

9.2.1 Experimental Methods

Murine fibroblastoid cells were as described before [41, 42]. Cells were grown in DMEM, 4.5 g/L glucose (Invitrogen, Karlsruhe, Germany) with 10% FCS (Sigma-Aldrich, Munich Germany), 2 mM glutamine, 1 mM sodium-pyruvate, 0.1 mM non-essential amino acids (Invitrogen) at 37°C and 7.5% CO_2. For visualization of the actin cytoskeleton cells were grown on acid-washed glass-coverslips. For the induction of lamellipodia cells were seeded sub-confluently onto glass-coverslips, serum-starved over night, and treated with DMEM alone or DMEM containing 10 ng/ml PDGF-BB (PDGF-BB; Sigma-Aldrich) for 5 min prior to fixation. Cells were then fixed with formaldehyde (4%) in PBS for 20 min, extracted with 0.1% Triton X-100 in 4% PFA for 1 min, and stained with Alexa dye-labelled phalloidin (Invitrogen) as described. Samples were analyzed on an inverted microscope (Axiovert 100TV; Zeiss, Jena Germany) using a 63x/1.4-numerical aperture plan-apochromatic objective and equipped for epifluorescence as described previously [43]. Images were acquired with a back-illuminated, cooled charge-coupled-device camera (TE-CCD 800PB; Princeton Scientific Instruments, Princeton, NJ, USA) driven by IPLab software (Scanalytics Inc., Fairfax, VA, USA). All microscopic images were further processed with Adobe Photoshop 7.0/CS software (Adobe Systems, Mountain View, CA, USA).

9.2.2 Model Details

In this chapter the theoretical model that is used is described. It is started with the derivation of the basic equations of motion for the membrane shape and density of membrane proteins. Then the linear stability analysis of the flat and round membrane shapes is described.

The geometries which will be explored in this work are only two-dimensional. In three dimensions there is the additional degree of freedom of the membrane proteins to re-orient themselves in response to changes in the local curvature tensor [44]. Their dynamics and distribution along the membrane will therefore be modified, and may well affect the long-time evolution of the membrane protrusions, while the short-time linear regime will not be very much affected by the dimensionality.

The first geometry is that of a round closed shape (Figure 1b), which can describe the outer contour of a flat cell that is adhered on a substrate. Such a geometry arises in many experiments where cells spread over a solid surface. In this geometry it is assumed that the cell has a preferred overall projected area (S_{target}), which it tries to maintain while its shape is evolving [38].

The second geometry describes a segment of the cell outer contour, where the membrane is initially straight (Figure 1c). This geometry can also describe a flat two-dimensional membrane, under the constraint of having undulations with translational symmetry. In this geometry we need to pin the membrane using an external harmonic force (F_{spring}, Eq.6), to prevent its drifting motion. Such an external force mimics the effects due to the adhesion of the rest of the cell to an external substrate.

Note that we do not explicitly describe the membrane shape along the cell thickness (Figure 1b). If the membrane curvature along the thickness is roughly constant, then it simply enters our calculation as a modified membrane tension and adhesion strength, as well as changing the value of S_{target} in Eq. 2 (see more details in Text S1).

Regarding the membrane proteins in our model, it is assumed that their overall number on the membrane is conserved, and that they are allowed to dynamically move along the fluid membrane.

In this model the adhesion and actin protrusive forces are described by two independent parameters [26] (α and A_{actin}, Figure 1d). In the cell the actin polymerization activity and adhesion are closely related [45, 46]; actin polymerization and treadmilling induces the initiation of focal adhesions [47], so the adhesion strength (α) increases with the actin polymerization activity (A_{actin}). On the other hand, the protrusive force on the membrane due to actin polymerization is found to depend in a biphasic manner on the adhesion strength [1, 2, 4]; it increases for low adhesions as the traction with the substrate increases, but eventually decreases as strong adhesion stalls the membrane. These two types of dependencies are indicated by the trajectories drawn in Figure 1d. Furthermore, the actin organization is different where adhesion or polymerization dominate; mature adhesion sites have internal actin stress fibers that have little direct contact with the membrane, while regions dominated by actin polymerization, such as filopodia and lamelipodia, have actin filaments that push actively the membrane surface. These two different scenarios lead us to describe these two forces by two independent parameters α and A_{actin}.

Inside cells the actin polymerization rate should also depend on the local membrane restoring force applied to the growing tips [48]. This relation is not well understood inside the cell, where the mechanism for polymerization depends on the membrane composition and type of actin nucleator. This effect can be implemented in our model by making the polymerization rate A_{actin} dependent on the membrane force, when this force opposes the actin protrusion [49]. This effect does not modify the (linear) stability of the system, but does change the shapes and dynamics of the resulting membrane undulations. In Text S1 an example of a calculation is given where it demonstrates the effects that this adds (Figure S1). For simplicity a constant (uniform) A_{actin} for the rest of this chapter is used. Note that it is easy to implement within our model a non-uniform actin polymerization rate or adhesion strength (A_{actin}, α) by making these parameters dependent on the local density of membrane proteins, membrane shape etc.

The actin polymerization in this model produces a force that is acting normally on the membrane, similar to an internal pressure force (see Eq.2). This may be a good description for Arp2/3-induced actin polymerization where a rather uniform network of filaments, with a distribution of angles, is protruding against the membrane surface. Actin polymerization that is induced by Formin-type proteins tends to be unidirectional [43], and the actin-membrane interactions of the resulting actin bundle can be strongly influenced by binding proteins and molecular motors [50, 51]. These additional effects are not explicitly treated here, and could be added in the future.

Another important point to note regarding this model, is that the adhesion is treated as a localized event on the membrane surface, while in adhering cells mature adhesion sites require stress-fibers that link adhesion domains on two distant locations on the membrane [45]. This non-local feature of adhesion is absent from the present model.

9.2.3 Equations of Motion

This model investigates the coupling between both the adhesion and the actin protrusion forces to the membrane curvature. We give below the free energy expression used in the model, from which we derive the equations of motion of the membrane shape and density distribution.

The continuum free energy for the model is based on the Helfrich form [52], including the membrane proteins interactions and entropy [26].

$$\mathcal{F} = \int \Big(\frac{1}{2}\kappa(H - \bar{H}n)^2 + (\sigma - \alpha n) + \gamma h^2 + k_B T n_s n(\log(n) - 1)) \\ - Jn^2 + \frac{J}{n_s}(\nabla n)^2\Big) ds, \quad (1)$$

where κ is the membrane bending rigidity, H is the local mean membrane curvature, $\bar{H}$ the intrinsic curvature of the membrane protein, n is the fractional area coverage of the membrane by the membrane proteins, n_s is the saturation density of membrane proteins on the membrane, σ is the membrane tension, α is a proportionality constant describing the effective adhesion interaction between the membrane proteins and the external substrate, γ is a restoring spring term, J is the direct binding interaction energy between the membrane proteins, and $ds = d.dl$ is an element of membrane area, where d is the thickness of membrane represented by our contour and dl is a line element along the membrane contour.

The first term in Eq.1 gives the curvature energy due to the mismatch between the membrane curvature and the spontaneous curvature of the membrane protein. The second term describes the negative contribution to the effective membrane tension, induced by the adhesion molecules. The third term describes an external harmonic potential that pins the membrane (in the flat geometry), representing the overall localization of the cell to the external matrix. The fourth term gives the entropic contribution due to the thermal motion of the membrane protein in the membrane. The fifth and sixth terms are the bulk and surface aggregation energies of the membrane protein.

Note that for simplicity in this model we have a single species of membrane protein complexes, described by the field n, which has the ability to both recruit actin polymerization and/or adhesion. In reality these two properties may exist on two (or more) independent membrane complexes, with different curvatures and interactions ($\bar{H}$ and J in Eq.1). Such an increased level of detail, and complexity, can be introduced in future elaborations of this model.

The equations of motions next are derived for a general contour in two dimensions, using the variation of the free energy (Eq.1) with respect to the membrane coordinate [53, 54] and membrane protein concentration [54], and adding the active forces due to actin polymerization which cannot be derived from the free energy [25, 26]. To take into account the drag force on the cell membrane due to viscous forces, it is assumed for simplicity only local friction forces [26, 54], with overall coefficient ξ. Note that the local friction coefficient for membrane motion contains also the effects of adhesion [26], so that: $\xi = \xi_0(1 + g(\alpha)n)$, where $g(\alpha)$ is some increasing function of the adhesion strength α, representing the stick-slip nature of the adhesive bonds [55, 56]. This term leads to non-linear effects, which do not modify the (linear) stability of the system or its qualitative dynamics, simply slows them down.

Before the equations of motion are given let us note again that we are interested in two geometries of the membrane; one is a closed, round shape which describes a whole cell, while the second is a flat membrane that describes a segment of the entire cell. For the round shape the equation of motion of the contour coordinate, $\vec{r}(s,t) = (x(s,t), y(s,t))$, is given by

$$\xi \frac{\partial \vec{r}}{\partial t} \cdot \vec{n} = -\frac{\delta F(s,t)}{\delta h} + A_{actin} n + K(S(t) - S_{target}) \qquad (2)$$

where s is the index along the contour length, A_{actin} is the proportionality factor representing the actin protrusive activity induced by the membrane proteins, K is an effective bulk modulus for the cell's projected area, $S(t)$ is the area enclosed by the contour, and $\vec{n}$ is a unit vector normal to the contour. The variation of the free energy is projected to give the forces normal to the membrane contour [53, 54]. The protrusive force of the actin is assumed to grow linearly with the local concentration of membrane protein [25, 26] and the area-preserving forces act as a global internal pressure. Both of these forces act normal to the membrane. Note that in this geometry we do not use the spring term in the free energy (γ=0 in Eq.1).

For the flat geometry the equation of motion of the membrane height $h(x)$ is given by

$$\xi \frac{\partial h}{\partial t} = -\frac{\delta F(s,t)}{\delta h} + A_{actin}(n - \langle n \rangle) \qquad (3)$$

where the average actin force is subtracted to prevent the membrane drift, and $\langle n \rangle$ is the average fractional area coverage of the membrane proteins along the membrane contour. Additionally, in the free energy a non-zero spring term is used to prevent an overall drift of the membrane. In this geometry only the projection of the force along the y-axis is taken, since the position of the membrane along the x-axis is fixed.

Now the forces (per unit area) derived from the variation of the free energy are listed (Eq.1) [53, 54]

$$F_{curv} = \kappa\left(-\nabla^2 H + \bar{H}\nabla^2 n + \frac{1}{2}n^2 \bar{H}^2 H - \frac{1}{2}H^3\right) \qquad (4)$$

$$F_{tension} = (\sigma - \alpha n)H \qquad (5)$$

$$F_{spring} = -2\gamma h \tag{6}$$

$$F_J = J\left(-n^2 + \frac{1}{n_s}(\nabla n)^2\right)H \tag{7}$$

where F_{curv} is the force due to the curvature energy mismatch between the membrane curvature and the spontaneous curvature of the membrane proteins, $F_{tension}$ is the membrane tension force, F_{spring} is the harmonic pinning force and F_J is the force due to the aggregation potential of the membrane proteins. There is in addition a force arising from the entropy of the membrane proteins in the membrane, which acts to expand the length of the contour, and has the form: $F_{entropy} = k_B T n_{sat}(n\log n - 1)H$. This force has been neglected in our calculations, since it is smaller than the other forces. All the derivatives are along the contour length (s).

In a cell the membrane area is finite and this leads to a non-linear form for the effective membrane tension [57]

$$\sigma = \sigma_0 \exp[\beta(L(t) - L_0)] \tag{8}$$

where $L(t)$ is the contour length, L_0 is the initial contour length and β is the factor that determines the length-scale at which the non-linear growth in the tension sets in. This restraint on the amplitude of membrane undulations also allowed us to avoid kinetically trapped configurations, by preventing the strong depletion of the membrane protein density which would have slowed down the evolution of the system, since it depends on the currents of membrane protein flowing in the membrane. Regions where the membrane protein density is highly depleted act as effective barriers for such flows [58]. A previous study [58] suggests that the steady-state of the system is unaffected by this limitation, only the rate at which the system is able to approach this steady-state.

Now the dynamics of the membrane protein density is calculated, using the following conservation equation (covariant version [59]) and the free energy F [60] (Eq.1)

$$\frac{\partial n}{\partial t} = -\nabla\cdot\vec{J} = \frac{\Lambda}{n_s}\nabla\left(n\nabla\frac{\delta F}{\delta n}\right) - \frac{n}{\sqrt{g}}\frac{\partial\sqrt{g}}{\partial t} \tag{9}$$

where $\wedge$ is the mobility of the proteins in the membrane and $\vec{J}$ is the total current of membrane proteins on the membrane, which includes the following terms

$$J_{curv} = \frac{\kappa\Lambda\bar{H}}{n_s} n\nabla H \tag{10}$$

$$J_{disp} = -\frac{\kappa\Lambda\bar{H}^2}{n_s} n\nabla n \tag{11}$$

$$J_{agg} = \frac{2J\Lambda}{n_s} n\nabla n + \frac{2J\Lambda}{n_s^2} n\nabla^3 n \tag{12}$$

$$J_{diff} = -D\nabla n \tag{13}$$

where J_{curv} is the flux resulting from the interaction between the membrane proteins through the membrane curvature, J_{disp} is the dispersion flux due to the membrane resistance to membrane protein aggregation due to their membrane bending effects, J_{agg} is the flux due to the direct membrane protein aggregation interactions, and Jdiff is the usual thermal diffusion flux, which depends on the diffusion coefficient, $D=\Lambda k_B T$.

The last term in Eq.(9) arises from the covariant derivative of the density with time on a contour whose length evolves with time [59]. In this term $\sqrt{g}$ is the matrix tensor, which in our one dimensional contour is simply the line element *dl*. This term ensures that the total number of membrane protein is conserved as the contour length changes.

Note that here a constant value for the mobility of the membrane proteins has been used (Λ), but this mobility is in reality diminished with increasing adhesion strength α. Furthermore, crowding effects in the membrane decrease the mobility with increasing local concentration of membrane proteins *n*. We checked that both of these effects do not qualitatively change the results that we present in this chapter, where for simplicity Λ is taken (and therefore also *D*) to be a constant, independent of α, *n* or the local shape of the membrane [61].

9.2.4 Linear Stability Analysis

Next a linear stability analysis of the model is performed, as was previously done for the flat case [26], in order to find the regions of instability of the system, and for the round case to calculate the equilibrium radius and membrane protein density. Note that we will consider only convex shape for the membrane proteins ($\bar{H}$ <0), in order to get instabilities in the dynamics of this system [25, 26].

9.2.5 Flat Geometry

In the flat geometry the contour is allowed to evolve only along one direction and the amplitude of the membrane height fluctuation is labeled as *h*(*x*) (Monge representation), where *x* is the coordinate along the initial contour length. In this representation the linearized curvature is: $H \simeq \nabla^2 h$, and the length element of the contour *dl* is given by: $dl = 1 + (\nabla h)^2 / 2$. Linearizing the equations of motion (Eqs.3–9), we then Fourier transform to get a 2×2 matrix whose eigenvalues $\omega(q)$ give the dynamic evolution of small fluctuations from the equilibrium uniform state. Both eigenvalues are real, and one of them is always negative and therefore represents only damped modes. The second solution can become positive in some a range of wavevectors and for certain parameters of the model, representing unstable modes that grow with time. The parameters of the model that represent the effects of actin polymerization and of adhesion are A_{actin} and α (Eqs.2,3,5).

Fixing all the other parameters, we plot in Figure 1d the phase-diagram from the linear stability analysis as a function of A_{actin} and α [26], where the three numbered points correspond to the adhesion strength used to calculate the dispersion relations given in Figure 1e. Below the solid line in Figure 1d the system is linearly stable and uniform, while above this line there are unstable modes and spontaneous patterns are initiated. The instability is driven by the positive feedback between the membrane shape and the density of membrane proteins [25, 26], due to their induction of protrusive forces and convex spontaneous curvature. Between the solid and dashed lines there is a growing range of wavevectors, $q_{c1}>q>q_{c2}$, which are unstable (Type-II). Above the dashed line this range extends to zero wavevectors ($q_{c2}=0$, Type-I). The most unstable mode (largest positive $\omega(q)$ is denoted by wavevector q_{max}. Along the solid line, the type-II instability first appears at a wavevector q_c^*, which increases as α (A_{actin}) increases (decreases).

9.2.6 Round Cell Geometry

For the round geometry, a similar linear stability analysis is followed, where the curvature by the following expansion is replaced (as a function of the angle θ)

$$H(\theta)=-\frac{1}{R_0}+\frac{h(\theta)+\ddot{h}(\theta)}{R_0^2}-\frac{2h(\theta)^2+\dot{h}(\theta)^2+4h(\theta)\ddot{h}(\theta)}{2R_0^3} \quad (14)$$

where R_0 is the initial radius and $h(\theta)$ is the deformation of the membrane in the radial direction, and the differentiation is with respect to θ. The line element dl, is also expanded up to quadratic order

$$dl=R_0+h(\theta)+\frac{\dot{h}(\theta)^2}{2R_0}. \quad (15)$$

The differences compared to the flat geometry is that now a spring energy term is not included, and the actin protrusive force does not have the mean force subtracted. Note also that the area-conserving term appearing in Eq.(2) does not contribute to the linear stability analysis. Following the same methods as described for the flat shape membrane, the equations of motion are linearized, and solved the dispersion relation. In this geometry there is a uniform force acting on the membrane, which vanishes for the initial equilibrium circular shape. This condition determines the initial radius R_0 and the initial uniform membrane protein area coverage which is defined by: $n_0=N_t/(2\pi n_s R_0)$, where N_t is the total number of membrane proteins on the initial contour. The dispersion relations for the circular geometry are very similar to those shown for the flat case in Figure 1e, while the wavevector q_0 has only integer values.

9.2.7 Numerical Simulations

The main results of this work are calculated using numerical simulations of the dynamics of our model system beyond the linear limit. For this purpose Eqs.(2–9) are solved using an explicit Euler method in Matlab. We checked for the convergence of our one-dimensional simulations, in space and time. Occasional re-discretization of the contour are used into equally spaced nodes, using the cubic "spline" routine, to prevent large changes in discretization density along the contour as it evolves with time. After such operation the membrane protein density is re-distributed among the new node locations using a linear interpolation algorithm, such that the total membrane protein number is conserved.

In the flat geometry the Monge representation *h(x)* is used, which leads to a complex curvature restoring force, but in order to simplify the numerics the curvature force only up to linear order are eventually kept. The boundary conditions on the flat geometry were taken to be periodic, for simplicity. In this geometry, the membrane moves only along the direction perpendicular to the initial flat state.

9.3 DISCUSSION

Now the results of this model are compared with observations on the shapes of cells. Before it is done we must be aware of the following complication; the membrane shape calculated in this model is dynamically evolving through coalescence of protrusions, and the time-scale for this evolution becomes very long as the critical values of adhesion and actin polymerization are approached (see Figure 2a,c). A living cell produces local actin or adhesion driven structures over time-scales that vary from several minutes to tens of hours, so when comparing to the calculated shapes it needs to be aware that the cellular shapes do not necessarily correspond to the steady-state shapes we predict at very long times. Cells also move around (even adhering cells), and divide and therefore drastically reorganize their cytoskeleton over time-scales that correspond to these two processes. Over such time-scales the cytoskeleton is "reset", and new features start growing from initial perturbations.

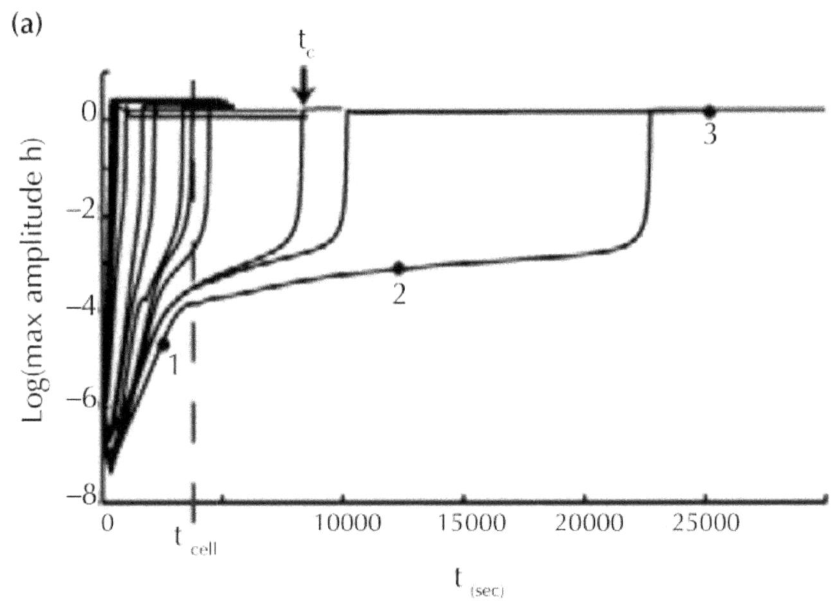

FIGURE 2 *(Continued)*

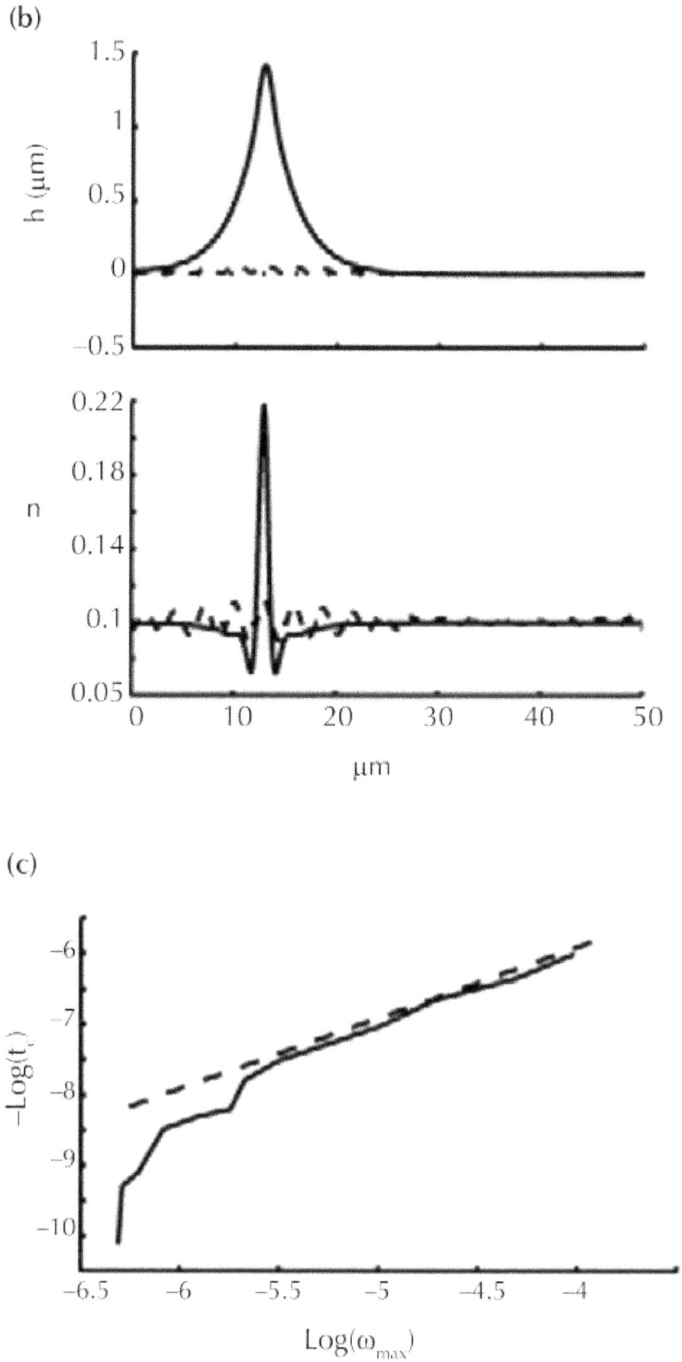

FIGURE 2 *(Continued)*

(d)

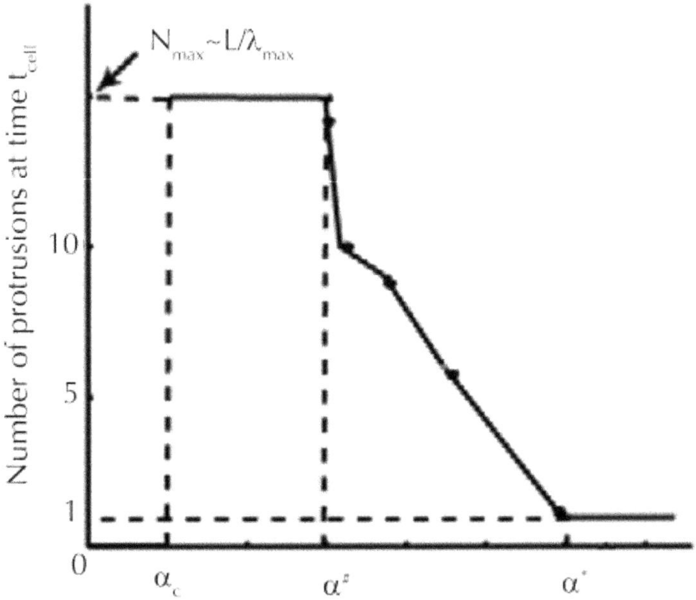

FIGURE 2 Dynamics of protrusion coalescence. (a) Plot of the maximum membrane amplitude (with respect to the average membrane shape) as a function of time, for decreasing values of the adhesive strength (left to right), approaching the critical values $\alpha_c = 0.01005\text{gr/sec}^2$ (for $A_{actin} = 0$, Figure 1d). The time when a single steady-state protrusion forms is denoted by t_c, shown for example for one of the trajectories. (b) Typical membrane shapes (h) and membrane protein density distributions (n) at the marked time points 1-3 in (a). Time point (1) is during the exponential growth of the most unstable mode at wavelength λ_{max} (dotted line), time point (2) is during the stalling in the dynamics as the non-linear tension stabilizes the undulations (dashed line), and time point (3) is at the steady-state of a single collapsed protrusion (solid line). (c) Log-log plot of the observed coalescence time t_c as a function of the maximal value of the dispersion relation ω_{max}. A linear relation at large values of ω_{max} (short t_c, dashed line) are found. A non-linear relation is observed for small values of ω_{max}, where t_c seems to diverge faster. (d) Plot of the total number of protrusions as a function of the adhesion strength x, measured at a time representing the cellular time scale t_{cell} and denoted in (a) by the vertical dashed line (chosen for illustration to be t = 3800 sec). The (arbitrary) threshold for a membrane protrusion to be counted is to be at least 20% of the maximum amplitude. Above α* all the simulations reach steady-state before t_{cell} and the total number of protrusions is 1. Within the range $\alpha^{\#} < \alpha < \alpha^{*}$ the number of peaks grows with decreasing adhesion strength, till it reaches a maximum number: $N_{max} = L/\lambda_{max}$. This value then remains unchanged until the critical value of the stability transition α_c, below which there are no protrusions forming and the total number of peaks collapses discontinuously to zero.

9.3.1 Cellular Shapes

The essential feature of this model is the feedback between the symmetry breaking of the membrane shape and the polarization of the cortical cytoskeleton, i.e. one cannot occur without the other. A recent study [62] demonstrates this property in a cell, where the shape was fixed by an external rigid confinement. It was found that when the confinement imposes a uniform shape, the polarization of the cytoskeleton (excited by an external signal) is transient and decayed rapidly to a uniform state. This observation strengthens the basic mechanism of this model.

Now the calculated steady-state shapes are compared with those of adhering cells on a two-dimensional surface, which are not very motile (Figure 3). On a qualitative level it is seen (Figure 3a,b) that regions of the cell that have strong adhesion (marked by stress fibers) tend to have a pointed tent-like protrusion, as it was calculated, while regions that are dominated by lamellipodia-like protrusions (marked by diffuse cortical actin) have a fan-like shape similar to those given by our model. Overlap between adhesion domains and lamellipodia can be seen in Figure 3b, where adhesion sites seem to serve as platforms for new lamellipodia or vice-versa. Such complex dynamics of overlapping structures is beyond the current version of our simple model.

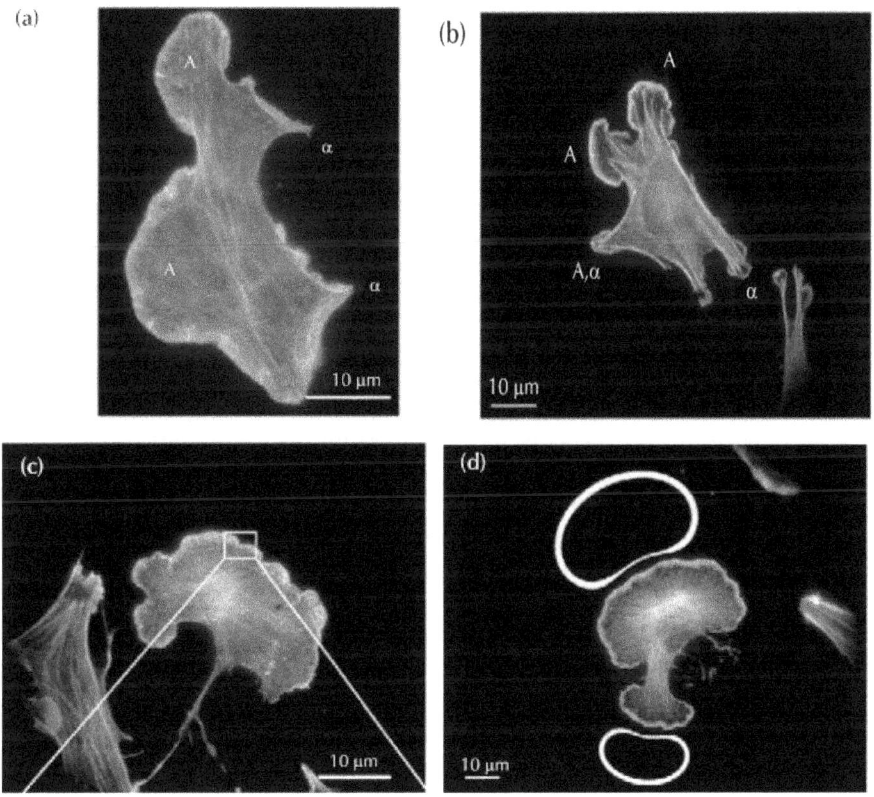

FIGURE 3 *(Continued)*

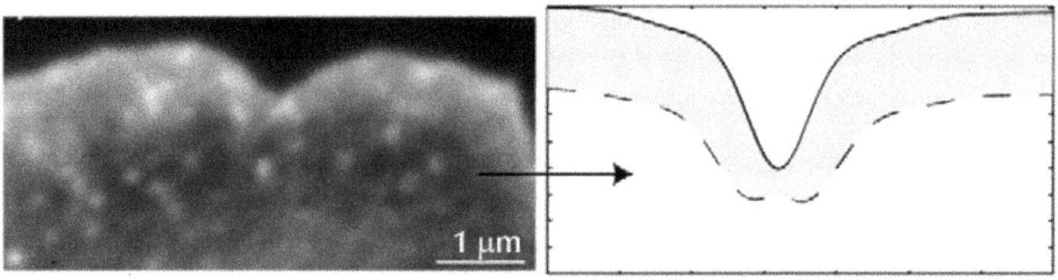

FIGURE 3 Qualitative comparison between observed and calculated cell shapes. (a,b) Shapes of adhering cells on a flat two-dimensional surface, with fluorescently labeled actin. Lamellipodia is denoted by "A" and adhesion domains by "α". (c) A cell dominated by lamellipodia, where a segment of the cell perimeter (lower panel) is shown and compared to the calculated (flat geometry) membrane shape (solid line) and cortical actin density (dashed line), assuming a constant rate of actin depolymerization behind the leading edge. (d) Comparison between cell shapes dominated by lamellipodia and the calculated cell shape and cortical actin density (round geometry).

The thickness of the observed cortical actin layer along the contour ($zw(s)$ can be related to the local membrane density of membrane protein that was calculated ($n(s)$, in the following way; assuming that the filaments have a constant rate of severing (depolymerization) k_{depol} after they are nucleated at the membrane, their number decay as a function of the distance z from the membrane according to: $n_{fil}(z)=n(s)\exp(-zk_{depol}/v)$, where v is the treadmilling velocity. The fluorescence signal is proportional to the number of filaments $n_{fil}(z)$, and the images show the signal above some threshold value n_{min}, which occurs at a distance: $zw(s)=(v/k_{depol})$ In $(n(s)/n_{min})$. As shown in Figure 3c, d (using: v/k_{depol}=0.2μm and n_{min} = 0.005) we indeed find that the actin density follows the membrane shape as it was calculated; actin is depleted where the membrane has concave curvature.

Another example for the transition in cell shapes from "spiky" (dominated by adhesion points) to "fan-shaped" (dominated by actin polymerization pressure) can be found in [63]. Especially intriguing are the shapes of cells where active Rac1 was expressed, leading to Arp2/3 recruitment to the membrane. These cells were predominantly in a shape similar to those shown in Figure 4d,e when actin polymerization is the dominant force. In these cells there is a broad fan-shape region, and a single concave depression, exactly as it was calculated. It is observed that Arp2/3 is absent from the membrane in the concave region. In this model the actin polymerization is absent from this region due to the local concave curvature, so based on this observation it is therefore expected that the Arp2/3 is activated by a membrane-bound complex with convex curvature. Candidates are the WASP-family proteins that have been shown to form complexes with convex proteins [23, 24].

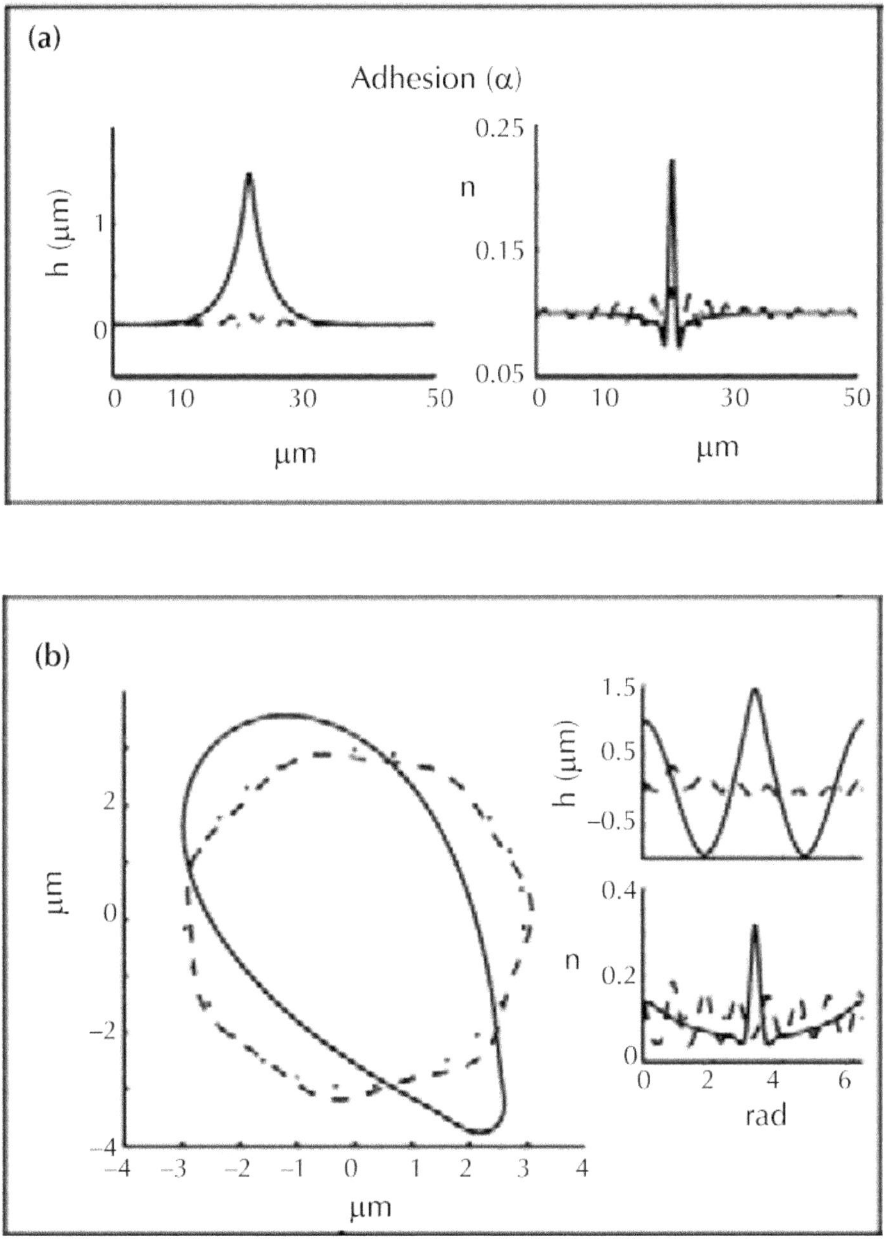

FIGURE 4 *(Continued)*

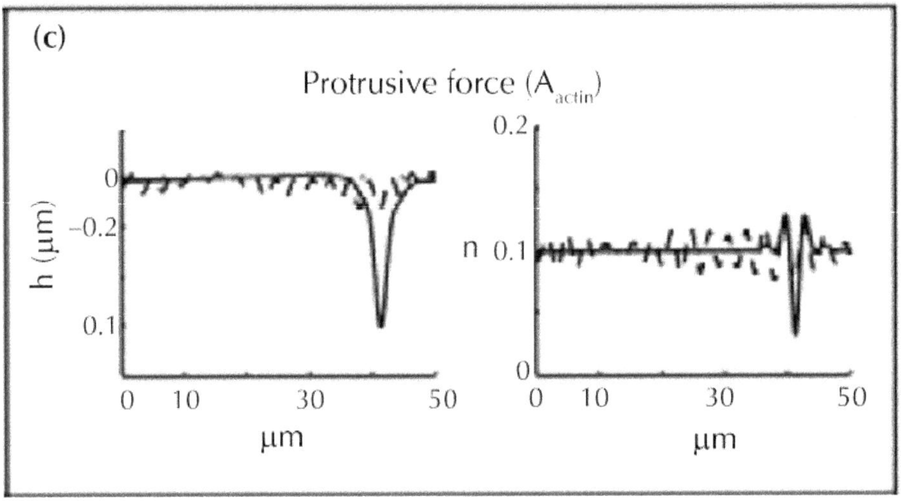

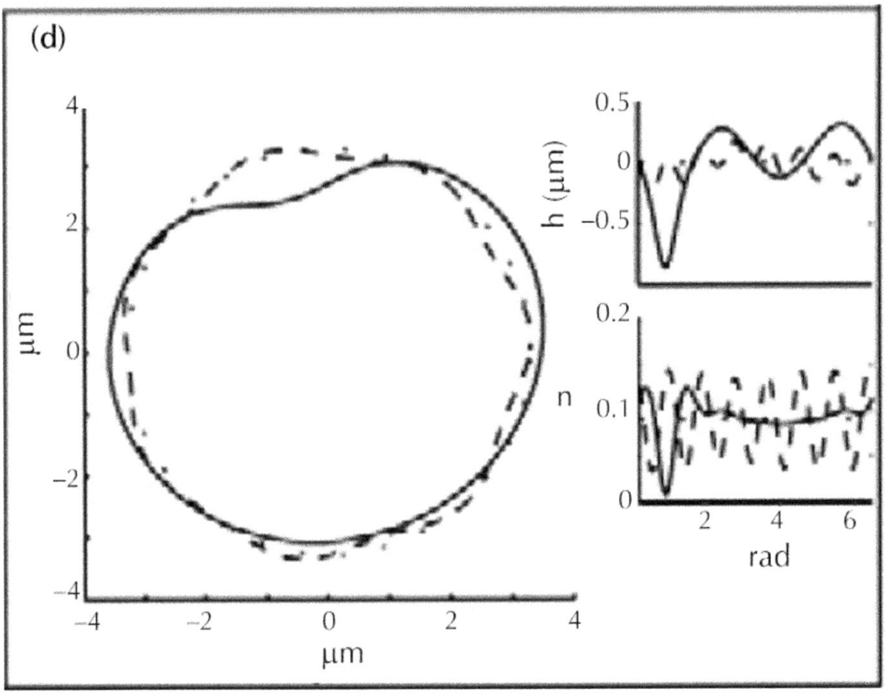

FIGURE 4 *(Continued)*

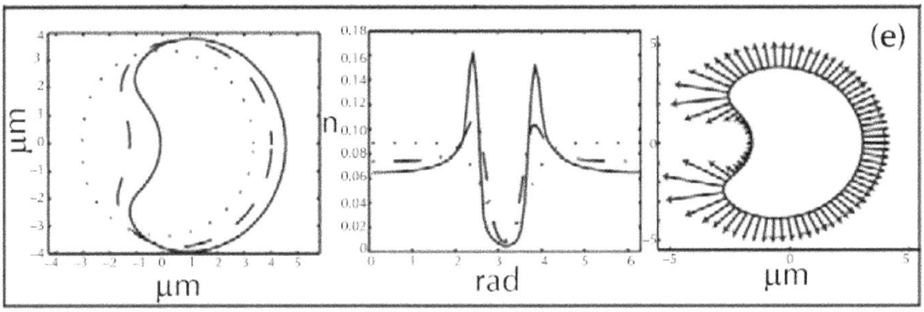

FIGURE 4 Cellular shapes driven by adhesion and actin protrusive forces. Numerical simulations of the evolution of the membrane shape (*h* in the flat geometry) and membrane protein distribution (*n*), for the flat (a) and round (b) geometries, driven by adhesion only (A_{actin}= 0): flat- α = 0.01gr/sec^2, round- α = 0.0089gr/sec^2. Dotted lines give the initial shape (uniform) and membrane protein distributions uniform with a 1% random noise). At an intermediate time equally spaced protrusions form (dashed lines- t = 4200 sec), which eventually coalesce to form a single protrusion at the final steady-state (solid lines- t = 5000). In (c,d) the evolution of the system is plotted for the case of only actin protrusive force (α = 0), using A_{actin}= 0.01grµm^{-1}sec^{-2}. All other conditions are as in (a,b) respectively. It can be seen again equally spaced protrusions at an intermediate time (*t* = 500 sec), which eventually coalesce to form a single protrusion at the final steady-state (solid lines- *t* = 2100 sec). The cell shape in the round geometry was centered at the origin. (e) Evolution of the system driven by actin protrusive force (A_{actin} = 0.0125grµm^{-1}sec^{-2}), for an initial condition of a highly concentrated Gaussian distribution of the membrane protein. The asymmetric distribution leads at first to a global motion of the cell (dashed line- *t* = 100 sec), which stops when the steady-state distribution is reached (solid line- *t* = 700 sec). The protrusive forces along the cell perimeter at the steady-state is illustrated by the arrows in the rightmost panel, which are proportional to the local density of membrane protein. The simulations corresponding to (b,d,e) are shown in supporting Videos S1, S2 and S3 respectively.

Another example for cell shapes that are dominated by actin polymerization comes from the study of spreading cells [34]. In this work it is shown that cells that normally spread in a roughly circular shape, become highly crescent when myosin-II is inhibited by a drug. This effect is attributed to the reduction in the contractile force due to myosin, which allows the protrusive forces of actin polymerization to effectively increase [64] and dramatically alter the cell shape. In our model the myosin activity is not taken explicitly into account, but it can be taken into account the reduction in contractility by increasing the effective protrusive force of actin (A_{actin} in Eq.2). In Figure 5 it is shown that as the actin protrusive force increases there seems to be an abrupt transition in the steady-state cell shape, from a circular shape with a small dip to a crescent-shape cell. In particular it is found that there is now essentially complete depletion of the actin nucleators from the membrane in the dip. Our results therefore provide a possible explanation for the dramatic shape transition reported in [34] when myosin-II was inhibited. Specifically, it was noted in [34] that there is complete absence of branched actin polymerization near the membrane in the dip region, as it is found in our model.

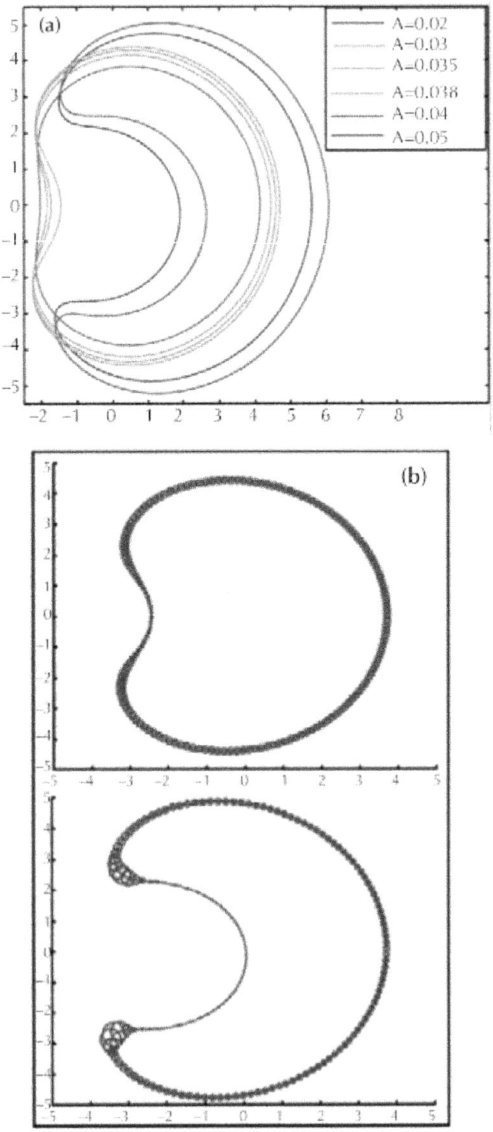

FIGURE 5 Crescent-cell shape transition at high actin polymerization levels. (a) Calculated steady-state shapes of cells driven by actin polymerization (no adhesion, $\alpha = 0$), for increasing levels of actin polymerization force parameter: $A_{actin} = 0.02, 0.03, 0.035, 0.038, 0.04, 0.05 gr\mu m^{-1} sec^{-2}$ (the non-linear tension parameter was taken to be β=0.1). The crescent shapes obtained above a threshold level of actin polymerization are similar to those observed in [34]. (b) Plot of the density distribution of the membrane proteins along the membrane for the two cases of A_{actin} = $0.038, 0.04 gr\mu m^{-1} sec^{-2}$ (top and bottom respectively), to demonstrate the strong depletion in the dip region that accompanies the formation of the crescent cell shape. The size of the circles that decorate the cell contour are proportional to the local concentration of the membrane proteins.

Finally, an example that comes from motile cells is given in [65], where it was observed that the localization of Ena/VASP to the leading edge is responsible for the formation of a cell with a single fan-shape (as it is found in Figure 4d,e). When the Ena/VASP was not localized to the edge, the cell assumed a more round and fluctuating shape. It is noted that Ena/VASP can be localized to the membrane by association with IRSp53 [66], which is precisely the type of linkage that our model proposes. In addition, when the cell membrane was forced to have a flat edge, it was observed that the localization of the Ena/VASP in this region disappeared, and reappeared only when the constraint was removed. This observation again points to the role of membrane shape in the localization of this protein at the leading edge, as it is proposed.

9.3.2 Evolution of Cellular Protrusions: Coalescence

While the coalescence phenomenon was previously discussed theoretically in [67, 68], our work is the first to our knowledge, to calculate this process explicitly for actin and adhesion driven protrusions on a membrane. The coalescence dynamics of the membrane protrusions that are calculated (Figure 2), allow to propose an explanation for a number of long-standing puzzles regarding the relation between cellular shapes and the properties of the surrounding matrix. This model predicts that initially the adhesion or actin-driven membrane undulations grow at the wavelength λ_{max}, so that the number of cellular protrusions along a membrane contour of length L is of order $N_{max}=L/\lambda_{max}$. The system then evolves by coalescence of the protrusions into progressively fewer and larger protrusions, until a single feature remains (Figure. 2b, 4).

It is hard to find much published data to compare with this general feature of this model. This type of dynamics was observed for adhering cells in [10], where the adhesion foci along the cell rim were initially numerous (10–20 after 4 hrs) and spread at rather uniform spacing, but later formed roughly two adhesion regions at the two opposing poles of the elongated cell (after 24 hrs). This model qualitatively captures these dynamics (Figure 4b), although it does not lead to a bi-polar distribution since it has not been the constraint imposed by the need to connect opposing adhesion regions by internal stress-fibers. Recently, the dynamics of cell morphology during spreading and adhesion was more closely investigated experimentally [39]. In this work one can observe some of the shape evolution that are calculated, such as the formation of regularly spaced protrusions from an initially circular cell (Figure 1 of [39]), as well as the later coalescence of such structures (Figure 3 of [39]).

Furthermore, this model predicts that as the phase transition line is approached (Figure 1d) the time-scale for coalescence of the protrusions becomes very long (Figure 2a,c). Since the cell has a typical time-scale (t_{cell}) over which it reorganizes its cytoskeleton (determined by external cues or division time, etc.), it is relevant to compare the calculated shapes at this particular time. It is plotted in Figure 2d the number of protrusions at this chosen time and find that it has the following non-linear behavior; for values of the adhesion or actin forces (α or A_{actin}) that are below the critical threshold, the system is stable (uniform) and the number of protrusions is therefore zero. Just above the critical line (either $\alpha > \alpha_c$ or $A_{actin} > A_{actin,c}$, Figure 1d) the time-scale for coalescence (t_c, Figure 2c) is much longer than t_{cell} and the observed number of protrusions is simply the maximal one $N_{max} = L/\lambda_{max}$ (note that near the transition qmax

$q_{max} \to q_c^*$, Figure 1e). As α (or A_{actin}) increase further the coalescence time becomes shorter and the protrusions begin to coalesce by the time t_{cell}, consequently reducing the number of protrusions that are counted. Above a certain value of the adhesion or actin parameters (α^* or A^*_{actin}), we arrive in a regime where: $t_c < t_{cell}$, and the protrusions have all coalesced to form a single feature. A membrane protrusion is counted if its amplitude is at least 20% that of the largest membrane protrusion. This value of the threshold was chosen arbitrarily, but does not change the qualitative behavior.

This unique prediction from our model suggests a possible explanation for the following puzzling observations; (i) In [9] neuronal cells adhering to a flat two-dimensional surface, have been shown to produce more (less) numerous and shorter (longer) protrusions, when the cells had less (more) activity of actin filament polymerization. This observation may correspond in our model to the cells having their actin force parameter vary within the region: $A_{actin,c} < A_{actin} < A_{actin}^*$. Note that the number of protrusions is stabilized (i.e. further coalescence is suppressed) in these cells when MTs invade the nascent protrusions along the cell edge. This invasion process sets the time scale t_{cell}. (ii) In [7], cells encapsulated in a three-dimensional matrix have been found to have more (less) numerous and shorter (longer) protrusions, when the surrounding gel was stiffer (softer) and therefore harder (easier) to degrade. The ability of cells can be mapped to degrade their surrounding matrix and protrude with the parameter describing the actin protrusion force, such that a stiffer gel corresponds to a smaller A_{actin}, and vice versa. The observations therefore suggest that the regime of stiffness explored in the experiments corresponds again to: $A_{actin,c} \leq A_{actin} \leq A_{actin}^*$. Note that in this experiment it was reported that when the stiffness increased above some threshold the number of protrusions collapsed to zero, as this model predicts (Figure 2d). A similar relation between the number of degradation-protrusions and substrate stiffness was also observed in [69], in the context of invadopodia produced by cancer cells during invasion of the ECM.

Most recently the number of cellular protrusions ("fingers") was measured as a function of the actin polymerization activity, using a drug [39]. As this model predicts (Figure 2d), the measurement shows that as the actin polymerization is inhibited the number of protrusions increases.

The effects of the adhesion strength of a cell on a flat two-dimensional surface, on the cell shape have been explored in a number of chapters [1, 70–72]. As the stiffness of the substrate is increased, so does the strength of the cell adhesion, and our model would therefore predict a decrease in the number of protrusions and more global cell polarization, as the stiffness increases. Indeed this is the observed trend.

The above discussion suggests that cells seem to naturally live in a parameter space near the transition lines between the stable (uniform) and unstable (protrusions) regime (Figure 1d). Such a location may allow cells to change their shape by only small changes to their cytoskeleton activity. This feature may explain the spontaneous cellular transitions observed in [11], from uniform round cells to cells covered by filopodia.

Note that there are several effects that may strongly suppress or delay the process of protrusions coalescence; long actin-driven protrusions, such as filopodia, can get anchored to the external substrate at their tips, and stabilize in such a way that any coalescence with neighboring filopodia is suppressed. Such dynamics are demonstrated in this model, where a small addition of adhesion stabilizes the actin-driven protrusions (Figure 6d-f). Additionally, strong adhesion may decrease the effective mobility and diffusion coefficient of the membrane proteins, again delaying or suppressing coalescence. This second process was not explicitly treated in the current model, but can be added in the future. These processes may result in cells retaining their "polygonal" (or "spiky") morphology, where by protrusions are separated by the typical wavelength λ_{max}.

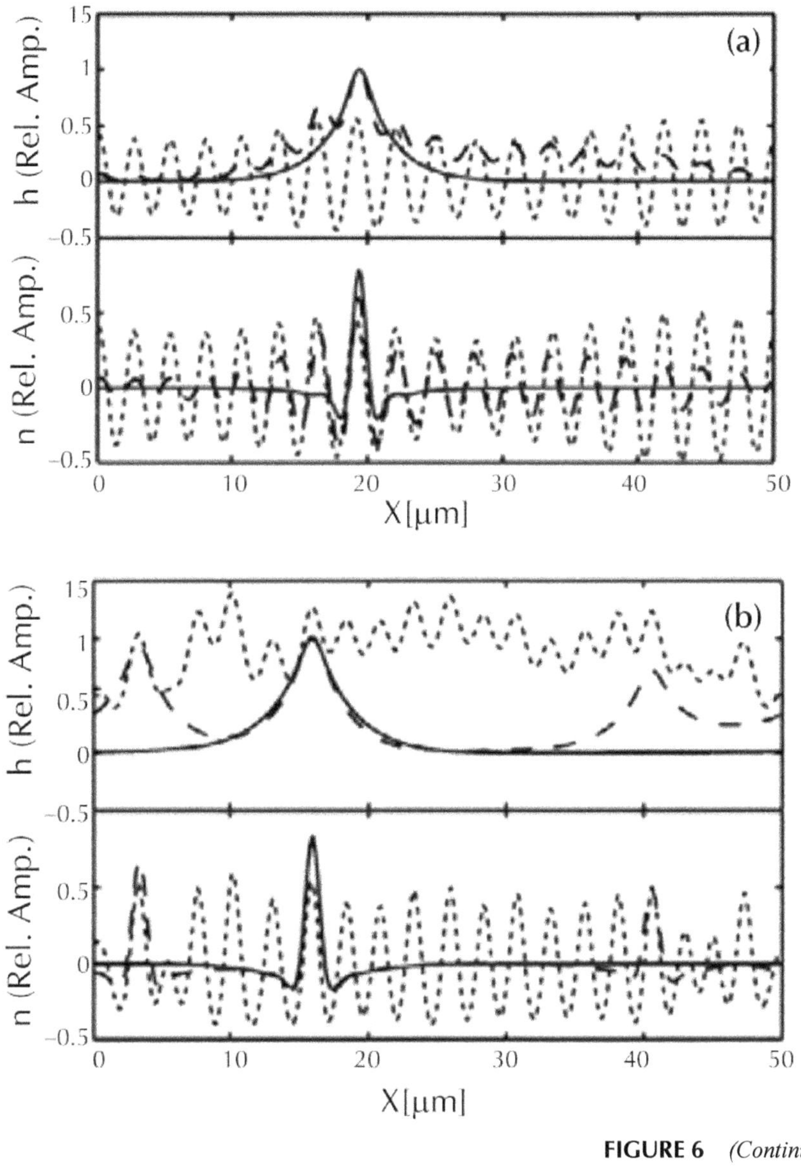

FIGURE 6 *(Continued)*

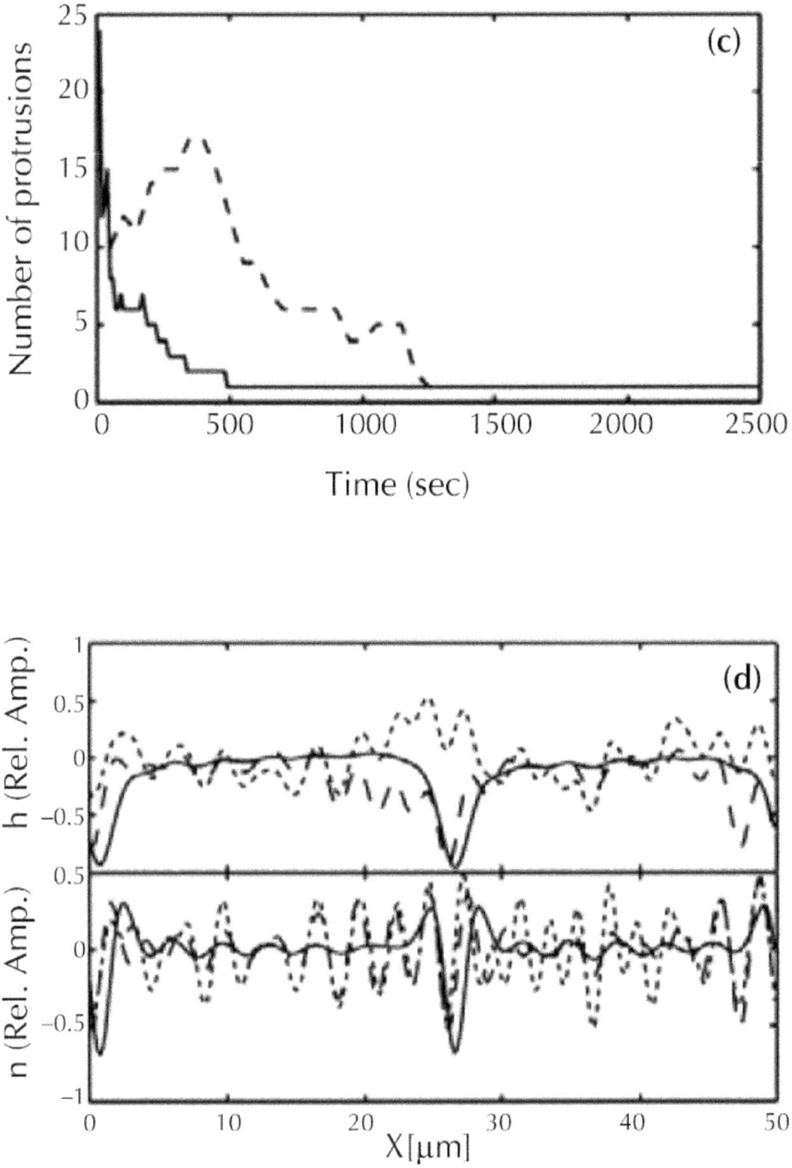

FIGURE 6 *(Continued)*

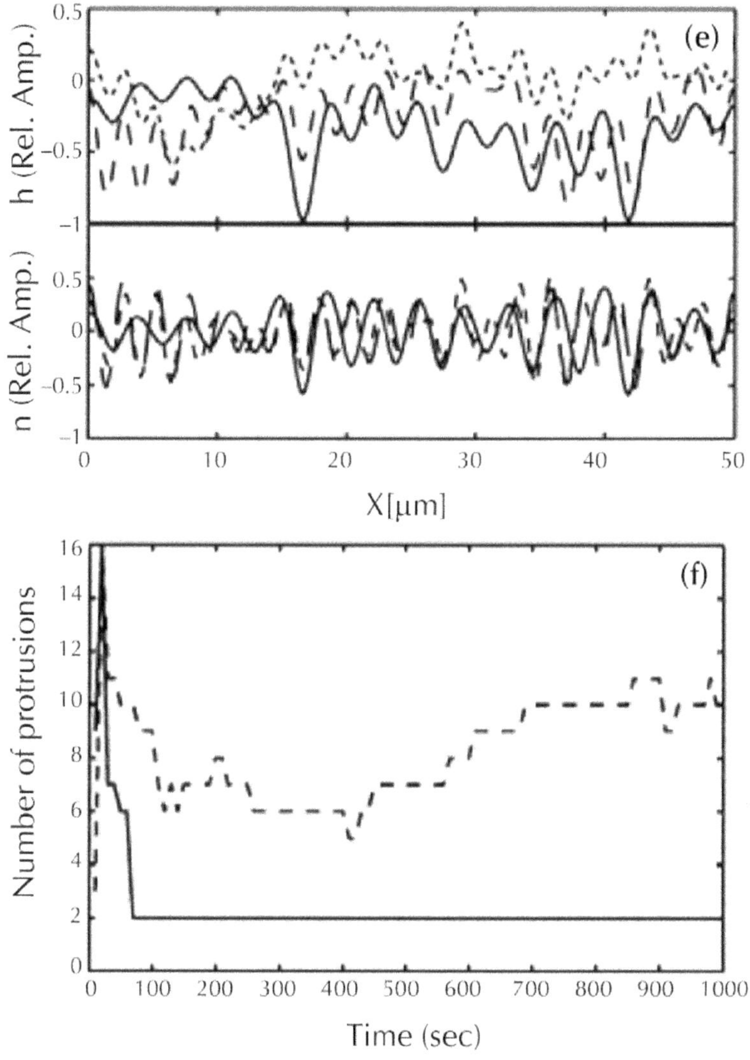

FIGURE 6 Dynamics driven by both actin polymerization and adhesion. (a, b) Evolution of the flat geometry dominated by adhesion (x=0.0105gr/sec2), alone (a) and with additional actin polymerization force (b) A_{actin} = 0.015grμm^{-1}sec^{-2}. The lines give snap-shots of the system at different times, with the dotted line at the earliest time, dashed line at intermediate time and the solid line at the final time. In (c) the number of protrusions is plotted as a function of time for the two simulations (dashed line for pure adhesion, solid line with additional actin polymerization force). (d, e) Evolution of the flat geometry dominated by actin polymerization force (A_{actin} = 0.015grμm^{-1}sec^{-2}), alone (d) and with additional adhesion force (e) α = 0.0008gr/sec^2. Lines give snap-shots of the system at different times, in the same scheme as in (a, b). In (f) the number is plotted of protrusions as a function of time for the two simulations (solid line for pure actin polymerization force, dashed line with additional adhesion force).

Finally, the processes that degrade cytoskeleton-membrane structures (protrusions and adhesion complexes) have not been discussed here. While in our model the overall number and activity of the membrane proteins is constant, in the cell each protein undergoes processes of degradation and deactivation. Such processes endow each cytoskeleton-membrane structure with a finite lifetime [73], which increases with the size of the protrusion and its protein content. The degradation process therefore further acts to inhibit coalescence as we approach the critical transition line where coalescence slows down (Figure 2). When the coalescence process is faster than the degradation, we will find cells that reach the steady-state shapes and attain global polarization, while slow coalescence will be further inhibited by the decay of the protrusions and the initiation of new ones (see for example [74]).

9.3.3 Cell Motility

Although we have been interested in cell shapes rather than cell motility, there are two features of this model that may be relevant for this problem as well, and suggests that curvature-driven feedback may play a role in cell motility as well:

1. In this model of the round cell driven only by adhesion (Figure 4b) there is spontaneous symmetry breaking due to the feedback between the convex membrane protein and their induced protrusive force. The protrusive force is driven by a local negative membrane tension which amounts to the continuous addition of membrane area at that location. Along the rest of the cell, the positive membrane tension pulls the cell rear inwards, and therefore an overall drift in the direction of the sharp feature ensues. This model may therefore be relevant for the study of amoeba-type cell motility observed by cells in a three-dimensional matrix [8], where localized adhesion to the matrix at the leading edge is the dominant feature, and cells often have the tear-drop shape we calculate.
2. In this model of the round cell driven only by actin protrusion (Figure 4d) there is overall symmetry breaking and the shape is very similar to that observed in highly motile cells, such as keratocytes [75], moving on a two-dimensional surface. Within this model there is a global cancelation of the protrusive forces such that the cell is stationary, and this cancelation arises due to the strong backward forces at the two highly curved corners, where the membrane shape is most convex (Figure 4e). These backward forces cancel the forward protrusive force along the fan-like front of the cell. In a real motile cell the actin seems to be prevented from pushing effectively the membrane backward at the cell back, due to myosin activity and polarization of the actin depolymerization processes [76, 77]. This model suggests that the localization and shape of the leading edge may be maintained by curved activators of actin [23], while additional symmetry breaking processes, such as retrograde flow and maturation of the adhesion contacts, are necessary to polarize the actin polymerization and result in overall cell motility [78–80].

9.4 RESULTS

The positive feedback between the protrusive forces (either due to adhesion or actin polymerization), the membrane shape and distribution of convex membrane proteins, leads to a dynamic instability that breaks the uniform configuration and produces membrane undulations where the membrane proteins are aggregated. At the linear regime this was explored in [26]. The main interest in this work is to follow the evolution of the cell contour shape in two dimensions (Figure 1b,c), beyond the regime of small perturbations which is captured by the linear stability analysis. In Figure 1d the stability phase diagram has been plotted for the system driven by only two types of forces; actin polymerization and adhesion. We wish to explore the long-time evolution and steady-state of the system when it is unstable.

In order to isolate the effects of actin polymerization and adhesion, this model is used to explore along the two axes shown in Figure 1d, such that we either took $A_{actin}=0, \alpha>0$ or $A_{actin}>0, \alpha=0$. In the real cell these two effects are closely linked, so such a complete separation is done here to better understand the dynamics when each of these factors is dominant. Additionally, in the simulations shown below we limited the amplitude of the membrane undulations to simplify the numerical analysis, but note that the model gives rise to highly elongated protrusions similar to filopodia, when the non-linear membrane tension is set to a small value. Finally, it is noted that the numerical values of the parameters used in the simulations (see Table 1) were not meant to fit any particular observation, but rather allow us to illustrate the qualitative features of the model.

TABLE 1 List of parameters used in the calculations.

Effective friction coefficient of membrane, ξ [$grsec^{-1}\mu m^{-2}$]	$8 \cdot 10^{-3}$
Diffusion coefficient of membrane protein in membrane, D [$\mu m^2 sec^{-1}$]	$2 \cdot 10^{-3}$
Mean area coverage of membrane protein, n_0	0.1
Saturating density of membrane protein, n_s [μm^{-2}]	10
Membrane bending rigidity, κ [$k_B T$]	100
Intrinsic membrane protein curvature, $\bar{H}$ [μm^{-1}]	-10
Membrane tension, σ [$grsec^{-2}$]	$1 \cdot 10^{-3}$
Spring constant, γ [$grsec^{-2}\mu m^{-2}$]	$4 \cdot 10^{-5}$
Membrane protein binding interaction, J [$grsec^{-2}$]	$3.5 \cdot 10^{-3}$
Cell effective bulk modulus, K [$gr\mu m^{-1}sec^{-2}$]	10^{-4}
Cell radius, r [μm]	3
Mobility of proteins, Λ	$D/(k_B T)$
Non-linear tension parameter, β [μm^{-1}]	0,0.1,1

doi:10.1371/journal.pcbi.1001127.t001

9.4.1 Shape Evolution Driven by Adhesion ($A_{actin} = 0, \alpha>0$)

In Figure 4a,b the evolution of the system driven by the adhesion forces is plotted, for the flat and round geometries respectively. The initial conditions in all cases are those of the uniform equilibrium state (flat or circular respectively), with a random perturbation in the membrane protein density distribution of maximal amplitude 1%. It is immediately clear that the system evolves initially according to the linear analysis, i.e. the most unstable mode from the dispersion relation (Figure 1e) grows the fastest and the system develops periodic undulations (protrusions) with wavelength $\lambda_{max} = 2\pi/q_{max}$.

At longer times the membrane shape and membrane protein density distribution no longer follow the linear behavior, and the coalescence of the protrusions into a single isolated feature is observed. In the flat geometry we end up with a single protrusion which has a sharp tent-like shape, and similarly in the round geometry we find a contour with a droplet shape. The density distribution of the membrane proteins in the steady-state ($n_{SS}(S)$), follows very closely the curvature of the membrane ($H(s)$). This arises due to the equality between the dominant currents in the steady-state, which are J_{curv} and J_{disp} (Eqs.10,11). Equating these currents, we find immediately that: $n_{SS}(s) \propto H(s)/H$ (Eq.S3 in Text S1). The distribution of membrane proteins has a sharp peak at the membrane peak, while depleted everywhere else, due to their convex spontaneous curvature. A further analysis of the steady-state shapes is described in Text S1.

In a real cell the stress-fibers that connect adhesion regions usually impose a bi-polar steady-state with adhesion localized at two opposing poles of the cell. In our model this non-local interaction is absent and therefore the adhesion region can collapse to a single localized domain (Figure 4a,b).

Note that in the round case, the adhesion forces along the whole membrane are not balanced (their sum does not vanish), since the highly concentrated membrane proteins at the sharp tip give an overall force pointing in that direction (local negative effective membrane tension), Eq.5, and leads to a global drift of the whole cell. This result arises due to the fact that the adhesion-induced protrusive force (Eq.5) depends on the local curvature of the membrane, since it appears as a negative membrane tension term. This means that the sharper the membrane shape at the peak, the stronger is the adhesion force due to both a larger concentration of membrane proteins (larger n at the peak) and a larger curvature, i.e. the force is proportional to $H(s)^2$. Its integral over the closed contour therefore does not vanish.

9.4.2 Shape Evolution Driven by Actin Polymerization (Aactin>0, $\alpha = 0$

In Figure 4c,d the evolution of the system driven by the actin protrusive forces is plotted, for the flat and round geometries respectively. The initial conditions are the same as for the case of adhesion-driven shapes (Figure 4a,b), and similarly the system evolves initially according to the linear analysis, where the most unstable mode from the dispersion relation (Figure 1e) grows the fastest and the system develops periodic undulations (protrusions) with wavelength λ_{max}.

At later times we again find that the protrusions coalesce, but instead of forming a single sharp peak, the system forms a broad and flat plateau, that is punctuated by a single sharp dip. In the round geometry the membrane develops a broad fan-like bulge, with a smaller concave dip. As for the adhesion-driven system, the density distribution of the membrane proteins again follows the membrane curvature, and is given by n_{SS} n_{SS} defined above. The membrane protein distribution is therefore rather flat, except for two peaks at the "shoulders" of the membrane dip, and are depleted from the dip itself. A further analysis of the steady-state shapes is described in Text S1.

The protrusive force due to actin on the closed membrane in the round geometry, sums up to zero at the steady-state. This is due to the linear relation between the actin force and the membrane protein density (Eqs.2,3), and that the steady-state membrane protein distribution (n_{SS}) is closely proportional to the curvature, while the integral over the change in the curvature vector vanishes along a closed contour. Until the steady-state shape settles, the forces can be unbalanced, and the whole shape drifts. This is clearly illustrated when we start with initial conditions, where the membrane proteins are localized (in a Gaussian shape) asymmetrically along the membrane contour (Figure 4e).

Note, that unlike the adhesion-induced protrusive force (Eq.5) whose strength depends on the local curvature of the membrane, the actin protrusive force acts as a local pressure term (Eqs.2,3).

9.4.3 Dynamics of the Approach to the Steady-State Shape: Coalescence

In Figure 2 the evolution of the flat system, driven by adhesion (a similar behavior is observed when actin drives the dynamics) is plotted, as one approaches the critical point the type-II dispersion vanishes ($\omega_{\max} \to 0$) and the system becomes linearly stable (Figure 1d,e). It is observed that initially the amplitude of the fluctuations grow exponentially as: $\exp(\omega_{max} t)$ (Figure 2a). As the amplitude grows, so does the restoring force due to the non-linear tension, which eventually, together with the spring force, stops the growth. The final amplitude h_{max} of the steady-state membrane peak (Figure 2b) is therefore given by the balance between the non-linear tension and the adhesion force of the steady-state shape.

When the coalescence of the protrusions is very slow, close to the instability transition line, non-linear tension is able to stop the growth of the most unstable mode, and the system is stalled with an approximately sinusoidal perturbation (point marked (2) in Figure 2a). The amplitude of the membrane undulation at this stage is much lower than the final steady-state amplitude, since in the sinusoidal case the membrane proteins are distributed among many peaks, rather than all of them concentrated at one single peak. Eventually small differences among the peak amplitudes and the slow diffusion of membrane proteins break this dead-lock, and allows the further growth of the single steady-state peak. These stages are illustrated in Figure 2a,b.

In Figure 2c the dependence of the final coalescence time t_c on ω_{max} is plotted, and find a simple linear relation: $t_c \propto \omega_{\max}^{-1}$ for large ω_{max}. As ω_{max} becomes smaller, this relation breaks down due to the partial stabilization of the system in the intermediate state ((2) in Figure 2a), by the non-linear tension.

9.4.4 Shape Evolution Driven by both Actin Polymerization and Adhesion

Although its simpler to analyze the effects of the two active forces separately, as shown above, both are present in a real cell. We now discuss how these two forces act in combination.

Since both adhesion and actin polymerization act to destabilize the membrane, when both forces act together the system is pushed deeper into the unstable regime. This means that the linear instability starts more quickly and with a larger number of protrusions.

However, the long-time non-linear evolution of the protrusions is very different in the adhesion or actin-dominated regimes; in the adhesion-dominated regime, the addition of actin forces results in faster coalescence of the protrusions, as is intuitively expected in a more unstable system. Since the actin acts to broaden the protrusions into fan-shapes, it speeds up local coalescence events. In Figure 6a,b typical evolutions of the system dominated by adhesion are plotted, with and without the addition of actin polymerization. The faster coalescence is shown for these two simulations in Figure 6c.

However, in the actin-dominated regime we find that additional adhesion forces result in stabilization of the small protrusions that form at the early stages, and suppression of their coalescence. This is due to the fact that the adhesion force stabilizes the pointed tips of the small protrusions, thereby slowing down their broadening and coalescence. This is demonstrated for typical simulations in Figure 6d-f.

9.5 CONCLUSIONS

To conclude, let us summarize the main findings:

- When protrusive forces, due to either actin polymerization and/or adhesion, are recruited by convex membrane proteins and exceed a threshold value, we find that protrusions spontaneously form, initially at regular spatial intervals. The protrusions evolve by a process of coalescence, leading to larger but fewer protrusions with time.
- The shape of the protrusions at long times differs significantly between the cases dominated by adhesion (pointed tent-like) or actin polymerization (broad fan-shaped), as observed in cells.
- The time-scale for the coalescence of protrusions diverges as the critical threshold of cytoskeleton activity is approached (from above). This means that the observed number of protrusions increases near the threshold value (from above) and vanishes below it. This type of dynamics can explain the puzzling observed dependencies of cell shapes on the properties of the surrounding matrix [7, 9].

Note that the process of pattern coarsening, which in our case is in the form of coalescence of protrusions, is a more general phenomenon than our specific model, and it appears in other biological systems as well [58]. Therefore the conclusions regarding the interplay between the cellular time-scales and observed patterns are also more general and may apply even if the underlying mechanisms for protrusion formation are different from those considered here.

This model not only recapitulates many features of observed cellular shapes, but also allows us to make predictions that await further measurements. It highlights the major role of curved membrane proteins that couple the membrane to the underlying cytoskeleton in determining cellular shapes [12, 23].

Let us emphasize again that the work presented here is just one step towards the understanding of the coupling of the cytoskeleton to the cell membrane. Our philosophy is to start with a simple model, which does not include all the complexity of the cell. We believe that the results we obtained are interesting and rich enough to encourage future extensions of this model, that will indeed add more details.

KEYWORDS

- **Actin Cytoskeleton**
- **Adhesive Forces**
- **Lamellipodia**
- **Membrane Protein**
- **Protrusive Forces**

ACKNOWLEDGMENTS

We thank useful discussions with Kinneret Keren.

REFERENCES

1. Engler, A. J., Sen, S., Sweeney, H. L., and Discher, D. E. Matrix elasticity directs stem cell lineage specification. *Cell* **126,** 677 (2006).
2. Discher, D. E., Janmey, P., Wang, Y. L. Tissue cells feel and respond to the stiffness of theirsubstrate. *Science* **310,** 1139 (2005).
3. Georges, P. C., Miller, W. J., Meaney, D. F, Sawyer, E. S, and Janmey, P. A. Matrices with compliance comparable to that of brain tissue select neuronal over glial growth in mixed cortical cultures. *Biophys J* **90,** 3012 (2006).
4. Yeung, T., Georges, P. C., Flanagan, L. A., Marg, B., Ortiz, M. et al. Effects of substrate stiffness on cell morphology, cytoskeletal structure, and adhesion. *Cell Motil Cytoskeleton* **60,** 24–34 (2005).
5. Giannone, G. and Sheetz, M. P. Substrate rigidity and force define form through tyrosine phosphatase and kinase pathways. *Trends Cell Biol* **16,** 213–223 (2006).
6. Martins, G. G. and Kolega, J. Endothelial cell protrusion and migration in three-dimensional collagen matrices. *Cell Motil Cytoskeleton* **63,** 101 (2006).
7. Dikovsky, D., Bianco-Peled, H., and Seliktar, D. Defining the role of matrix compliance and proteolysis in three-dimensional cell spreading and remodeling. *Biophys J* **94,** 2914–2925 (2008).
8. Fraley, S. I., Feng, Y., Krishnamurthy, R., Kim, D. H., Celedon, A. et al. A distinctive role for focal adhesion proteins in three-dimensional cell motility. *Nat Cell Biol* **12,** 598 (2010).
9. Korobova, F. and Svitkina, T. Arp2/3 complex is important for filopodia formation, growth cone motility, and neuritogenesis in neuronal cells. *Mol Biol Cell* **19,** 1561–1574 (2008).
10. Cavalcanti-Adam, E. A., Volberg, T., Micoulet, A., Kessler, H., Geiger, B. et al. Cell spreading and focal adhesion dynamics are regulated by spacing of integrin ligands. *Biophys J* **92,** 2964–2974 (2007).
11. Applewhite, D. A., Barzik, M., Kojima, S. I., Svitkina, T. M., Gertler, F. B. et al. Ena/vasp proteins have an anti-capping independent function in filopodia formation. *Mol Biol Cell* **18,** 2579 (2007).
12. Scita, G., Confalonieri, S., Lappalainen, P., and Suetsugu, S. Irsp53: crossing the road of membrane and actin dynamics in the formation of membrane protrusions. *Trends Cell Biol* **18,** 52–60 (2008).

13. Peter, B. J., Kent, H. M., Mills, I. G., Vallis, Y., Butler, P. J. et al. Bar domains as sensors of membrane curvature: the amphiphysin bar structure. *Science* **303,** 495–499 (2003).
14. Itoh, T. and Camilli, P. D. Bar, f-bar (efc) and enth/anth domains in the regulation of membrane-cytosol interfaces and membrane curvature. *Biochim Biophys Acta* **1761,** 897 (2006).
15. Heath, R. J. W. and Insall, R. H. F-bar domains: multifunctional regulators of membrane curvature. *J Cell Sci* **121,** 1951–1954 (2008).
16. Mattila, P. K., Pykalainen, A., Saarikangas, J., Paavilainen, V. O., Vihinen, H. et al. Missing-in metastasis and irsp53 deform pi(4,5)p2-rich membranes by an inverse bar domain-like mechanism. *J Cell Biol* **176,** 953 (2007).
17. Takenawa, T. and Suetsugu, S. The wasp-wave protein network: connecting the membrane to the cytoskeleton. *Nat Rev Mol Cell Biol* **8,** 37–48 (2007).
18. Takano, K., Toyooka, K., and Suetsugu, S. Efc/f-bar proteins and the n-wasp-wip complex induce membrane curvature-dependent actin polymerization. *EMBO J* **27,** 2817–2828 (2008).
19. Mattila, P. K. and Lappalainen, P. Filopodia: molecular architecture and cellular functions. *Nat Rev Mol Cell Biol* **9,** 446–454 (2008).
20. Saarikangas, J., Hakanen, J., Mattila, P. K., Grumet, M., Salminen, M., et al. Abba regulates plasma-membrane and actin dynamics to promote radial glia extension. *J Cell Sci* **121,** 1444–1454 (2008).
21. Yang, C., Hoelzle, M., Disanza, A., Scita, G, and Svitkina, T. Coordination of membrane and actin cytoskeleton dynamics during filopodia protrusion. *PLoS ONE* **4,** e5678 (2009).
22. Zhao, H., Pykäläinen, A., and Lappalainen, P. I-bar domain proteins: linking actin and plasma membrane dynamics. *Curr Opin Cell Biol* **23,** 1–8 (2010).
23. Suetsugu, S. The proposed functions of membrane curvatures mediated by the bar domain superfamily proteins. *J Biochem* **148,** 1–12 (2010).
24. Nakagawa, H., Miki, H., Nozumi, M., Takenawa, T., Miyamoto, S., et al. Irsp53 is colocalised with wave2 at the tips of protruding lamellipodia and filopodia independently of mena. *J Cell Sci* **116,** 2577–2583 (2003).
25. Gov, N. S. and Gopinathan, A. Dynamics of membranes driven by actin polymerization. *Biophys J* **90,** 454 (2006).
26. Veksler, A. and Gov, N. S. Phase transitions of the coupled membrane-cytoskeleton modify cellular shape. *Biophys J* **93,** 3798–3810 (2007).
27. Jiang, G., Huang, A. H., Cai, Y., Tanase, M., and Sheetz, M. P. Rigidity sensing at the leading edge through $\alpha_v\beta_3$ integrins and *rptpα*. *Biophys J* **90,** 1804 (2006).
28. Hall, A. Rho gtpases and the control of cell behaviour. *Biochem Soc Trans* **33,** 891 (2005).
29. von Andrian, U. H., Hasslen, S. R., Nelson, R. D., Erlandsen, S. L., and Butcher, E. C. A central role for microvillous receptor presentation in leukocyte adhesion under flow. *Cell* **82,** 989–999 (1995).
30. Abitorabi, M. A., Pachynski, R. K., Ferrando, R. E., Tidswell, M., and Erle, D. J. Presentation of integrins on leukocyte microvilli: a role for the extracellular domain in determining membrane localization. *J Cell Biol* **139,** 2563–2571 (1997).
31. van Buul, J. D., van Rijssel, J., van Alphen, F. P., Hoogenboezem, M., Tol, S. et al. Inside-out regulation of icam-1 dynamics in tnf-alpha-activated endothelium. *PLoS One* **5,** e11336 (2010).
32. Raucher, D. and Sheetz, M. P. Cell spreading and lamellipodial extension rate is regulated by membrane tension. *J Cell Biol* **148,** 127–136 (2000).
33. Wakatsuki, T., Wysolmerski, R. B., Elson, E. L. Mechanics of cell spreading: role of myosin ii. *J Cell Sci* **116,** 1617–1625 (2003).
34. Cai, Y., Rossier, O., Gauthier, N. C., Biais, N., Fardin, M. A. et al. Cytoskeletal coherence requires myosin-iia contractility. *J Cell Sci* **123,** 413–423 (2010).
35. Vicente-Manzanares, M., Ma, X., Adelstein, R. S., and Horwitz, A. R. Non-muscle myosin ii takes centre stage in cell adhesion and migration. *Nat Rev Mol Cell Bio* **10,** 778–790 (2009).
36. Vicente-Manzanares, M., Ma, X., Adelstein, R. S., and Horwitz, A. R. Non-muscle myosin ii takes centre stage in cell adhesion and migration. *Nat Rev Mol Cell Bio* **10,** 778–790 (2009).

37. Enculescu, M., Sabouri-Ghomi, M., Danuser, G., and Falcke, M. Modeling of protrusion phenotypes driven by the actin-membrane interaction. *Biophys J* **98,** 1571–1581 (2010).
38. Shao, D., Rappel, W. J., and Levine, H. Computational model for cell morphodynamics. *Phys Rev Lett* **105,** 108104 (2010).
39. Fardin, M. A., Rossier, O. M., Rangamani, P., Avigan, P. D., Gauthier, N. C. et al. Cell spreading as a hydrodynamic process. *Soft Matter* **6,** 4788–4799 (2010).
40. Satulovsky, J., Lui, R., and l Wang, Y. Exploring the control circuit of cell migration by mathematical modeling. *Biophys J* **94,** 3671–3683 (2008).
41. Bosse, T., Ehinger, J., Czuchra, A., Benesch, S., Steffen, A. et al. Cdc42 and phosphoinositide 3-kinase drive rac-mediated actin polymerization downstream of c-met in distinct and common pathways. *Mol Cell Biol* **27,** 6615–6628 (2007).
42. Czuchra, A., Wu, X., Meyer, H., van Hengel, J., Schroeder, T. et al. Cdc42 is not essential for filopodium formation, directed migration, cell polarization, and mitosis in fibroblastoid cells. *Mol Biol Cell* **16,** 4473–4484 (2005).
43. Steffen, A., Faix, J., Resch, G. P., Linkner, J., Wehland, J. et al. Filopodia formation in the absence of functional wave- and arp2/3-complexes. *Mol Biol Cell* **17,** 2581–2591 (2006).
44. Iglic, A., Hägerstrand, H., Veranic, P., Plemenitas, A., and Kralj-Iglic, V. Mcurvature-induced accumulation of anisotropic membrane components and raft formation in cylindrical membrane protrusions. *J Theor Biol* **240,** 368–373 (2006).
45. Clainche, C. L. and Carlier, M. F. Regulation of actin assembly associated with protrusion and adhesion in cell migration. *Physiol Rev* **88,** 489–513 (2008).
46. Locka, J. G., Wehrle-Hallerb, B., and Strömblad, S. Cellmatrix adhesion complexes: master control machinery of cell migration. *Semin Cancer Biol* **18,** 65–76 (2008).
47. Brown, C. M., Hebert, B., Kolin, D. L., Zareno, J., Whitmore, L. et al. Probing the integrin-actin linkage using high-resolution protein velocity mapping. *J Cell Sci* **119,** 5204–5214 (2006).
48. Mogilner, A. On the edge: modeling protrusion. *Curr Opin Cell Biol* **18,** 32–39 (2006).
49. Mogilner, A. and Rubinstein, B. Actin disassembly clock and membrane tension determine cell shape and turning: a mathematical model. *J Phys Condens Matter* **22,** 194118 (2010).
50. Faix, J. and Rottner, K. The making of filopodia. *Curr Opin Cell Biol* **18,** 18–25 (2006).
51. Nambiar, R., McConnell, R. E., and Tyska, M. J. Myosin motor function: the ins and outs of actin-based membrane protrusions. *Cell Mol Life Sci* **67,** 1239–1254 (2010).
52. Helfrich, W. Elastic properties of lipid bilayerstheory and possible experiments. *Z Naturforsch C* **28,** 693–703 (1973).
53. Mark, S., Shlomovitz, R., Gov, N. S., Poujade, M., Grasland-Mongrain, E. et al. Physical model of the dynamic instability in an expanding cell culture. *Biophys J* **98,** 361–370 (2010).
54. Kabaso, D., Shlomovitz, R., Auth, T., Lew, V. L., and Gov, N. S. Curling and local shape changes of red blood cell membranes driven by cytoskeletal reorganization. *Biophys J* **99,** 1 (2010).
55. Walcott, S. and Sun, S. X. A mechanical model of actin stress fiber formation and substrate elasticity sensing in adherent cells. *Proc Natl Acad Sci U S A* **107,** 7757–7762 (2010).
56. Gerbal, F., Chaikin, P., Rabin, Y., and Prost, J. An elastic analysis of listeria monocytogenes propulsion. *Biophys J* **79,** 2259–2275 (2000).
57. Sens, P. and Safran, S. A. Pore formation and area exchange in tense membranes. *Europhys Lett* **43,** 95 (1998).
58. Shlomovitz, R. and Gov, N. S. Membrane-mediated interactions drive the condensation and coalescence of ftsz rings. *Phys Biol* **6,** 046017 (2009).
59. Cai, W. and Lubensky, T. C. Covariant hydrodynamics of fluid membranes. *Phys Rev Lett* **73,** 1186 (1994).
60. Ramaswamy, S., Toner, J., and Prost, J. Nonequilibrium fluctuations, traveling waves, and instabilities in active membranes. *Phys Rev Lett* **84,** 3494 (2000).
61. Gov, N. Diffusion in curved fluid membranes. *Phys Rev E* **73,** 041918 (2006).
62. Kandere-Grzybowska, K., Soh, S., Mahmud, G., Komarova, Y., Pilans, D., et al. Short-term molecular polarization of cells on symmetric and asymmetric micropatterns. *Soft Matter* **6,** 3257–3268 (2010).

63. Johnston, S. A., Bramble, J. P., Yeung, C. L., Mendes, P. M., and Machesky, L. M. Arp2/3 complex activity in filopodia of spreading cells. *BMC Cell Biol* **9,** 65 (2008).
64. Medeiros, N. A., Burnette, D. T., and Forscher, P. Myosin ii functions in actin-bundle turnover in neuronal growth cones. *Nat Cell Biol* **8,** 216–226 (2006).
65. Lacayo, C. I., Pincus, Z., VanDuijn, M. M., Wilson, C. A., Fletcher, D. A., et al. Emergence of largescale cell morphology and movement from local actin filament growth dynamics. *PLoS Biol* **5,** e233 (2007).
66. Ahmed, S., Goh, W. I., and Bu, W. I-bar domains, irsp53 and filopodium formation. *Semin Cell Dev Biol* **21,** 350–356 (2010).
67. Atilgan, E., Wirtz, D., and Sun, S. X. Mechanics and dynamics of actin-driven thin membrane protrusions. *Biophys J* **90,** 65–76 (2006).
68. Gov, N. S. Dynamics and morphology of microvilli driven by actin polymerization. *Phys Rev Lett* **97,** 018101 (2006).
69. Alexander, N. R., Branch, K. M., Parekh, A., Clark, E. S., Iwueke, I. C. et al. Extracellular matrix rigidity promotes invadopodia activity. *Curr Biol* **18,** 1295–1299 (2008).
70. Ahmed, I., Ponery, A. S., Nur-E-Kamal, A., Kamal, J., Meshel, A. S. et al. Morphology, cytoskeletal organization, and myosin dynamics of mouse embryonic fibroblasts cultured on nanofibrillar surfaces. *Mol Cell Biochem* **301,** 241 (2007).
71. Ghosh, K., Pan, Z., Guan, E., Ge, S., Liu, Y. et al. Cell adaptation to a physiologically relevant ecm mimic with different viscoelastic properties. *Biomaterials* **28,** 671–679 (2007).
72. Ingber, D. E., Prusty, D., Sun, Z., Betensky, H., and Wang, N. Cell shape, cytoskeletal mechanics, and cell cycle control in angiogenesis. *J Biomech* **28,** 1471–1484 (1995).
73. Zhuravlev, P. I. and Papoian, G. A. Molecular noise of capping protein binding induces macroscopic instability in filopodial dynamics. *Proc Natl Acad Sci U S A* **106,** 11570–11575 (2009).
74. Enderling, H., Alexander, N. R., Clark, E. S., Branch, K. M., Estrada, L. et al. Dependence of invadopodia function on collagen fiber spacing and cross-linking: computational modeling and experimental evidence. *Biophys J* **95,** 2203–2218 (2008).
75. Mogilner, A. and Keren, K. The shape of motile cells. *Curr Biol* **19,** R762–R771 (2009).
76. Keren, K., Pincus, Z., Allen, G. M., Barnhart, E. L., Marriott, G. et al. The shape of motile cells. *Nature* **453,** 475–480 (2008).
77. Wilson, C. A., Tsuchida, M. A., Allen, G. M., Barnhart, E. L., Applegate, K. T. et al. Myosin ii contributes to cell-scale actin network treadmilling through network disassembly. *Nature* **465,** 373–377 (2010).
78. Marée, A. F., Jilkine, A., Dawes, A., Grieneisen, V. A., and Edelstein-Keshet, L. Polarization and movement of keratocytes: a multiscale modelling approach. *Bull Math Biol* **68,** 1169–1211 (2006).
79. Rubinstein, B., Fournier, M. F., Jacobson, K., Verkhovsky, A. B., and Mogilner, A. Actin-myosin viscoelastic flow in the keratocyte lamellipod. *Biophys J* **97,** 1853–1863 (2009).
80. Fournier, M. F., Sauser, R., Ambrosi, D., Meister, J. J., and Verkhovsky, A. B. Force transmission in migrating cells. *J Cell Biol* **188,** 287–297 (2010).

10 Amoeboid Cells' Protrusions

Peter J. M. Van Haastert

CONTENTS

10.1 Introduction 193
10.2 Methods 195
10.3 Results and Discussion 195
10.3.1 Walking of Cells on a Solid Support 195
10.3.2 Traveling Waves of Convexities in Tail-Attached Cells 197
10.3.3 Swimming Cells Use Pseudopods that Convert to Paddles 197
10.3.4 How Fast can Cells Glide, Walk and Swim? 200
10.4 Conclusion 201
Keywords 202
References 203

10.1 INTRODUCTION

Amoeboid cells crawl using pseudopods, which are convex extensions of the cell surface. In many laboratory experiments, cells move on a smooth substrate, but in the wild cells may experience obstacles of other cells or dead material, or may even move in liquid. To understand how cells cope with heterogeneous environments we have investigated the pseudopod life cycle of wild type and mutant cells moving on a substrate and when suspended in liquid. It is shown that the same pseudopod cycle can provide three types of movement that we address as walking, gliding and swimming. In walking, the extending pseudopod will adhere firmly to the substrate, which allows cells to generate forces to bypass obstacles. Mutant cells with compromised adhesion can move much faster than wild type cells on a smooth substrate (gliding), but cannot move effectively against obstacles that provide resistance. In a liquid, when swimming, the extending pseudopods convert to side-bumps that move rapidly to the rear of the cells. Calculations suggest that these bumps provide sufficient drag force to mediate the observed forward swimming of the cell.

Many cells have a mode of migration known as amoeboid movement that is characterized by frequent changes in cell shape due to the extension of protrusions [1, 2]. The protrusions of an amoeboid cell, often termed pseudopods or lamellipods, are crucial for cell movement, as they determine the speed, direction and trajectory of the cell. An important aspect of cell motility is the ability of cells to respond to directional cues with oriented movement. Gradients of diffuse chemicals give rise to chemotaxis [3, 4]. Other directional cues that can induce oriented movement are temperature gradients or electric fields [5, 6]. These signals somehow modulate the direction of pseudopods such that, on average, cells move in the direction of the positional cues.

Recently, cell migration has been investigated using a 'pseudopod-centred' approach, in which large data sets are collected on the spatiotemporal properties of pseudopods that are extended by cells in the absence or presence of directional cues [7–13]. In the absence of external cues, cells are more likely to extend a new pseudopod in the direction of the existing pseudopods, and their directions are alternating to the left and right. With these characteristics cells move with strong persistence to cover a large distance in a short period [11, 14]. A shallow gradient of chemoattractant induces a small positional bias, such that cells are more likely to start a new pseudopod at the side of the cell facing the highest chemoattractant concentration [8, 9]. In addition, cells with multiple pseudopods often retract the pseudopod that is positioned in the worse direction relative to the cAMP gradient [7]. As pseudopods are generally extended perpendicular to the cell surface, this positional bias where pseudopods begin will direct the cell towards the attractant [15]. Cell migration and chemotaxis generally are studied in two dimensions as the cells crawl over various solid substrates. However, in vivo, cells move in a complex three dimensional environment [16, 17]. Such cells may experience obstacles such as other cells, soil particles, cavities, or liquids. Movement in a complex environment may require the ability to generate substantial force to resist obstacles, as well as the ability to swim [18, 19].

It has been suggested that pseudopods are self-organizing structures, which means that their organization is largely intrinsically controlled; although external signals can trigger the formation and location of a pseudopod, the pseudopod otherwise follows a typical life cycle [20]. In our studies on the movement in *Dictyostelium* cells we observed that wild type and many mutant cells extend a new pseudopod every ~15 s. After ~12 s, the pseudopod suddenly stops growing. In wild type cells on a solid support about 75% of the pseudopods make contact with the substrate, followed by outward expansion of the pseudopodia thereby contributing to the forward movement of the cell [9, 14]. Well after the pseudopodia cease to expand, they remain recognizable as convex extensions at the side of the cell. We have made some investigations to such "side-bumps", because likely they are places of attachment of the cell to the substrate [21–24] and therefore may contribute to movement in an environment with obstacles providing enhanced resistance. Surprisingly, cells in suspension have side bumps that move to the rear of the cell, potentially providing a drag force that may contribute to the forward movement of cells in suspension. Previously it has been suggested that traveling waves of surface deformations may explain the movement of cyanobacteria [25]. Here I report on the conversion of pseudopods to side-bumps and their potential role in movement of amoeboid cells on solid supports and in suspension.

10.2 METHODS

The strains used are wild type AX3 and *gbpD*-null cells lacking gene encoding a Rap-GEF [28]. Cells were grown in HG5 medium (contains per liter: 14.3 g oxoid peptone, 7.15 g bacto yeast extract, 1.36 g $Na_2HPO_4 \cdot 12H_2O$, 0.49 g KH_2PO_4, 10.0 g glucose), harvested in PB (10 mM KH2PO4/Na2HPO4, pH 6.5), and allowed to develop in 1 ml PB for 5 hr in a well of a 6-wells plate (Nunc). Movies of starved cells in PB on an objective glass were recorded with an inverted light microscope (Olympus Type CK40 with 20x objective) and images were captured at a rate of 1 frame/s with a JVC CCD camera. For movement under agar, cells were covered with an approximately cubic (length 3 mm) block of 1.5% agar in PB. Excess buffer was removed and movies were recorded as described above.

Images were analyzed with the fully automatic algorithm Quimp3 [26]. In short, the program uses an active contour analysis to identify the outline of the cell as ~150 nodes [34]. The local curvature of the cell outline is defined as the angle of the line segments pointing from a given node to its two neighbors, and the program identifies the central node of convex regions. With the convexity and area change of the nodes, extending pseudopodia were identified that fulfill the requirement of used-defined minimal number of adjacent convex nodes and minimal area change. The x,y coordinates of the central convex node of the convex area were recorded from start to end of pseudopod growth as described [26]; the position of this node was followed till the node disappeared by retraction.

The data are presented as the means and standard deviation (SD), where *n* represents the number of pseudopodia or number of cells analyzed, as indicated. The number of bumps in swimming and tail-attached cells is based on the observed number bumps at the lateral and upper sides of the cell, and multiplied by 4/3 to account for the invisible bumps in the lower side of the cell.

10.3 RESULTS AND DISCUSSION

10.3.1 Walking of Cells on a Solid Support

The algorithm Quimp for pseudopod analysis describes the cell boundary as a polygon of ~100 nodes [26]. Each node has an address, and therefore its speed and convexity can be determined in subsequent images. Pseudopods are identified as extending convex areas, with the central node of this convex area assigned as the tip of the pseudopod. We followed the tip node during and after pseudopod extension. The speed of the pseudopod tip relative to the substrate during the extending phase of the pseudopod is very high (~30 μm/min). At the end of the extension period, the speed of the tip node declines abruptly to nearly zero (Figure 1A). Interestingly, these stationary tip nodes are still present in convex areas, as identified by the computer algorithm [27] and by visual inspection of individual cells (Figure 1B). It thus appears that pseudopods frequently convert to convex bump at the side of the cell. Since the cell moves forward, the stationary bumps are found after about 1 min in the rear of the cell, where they are retracted.

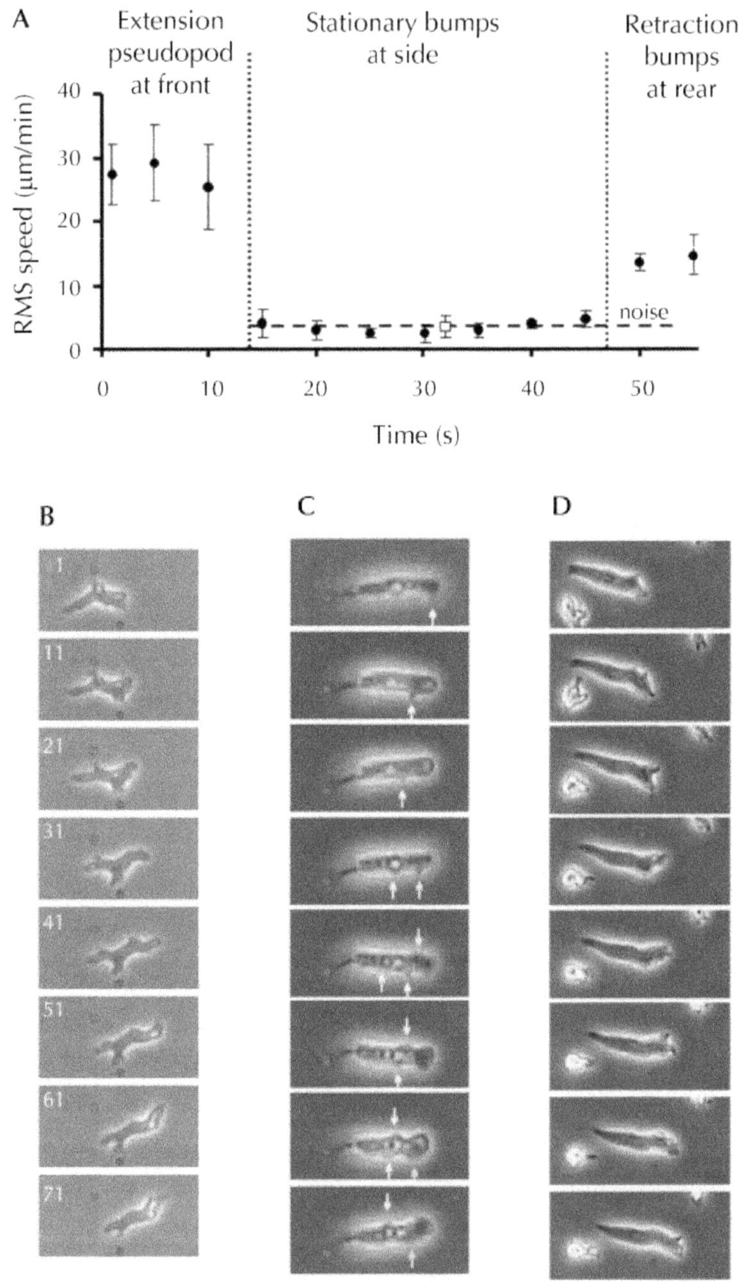

FIGURE 1 Side bumps on walking cells. A. Presented is the root-mean-square speed relative to the substrate of the tip of 20 pseudopods.. B. Images of wild-type walking cells. C. Images of tail-attached wild-type cells. D. Images of gliding gbpD-null cells. In the three cases the frames are static and the dots are placed at fixed positions; the arrows point to moving bumps. Numbers indicate time in seconds.

Pseudopods are formed at the front alternating to the right and left [11, 14], and therefore the sideway extensions are also alternating right/left from front to tail (Figure 1B). Previous experiments suggest that positions where pseudopods convert to side bumps are places of cell adhesion to the substrate [21–24]. Therefore, cell movement on a solid medium has the appearance of walking, because cells have sideway stationary "foots"; cells retract these foots in the rear while cells extend pseudopod at the front that become new foots.

10.3.2 Traveling Waves of Convexities in Tail-Attached Cells

Dictyostelium cells occasionally appear to be attached to the substrate exclusively at the tail region, whereas other parts of the cell move freely in suspension (Figure 1C). This interpretation is based on the observations that i) the tail touches the surface in the focal plane of the tail ii) the position of the tail does not change (thus the tail does not move), and iii) the rest of the cell can move; sometimes the cell is relatively stationary as in figure 1C (and therefore pseudopod and bump data can be easily collected for a few minutes) and sometimes cells move actively to the right and left ("waving"). In wild type cells, about 5–10% of the cells show this behavior. In some mutants, such as talinB-null cells, tail-attached movement is very common (83% of the cells). Tail-attached cells extend pseudopods at nearly the same frequency compared to surface-attached cells (Table S1). However, pseudopods grow for a shorter period and are also smaller. As in cell body-attached cells, the convex pseudopod tips frequently convert to side bumps. Interestingly, these side bumps travel to the rear of the cell (Figure 1C) at a speed of about -13 μm/min; the minus sign is to indicate that side-bumps travel in a direction opposite to extending pseudopods. After approximately 1 min the side bumps have arrived at the tail of the cell.

10.3.3 Swimming Cells Use Pseudopods that Convert to Paddles

Occasionally the tail-attached cells detach from the substrate. Although such cells will slowly sink, they can be followed while completely free in suspension during a few minutes (Figure 2A). We followed 8 tail-attached cells after they detach from the substrate. Such cells continue to form pseudopods and side bumps that travel to the rear of the cell, with essentially the same properties as tail-attached cells. The swimming cells move forward (i.e. in the direction of the extended pseudopods) at a slow speed of ~ 3 μm/min (Table S1). The trajectories of swimming cells reveal persistent directional movement (Figure 2C).

Gliding of adhesion-defective mutant cells on a smooth surface

The GTP-binding protein Rap1 has been shown to be involved in cell adhesion [28]. Cells lacking GbpD, a Rap-GEF, exhibit strongly reduced adhesion (Figure 3A). We analyzed pseudopod formation ingbpD-null cells (Table S1). Pseudopods are formed at a frequency and speed that are similar compared to wild type cells, but they grow longer and are therefore larger. After the pseudopod tip stops outward expansion, the pseudopod rarely continues as convex extension at the side of the cell body (Figure 1D). Instead, the cytosol flows into the pseudopod and the old pseudopod tip merges with the cell body. Interestingly, these cells have enhanced speed of movement, 17 μm/min, compared to ~10 μm/min for wild type cells. The enhanced speed was also observed for other mutants with reduced adhesion: talinA-null cells with defects in the cytoskeleton [29] and sadA-null cells with defects in adhesion molecules [30]. The adhesion-defective mutant cells move nearly with the speed of the forward moving pseudopods, suggesting that cells in the absence of strong adhesion glide over the substrate.

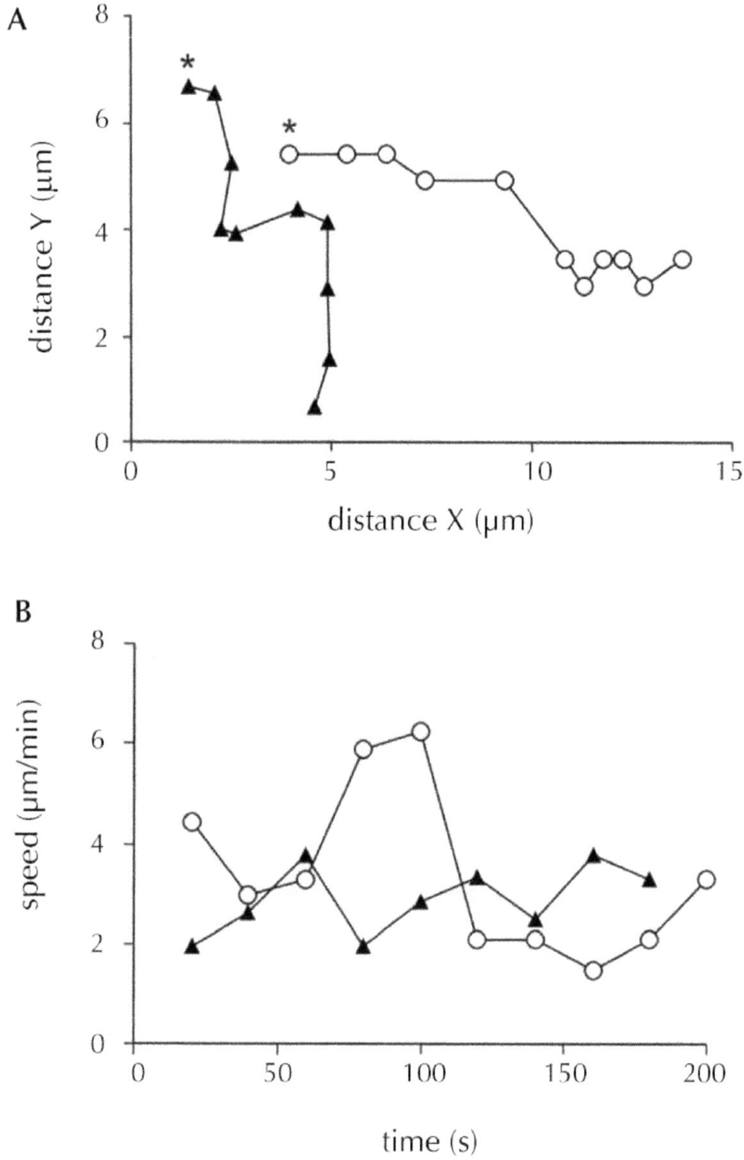

FIGURE 2 *(Continued)*

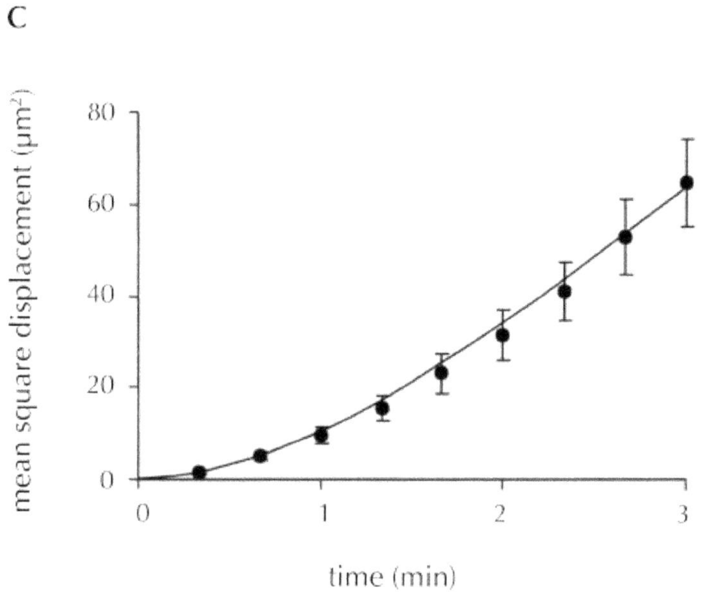

FIGURE 2 Swimming cells. The trajectory of eight tail-attached cells was followed after detachment from the surface. A. Tracks of two swimming cells at 20 s interval; * indicate the start. B. Speed of the same two cells. C. Mean square displacement $\langle D^2 \rangle$ of eight swimming cells. The parameters of the equation for a persistent random walk in three dimensions $\langle D^2 \rangle = 3S^2[Pt-P^2(1-e^{-t/p})]$ were fitted to the data points; the line presents the optimal fit with speed S = 3 μm/min and persistence time P = 1.3 min.

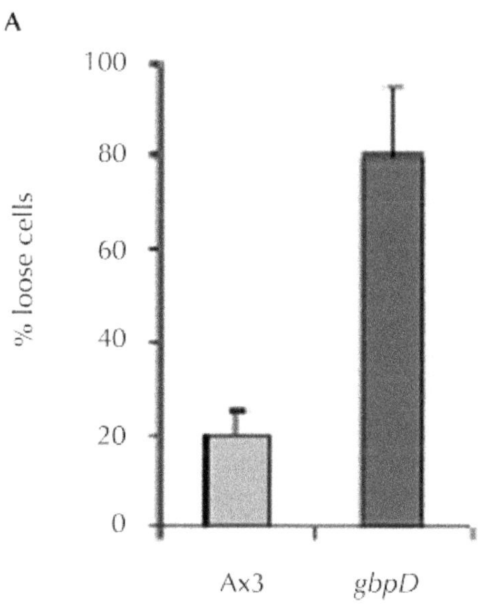

FIGURE 3 *(Continued)*

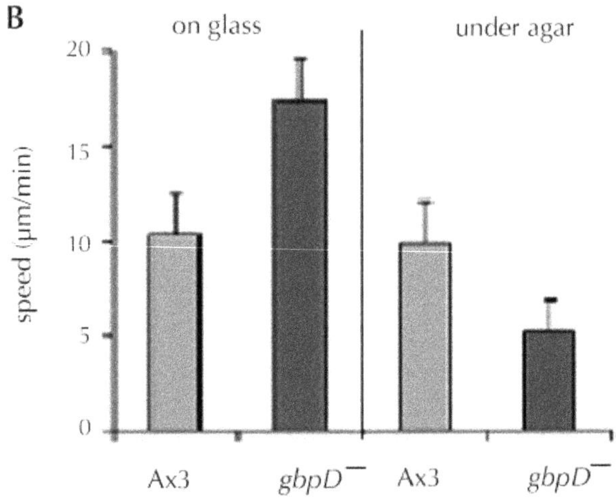

FIGURE 3 Gliding cells. A. Adhesion of wild type and gbpD-null cells, expressed as the fraction of cells that detach from a plastic surface after shaking in buffer for 1 hr [28]. B. Speed of wild-type and mutant cells on a solid support and under a block of agar.

Can cells with reduced adhesion move against obstacles? Wild-type and *gbpD*-null cells were covered with a block of agar to provide resistance. Whereas movement of wild type cells under agar is only slightly slower than movement on agar, the speed of *gbpD*-null cells is strongly reduced from ~17 μm/min to ~5 μm/min under agar.

10.3.4 How Fast can Cells Glide, Walk and Swim?

The maximal speed of walking or gliding cells on a substrate depends on the frequency F and size l_p of extending pseudopods according to

$$v_c = Fl_p ab \cos \alpha \tag{1}$$

where a = fraction of pseudopods that contribute to movement, b = positional overlap of pseudopods, and α = the mean angle between pseudopod and forward cell movement. The experimental data (Table S2) predict that wild type cells can walk at a maximal speed of 11.4 μm/min, which is close to the observed speed of 10.4 μm/min.

Gliding *gbpD*-null cells extend pseudopods at nearly the same frequency as wild type cells, but with some subtle differences: pseudopods are larger and rarely retracted, thereby providing more forward movement to the cells. Using these measured pseudopod properties, the predicted speed of gliding cells is 17.8 μm/min, which is close to the observed speed of 17.3 μm/min.

Apparently, the lower speed of wild type cells is attributed to the shorter pseudopods and the more frequent retraction of new pseudopods. Why? Movement of substrate-attached cells can be regarded as the transport of material from behind the attachment zone at the rear of the cell to before the attachment zone at the front of the cell. Materials to be transported are cytoplasm with organelles and plasma membrane.

Cytoplasm may flow freely to the front by hydrostatic pressure [31], but transport of membrane may be restricted by adhesion of the cell to the surface. The reduced flow of membrane may lead to increased membrane tension at the front of the cell, which could impair pseudopod growth and induce retraction of newly formed pseudopods, as is observed in wild type cells.

How fast can cells swim in liquid? An object moving in a fluid will experience a drag force that, according to Newton›s third law, will induce an equal counterforce on the object. For small objects the drag occurs at low Reynolds numbers [32] and therefore the drag force F_D is given by the Stokes equation: $F_D = -6\pi\rho Rv$, where ρ is the viscosity of the fluid, R is the radius of the frontal cross sectional area of the object and v is the speed of the object. A swimming *Dictyostelium* cell may experience three forces: the drag induced by the extending pseudopod, the drag induced by the rearward moving bumps, and the drag induced by the total movement of the cell. Assuming that these three drag forces experience the same viscosity, the balanced force equation yields:

$$v_c R_c + n_b R_b \vec{v}_b + n_p R_p \vec{v}_p = 0 \tag{2}$$

where the subscripts indicate cells (c) bumps (b) or pseudopods (p), $\vec{v}$ indicates the mean speed of the object in the direction of the cell and n indicates the mean number of moving bumps or pseudopods. We measured these properties for cells in suspension (Table S2), predicting an average swimming rate of ~ 3.5 μm/min, which is close to the observed rate of 3 μm/min for freely swimming Dictyostelium cells.

10.4 CONCLUSION

Polarized amoeboid cells may move by three modes, walking, gliding and swimming (Figure 4). Investigations to swimming cells are complicated because cells in suspension will sink. In isodense suspensions of ficol cells can be observed easily [18], but ficol may induce an osmotic response or it may provide some form of structure to move on, as ficol may not be a Newtonian solution [33]. We investigated cells that detached from the surface by which they move in suspension, and obtained a swimming speed of ~3 μm/min, about 4-times slower that the crawling speed of ~12 μm/min. In ficol, cells swim and crawl at about the same speed of 4.2 and 3.8 μm/min, respectively [18]. The current observations suggest that the three moving modes, walking, gliding and swimming, are all based on pseudopods, which are extending convex areas of the cell boundary. At the end of the extension period the convex pseudopods often convert to convex bumps at the side of the cell, which move in about 1 min to the rear of cells in suspension. The ~3 rearward moving bumps provide sufficient drag force to explain the observed forward movement of swimming cell. Pseudopods of cells attached to a substrate also convert to side bumps that end up in the rear of the cell after about 1 min. However, these bumps do not move, but are fixed to the substrate, presumably because they form attachment sides of the cell to the substrate [21–24]. Surprisingly, adhesion-defective mutant cells move at a much higher speed than wild type cells on a smooth surface, but exhibit a much lower speed then wild type cells when exposed to strong resistance, such as movement under a block of agar. *Dictyostelium* cells live in a heterogeneous environment composed of soil particles and surrounding liquid. Cells move probably most of the time on 2D surfaces of soil particles, but may also experience clefts and obstacles. Cells have the ability to walk on these complicated surfaces with stronger adhesion, and to swim in water, all using essentially the same cycle of pseudopod formation with conversion to sideway extensions. This may allow the cells to effectively move optimally in its physically complex habitat.

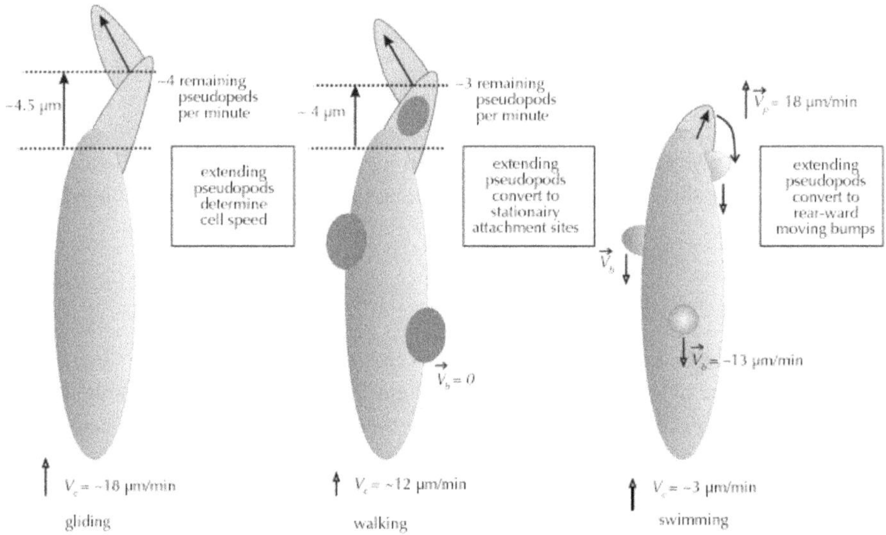

FIGURE 4 Model of gliding, walking and swimming. All cells extend pseudopods at a frequency of ~4/min. In gliding cells, pseudopods are large and all contribute to forward moving; side bumps are rare. In walking cells, pseudopods are smaller and only ~75% of pseudopods contribute to forward movement; these pseudopods convert to side-bumps that are stationary relative to the substrate. In swimming cells, pseudopods are small and convert to side-bumps that move to the rear of the cell. On average walking and swimming cells have ~3 side-bumps (see table S1)

KEYWORDS

- Amoeboid cells
- Chemoattractant concentration
- Chemotaxis
- *Dictyostelium* cells
- Pseudopod life cycle
- Pseudopod-centred' approach
- Tail-attached cells

REFERENCES

1. Friedl, P. and Wolf, K. Plasticity of cell migration: a multiscale tuning model. *J Cell Biol*, **188**, 1119 (2009).
2. Lammermann, T. and Sixt, M. Mechanical modes of 'amoeboid' cell migration. *Curr Opin Cell Biol*, **21**, 636–644 (2009).
3. Weiner, O. D. Regulation of cell polarity during eukaryotic chemotaxis: the chemotactic compass. *Curr Opin Cell Biol*, **14**, 196–202 (2002).
4. Hoeller, O. and Kay, R. Chemotaxis in the absence of PIP3 gradients. *Current Biology*, **17**, 813–817 (2007).
5. Bahat, A. and Eisenbach, M. Sperm thermotaxis. *Mol Cell Endocrinol*, **252**, 115–119 (2006).
6. Zhao, M. Electrical fields in wound healing-An overriding signal that directs cell migration. *Semin Cell Dev Biol*, **20**, 674–682 (2009).
7. Andrew, N. and Insall, R. H. Chemotaxis in shallow gradients is mediated independently of PtdIns 3-kinase by biased choices between random protrusions. *Nat Cell Biol*, **9**, 193–200 (2007).
8. Arrieumerlou, C. and Meyer, T. A local coupling model and compass parameter for eukaryotic chemotaxis. *Dev Cell*, **8**, 215–227 (2005).
9. Bosgraaf, L. and Van Haastert, P. J. M. Navigation of chemotactic cells by parallel signaling to pseudopod persistence and orientation. *PLoS ONE*, **4**, e6842 (2009).
10. Maeda, Y. T., Inose, J., Matsuo, M. Y., Iwaya, S., and Sano, M. Ordered patterns of cell shape and orientational correlation during spontaneous cell migration. *PLoS ONE*, **3**, e3734 (2008).
11. Li, L., Norrelykke, S. F., and Cox, E. C. Persistent cell motion in the absence of external signals: a search strategy for eukaryotic cells. *PLoS ONE*, **3**, e2093 (2008).
12. Takagi, H., Sato, M. J., Yanagida, T., and Ueda, M. Functional analysis of spontaneous cell movement under different physiological conditions. *PLoS ONE*, **3**, e2648 (2008).
13. Insall, R. H. Understanding eukaryotic chemotaxis: a pseudopod-centred view. *Nat Rev Mol Cell Biol*, **11**, 453–458 (2010).
14. Bosgraaf, L. and Van Haastert, P. J. M. The Ordered Extension of Pseudopodia by Amoeboid Cell in the Absence of external Cues. *PLoS ONE*, **4**, e5253 (2009).
15. Van Haastert, P. J. M. and Bosgraaf, L. The local cell curvature guides pseudopodia towards chemoattractants. *Hfsp J*, **3**, 282–286 (2009).
16. Franck, C., Maskarinec, S. A., and Tirrell, D. A., and Ravichandran, G. Three-dimensional traction force microscopy: a new tool for quantifying cell-matrix interactions *PLoS ONE*, **6**, e17833 (2011).
17. Provenzano, P. P., Eliceiri, K. W., and Keely, P. J. Shining new light on 3D cell motility and the metastatic process. *Trends Cell Biol*, **19**, 638–648 (2009).
18. Barry, N. P. and Bretscher, M. S. Dictyostelium amoebae and neutrophils can swim. *Proc Natl Acad Sci USA*, **107**, 11376–11380 (2010).
19. Bae, A. J., and Bodenschatz, E. On the swimming of Dictyostelium amoebae. *Proc Natl Acad Sci USA*, **107**, E165–166 (2010).
20. Karsenti, E. Self-organization in cell biology: a brief history. *Nat Rev Mol Cell Biol*, **9**, 255–262 (2008).
21. Weber, I. Is there a pilot in a pseudopod? *Eur J Cell Biol*, **85**, 915–924 (2006).
22. Uchida, K. S. and Yumura, S. Dynamics of novel feet of Dictyostelium cells during migration. *J Cell Sci*, **117**, 1443–1455 (2004).
23. Fukui, Y. and Inoue, S. Amoeboid movement anchored by eupodia, new actin-rich knobby feet in Dictyostelium. *Cell Motil Cytoskeleton*, **36**, 339–354 (1997).
24. Iwadate, Y. and Yumura, S. Actin-based propulsive forces and myosin-II-based contractile forces in migrating Dictyostelium cells. *J Cell Sci*, **121**, 1314–1324 (2008).
25. Stone, H. and Samuel, A. Propulsion of Microorganisms by Surface Distortions. *Phys Rev Lett*, **77**, 4102–4104 (1996).
26. Bosgraaf, L. and Van Haastert, P. J. M. Quimp3, an automated pseudopod-tracking algorithm. *Cell Adh Migr*, **4**, 1 (2009).

27. Bosgraaf, L. and Van Haastert, P. J. M., and Bretschneider, T. Analysis of cell movement by simultaneous quantification of local membrane displacement and fluorescent intensities using Quimp2. *Cell Motil Cytoskeleton*, **66**, 156–165 (2009).
28. Kortholt, A., Rehmann, H., Kae, H., Bosgraaf, L., Keizer-Gunnink, I., et al. Characterization of the GbpD-activated Rap1 pathway regulating adhesion and cell polarity in Dictyostelium discoideum. *J Biol Chem*, **281**, 23367–23376 (2006).
29. Niewohner, J., Weber, I., Maniak, M., Muller-Taubenberger, A., and Gerisch, G. Talin-null cells of Dictyostelium are strongly defective in adhesion to particle and substrate surfaces and slightly impaired in cytokinesis. *J Cell Biol*, **138**, 349–361 (1997).
30. Fey, P., Stephens, S., Titus, M. A., and Chisholm, R. L. SadA, a novel adhesion receptor in Dictyostelium. *J Cell Biol*, **159**, 1109–1119 (2002).
31. Keren, K., Yam, P. T., Kinkhabwala, A., Mogilner, A., and Theriot, J. A. Intracellular fluid flow in rapidly moving cells. *Nat Cell Biol*, **11**, 1219–1224 (2009).
32. Dusenbery, D. *Living at Micro Scale: the Unexpected Physics of Being Small*. Harvard University Press, Cambridge, Mass ISBN 978-0-674-03116-6 (2009).
33. Winet, H. Ciliary propulsion of objects in tubes: wall drag on swimming Tetrahymena (Ciliata) in the presence of mucin and other long-chain polymers. *J Exp Biol*, **64**, 283–302 (1976).
34. Bosgraaf, L., Van Haastert, P. J. M., and Bretschneider, T. Analysis of cell movement by simultaneous quantification of local membrane displacement and fluorescent intensities using Quimp2. *Cell Motil Cytoskel*, **66**, 156–165 (2009).

11 Collective Cell Migration

Masataka Yamao, Honda Naoki, and Shin Ishii

CONTENTS

11.1 Introduction 205
11.2 Materials and Methods 207
11.2.1 Non-Dimensionalization 207
11.2.2 Effective Interaction Identification 208
11.3 Discussion 212
11.4 Results 219
11.4.1 Model of Multi-Cellular Migration 219
11.4.2 Effect of Single Cellular Stochasticity on Multi-Cellular Transportation 222
11.4.3 Interaction Between Migratory Cells 225
11.4.4 Other Properties of Multi-Cellular Logistics 230
11.5 Conclusion 233
Keywords 233
Author Contributions 234
Acknowledgments 234
References 234

11.1 INTRODUCTION

During development, the formation of biological networks (such as organs and neuronal networks) is controlled by multicellular transportation phenomena based on cell migration. In multi-cellular systems, cellular locomotion is restricted by physical interactions with other cells in a crowded space, similar to passengers pushing others out of their way on a packed train. The motion of individual cells is intrinsically stochastic and may be viewed as a type of random walk. However, this walk takes place in a noisy environment because the cell interacts with its randomly moving neighbors. Despite this randomness and complexity, development is highly orchestrated and precisely regulated, following genetic (and even epigenetic) blueprints. Although individual cell migration has long been studied, the manner in which stochasticity affects multicellular transportation within the precisely controlled process of development remains largely unknown. To explore the general principles underlying multicellular migration, we focus on the migration of neural crest cells, which migrate collectively and form streams. We introduce a mechanical model of multi-cellular migration. Simulations based on the model show that the migration mode depends on the relative strengths of the noise from migratory and non-migratory cells. Strong noise from migratory cells and weak noise from surrounding cells causes "collective migration," whereas strong noise from non-migratory cells causes "dispersive migration." Moreover, theoretical analyses reveal that migratory cells attract each other over long distances, even without direct mechanical contacts. This effective interaction depends on the stochasticity of the migratory and non-migratory cells. On the basis of these findings, we propose that stochastic behavior at the single-cell level works effectively and precisely to achieve collective migration in multi-cellular systems. Movements of various cell groups are ubiquitous during development. The extent and speed of migrations must be well-controlled to achieve precise axon placement in the wiring of neuronal networks and to ensure the appropriate morphogenesis of tissues and organs [1]. In this chapter, we focus on multi-cellular collective migration, which can be observed in the behaviors of cranial neural crest cells during embryonic development, as a model system for understanding how the system-level control of cellular transportation is achieved; such system-level control is called "logistics". This transportation is accompanied by cell migration that is directed by extra-cellular signaling molecules working as chemo-attractants or repellants. In multi-cellular systems, cellular locomotion is restricted by physical interactions with other cells in a crowded space, similarly to passengers pushing others out of their way on a packed train. The mechanisms underlying multi-cellular logistics in these crowded space remain largely unknown.

At the level of individual cells and neuronal growth cones, migratory behavior is often stochastic rather than deterministic, due largely to the small number of signaling molecules within such cells, which perform biased random walks along chemo-attractant gradients [2], [3]. Nevertheless, the developmental process remains consistent across different embryos, even though the stochastic behavior of individual cells might seem to make precise and consistent control difficult. There must be a homeostasis (stability) mechanism at the multi-cellular systems level that absorbs the stochastic behavior. Also, developmental processes need to be variable enough to construct a variety of biological patterns starting from a single fertilized egg cell, while being stable enough to maintain the consistency of the patterns; this requirement is a typical plasticity-stability dilemma [4]. Therefore, the relationship between microscopic properties of individual cell migration and macroscopic multi-cellular migration patterns needs to be clarified.

Multi-cellular migration employs various modes of transportation, depending on the cell type and the developmental stage. These modes can be classified into two main categories, individual and collective migration [5]. Individual migration is dispersive and enables cells to cover a local area, as can be seen in immune cell trafficking [6]. Collective migration consists of multi-cellular units and is used mainly to build complex tissues. Typically, neural crest cells migrate together by a forming "stream" [7], and neural precursor cells sometimes migrate along a single dimension in a "chain"-like manner [8]. Understanding how these modes of migration emerge is important for understanding the mechanisms of multi-cellular development.

It has recently been shown that pattern modes can be experimentally inter-converted by manipulating the expression of proteins involved in cellular mechanics; up-regulating a cell adhesion molecule (CAM) in individually migratory cells leads to collective migration [9], whereas down-regulating a CAM in collectively migrating cohorts leads to individual migration [10], [11]. These observations suggest that the various transport pattern modes are not achieved simply by system-specific molecular regulations. In addition, it has been suggested that the pattern modes can be controlled through altering physical parameters in cell migration such as driving force, cellular stiffness, and the randomness of the migration [12]. However, the mechanisms by which microscopic mechanical parameters at the level of single cells contribute to the macroscopic pattern of multi-cellular migration remains elusive.

In this chapter, we studied the multi-cellular logistics of biological systems with a special emphasis on collective migration in a crowded environment. To this end, we focused on "neural crest migration", because even without the guidance of extra-cellular signals, neural crest cells collectively migrate from rhombomeres to branchial arches along stream [13]. We constructed a bio-physical model of a multicellular system in which cells migrate through crowded cell population using their chemotactic abilities. Note that neural crest cell migration is driven by both chemotactic abilities and population pressure due to proliferation [14], [15]. In this chapter, we particularly examined cell migration phenomena in a crowded situation. We then performed a computer simulation, which led us to hypothesize that migratory cells exploit the stochasticity within multi-cellular systems to collectively and efficiently migrate using an autonomously emerging stream. Theoretical analysis could shed light on the mechanisms that govern various migration pattern modes. Moreover, we discuss the properties of multi-cellular logistics on the basis of simulation.

11.2 MATERIALS AND METHODS

11.2.1 Non-Dimensionalization

The original bio-physical model given by Equations (13–17) is non-dimensionalized by setting a common cellular radius for all of the cells, i.e., $R_i = R$ for all i, as follows:

$$\frac{d\left(\frac{1}{R}\mathbf{r}_i\right)}{d\left(\frac{F_m}{\mu R}t\right)} = \sum_{j\in N_i}\mathbf{F}'_{\mathrm{rep}_{ij}} + \sum_{j\in M_i}\mathbf{F}'_{\mathrm{mig}_{ij}} + \mathbf{F}'_{\mathrm{flu}_i} \tag{1}$$

$$\mathbf{F}'_{\mathrm{rep}_{ij}} = \frac{kR}{F_m}\left\{2 - \left\|\frac{\mathbf{r}_i}{R} - \frac{\mathbf{r}_j}{R}\right\|\right\}\frac{\mathbf{r}_i/R - \mathbf{r}_j/R}{\|\mathbf{r}_i/R - \mathbf{r}_j/R\|} \tag{2}$$

$$\mathbf{F}'_{\mathrm{mig}_{ij}} = \begin{pmatrix} F_{\mathrm{mig}_x} \\ F_{\mathrm{mig}_y} \end{pmatrix} = s(i)\begin{pmatrix} \dfrac{\|\Delta y_{ij}/R\|}{\|\mathbf{r}_i/R - \mathbf{r}_j/R\|} \\ -\dfrac{\|\Delta y_{ij}/R\|}{\Delta y_{ij}/R}\dfrac{\Delta x_{ij}/R}{\|\mathbf{r}_i/R - \mathbf{r}_j/R\|} \end{pmatrix} \tag{3}$$

$$\mathbf{F}'_{\mathrm{flu}_i} = \frac{\sigma_i}{F_{\mathrm{m}}}\begin{pmatrix}\xi_{xi}\\ \xi_{yi}\end{pmatrix} \tag{4}$$

$$s(i) = \begin{pmatrix} 1 & \text{if cell i is migratory} \\ -1 & \text{if cell i is non-migratory} \end{pmatrix} \tag{5}$$

We have defined the dimension-less variables as $\mathbf{r}' = \frac{\mathbf{r}}{R}$, $t' = \frac{F_{\mathrm{m}}}{\mu R}t$, $k' = \frac{kR}{F_{\mathrm{m}}}$, and $\sigma_i' = \frac{\sigma_i}{F_{\mathrm{m}}}$ here. After this non-dimensionalization, the free parameters of the model are reduced to k' and σ_i', implying that the noise intensity differs between the migratory and non-migratory cells.

The adhesive force in the model is also non-dimensionalized as follows:

$$\mathbf{F}'_{\mathrm{adh}_{ij}} = -\frac{aR}{F_{\mathrm{m}}}\left\{\frac{L}{R} + 2 - \left\|\frac{\mathbf{r}_i}{R} - \frac{\mathbf{r}_j}{R}\right\|\right\}\frac{\mathbf{r}_i/R - \mathbf{r}_j/R}{\|\mathbf{r}_i/R - \mathbf{r}_j/R\|}. \tag{6}$$

Then, we set $a' = \frac{aR}{F_{\mathrm{m}}}$ and $L' = \frac{L}{R}$.

11.2.2 Effective Interaction Identification

We proposed a method for identifying the effective potential U, which is defined in Equation (19). Because the non-parametric (sample-based) estimation of the velocity vector field, $v = (v_x, v_y)_T$, can be rough due to the lack of continuous constraint, the potential was modeled as a continuous parametric polynomial function:

$$U(x,y) = \sum_{i,j} a_{ij} x^i y^j, \tag{7}$$

where a_{ij} is the coefficient of $x^i y^j$ and $x = (x, y)_T$ denotes the relative coordinates in two-dimensional space. Because a_{00} does not affect the dynamics (Equation (19)), it was simply set to 0. The relative velocity of the migrating cell is then re-expressed by

$$v_x = -\frac{\partial U}{\partial x} = a_{10} + 2a_{20}x + a_{11}y + \ldots \tag{8}$$

$$v_y = -\frac{\partial U}{\partial y} = a_{01} + a_{11}x + a_{02}y + \ldots . \tag{9}$$

These coefficients were estimated by a least-squares regression on the basis of the vector field sampled by the simulations (Figure 1D):

$$\begin{pmatrix} a_{10} \\ a_{01} \\ a_{20} \\ a_{11} \\ \vdots \end{pmatrix} = - \begin{pmatrix} 1 & 0 & 2x_1 & y_1 & \cdots \\ 0 & 1 & 0 & x_1 & \cdots \\ 1 & 0 & 2x_2 & y_2 & \cdots \\ 0 & 1 & 0 & x_2 & \cdots \\ \vdots & \vdots & \vdots & \vdots & \ddots \end{pmatrix}^{\dagger} \begin{pmatrix} v_{x1} \\ v_{y1} \\ v_{x2} \\ v_{y2} \\ \vdots \end{pmatrix} \quad (10)$$

A

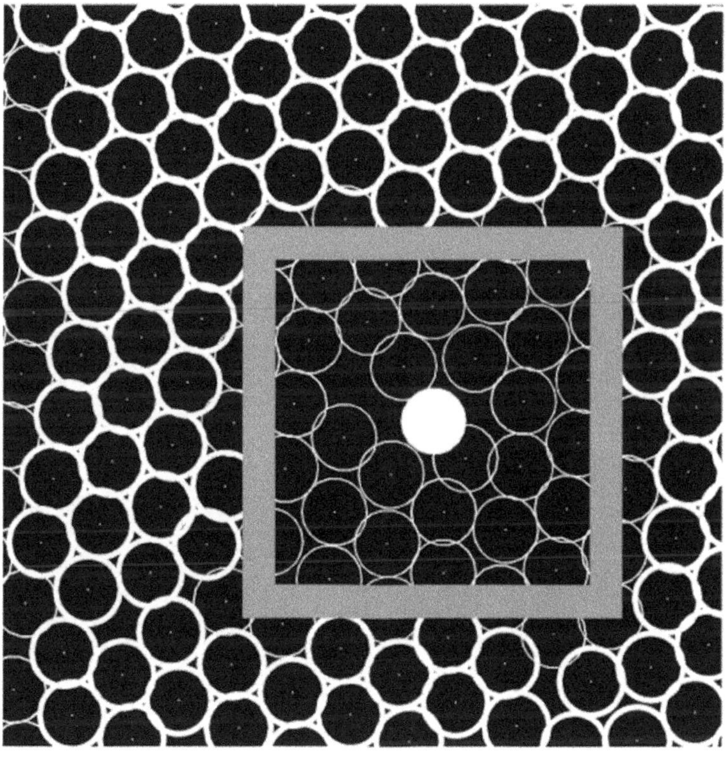

FIGURE 1 *(Continued)*

B

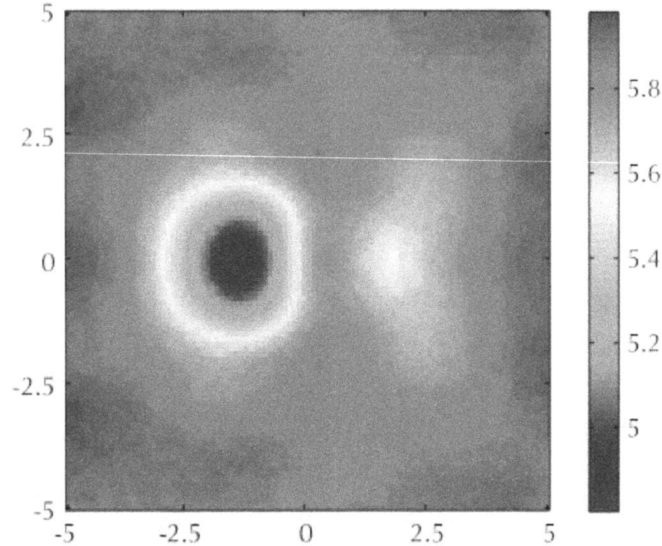

C

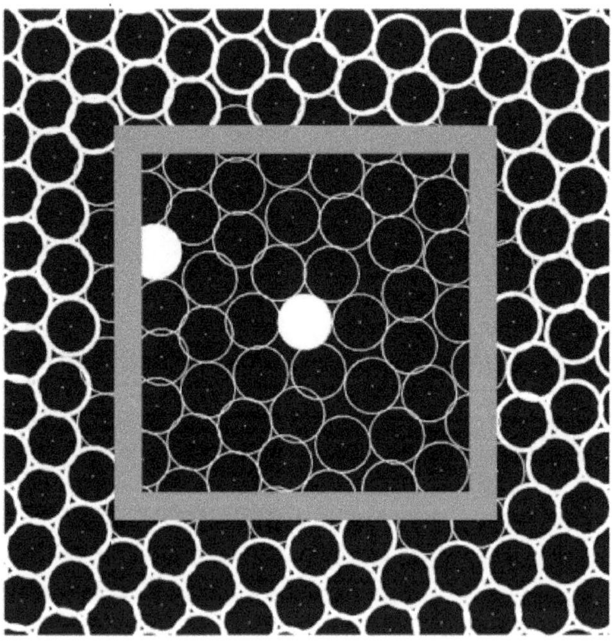

FIGURE 1 *(Continued)*

D

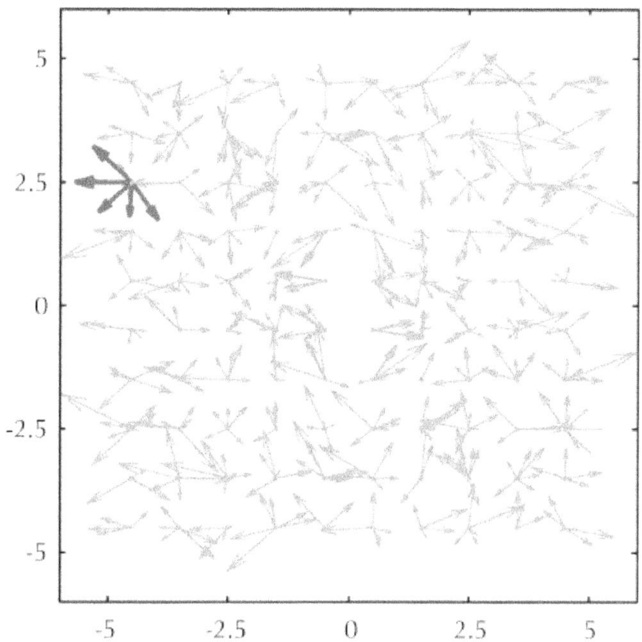

FIGURE 1 An illustration of the method for estimating cell density and the effective potential field around a migratory cell. The simulations were performed after placing a single migratory cell (A) or two migratory cells (C) to be surrounded by non-migratory cells. The white and black circles indicate migratory and non-migratory cells, respectively. (B) The average density of the non-migratory cells was estimated relative to the position of the migratory cell at the origin (i.e., (A)). This density is estimated by kernel density estimation with Gaussian kernel functions with variances equal to the cellular radius. The square region shown in this panel corresponds to the cyan square in (B). (D) The sample-based velocity vector field. We performed a short-term (0.5 sec.) simulation after placing a migratory cell on each grid point, and the vector differences of each migratory cell in its position are displayed at the each grid points. The square region shown in this panel corresponds to the cyan square in (C).

where † denotes the Moore-Penrose pseudo-inverse. Because the effective noise intensity along the vertical axis is important for collectivity, σ_y was defined as the mean-squared error between the sampled velocity v_{yi} at (x_i, y_i) and its expected value, $-\partial U_{est}/\partial y_i$, i.e., $\|v_{yi} - (-\partial U_{est}/\partial y_i)\|^2$. To smoothly estimate the position-based effective variance, we again used a polynomial fitting:

$$\sigma_y^2(x,y) = \sum_{i,j} b_{ij} x^i y^j, \tag{11}$$

the coefficients of which, b_{ij}, were identified using a least-squares regression on the simulation samples:

$$\begin{pmatrix} b_{00} \\ b_{10} \\ b_{01} \\ b_{20} \\ b_{11} \\ \vdots \end{pmatrix} = \begin{pmatrix} 1 & x_1 & y_1 & x_1^2 & x_1 y_1 & \\ 1 & x_2 & y_2 & x_2^2 & x_2 y_2 & \\ 1 & x_3 & y_3 & x_3^2 & x_3 y_3 & \\ \vdots & \vdots & \vdots & \vdots & \vdots & \ddots \end{pmatrix}^{\dagger} \begin{pmatrix} \left\| v_{y1} - \left(-\frac{\partial U_{est}}{\partial y_1} \right) \right\|^2 \\ \left\| v_{y2} - \left(-\frac{\partial U_{est}}{\partial y_2} \right) \right\|^2 \\ \left\| v_{y3} - \left(-\frac{\partial U_{est}}{\partial y_3} \right) \right\|^2 \\ \vdots \end{pmatrix} \tag{12}$$

11.3 DISCUSSION

During multi-cellular development, in addition to intra-cellular biochemical features, mechanical cellular features evoked by direct physical interactions with neighboring cells become dominant. By simulating multi-cellular migration using simple mechanical cells, we have shown that microscopic stochasticity plays a significant role in the emergence of population migration patterns and their logistics.

Model can explain the collective migration of neural crest cells, which occurs through the autonomous formation of a stream (Figure 2A, upper panel). Stream formation has been hypothesized to be extra-cellularly regulated by repulsive cue molecules [22]. Recently, however, it has been reported that down-regulation of the repulsive cue molecule neurophilin-1 does not affect the collective migration of neural crest cells [7]. Model has indicated that the combination of strongly fluctuating migratory cells and weakly fluctuating non-migratory cells lead to collective migration with autonomously stream formation (point α in Figure 3D). Although the mechanisms by which chemotactic cells manage to suppress the intrinsic stochasticity of signal transduction have been previously discussed [3], the ways in which multicellular functions are implemented in the context of the stochastic migration of individual cells have not been examined.

Several theoretical models have been proposed for multi-cellular migration [23], [24]. The dynamics of the cell population have often been modeled by reaction-diffusion systems [25]. Although approaches based on reaction-diffusion systems likely ignore the detailed dynamics at the single-cell level, they are still useful for providing insight into macroscopic mechanisms. A previous theoretical study addressed the migration of neural crest cells in the intestine, thereby highlighting the biological significance of cell proliferation [14]. Proliferative activity was also found in the cranial neural crest cells that we addressed in this chapter [15]. Therefore developing a model that includes this proliferation is an important step towards reproducing multi-cellular migration more realistically. Proliferation was easily introduced into model by adding a new cell near the existing cell, as in an existing computational model [26]. An alternative approach is the phase-field model, in which cellular morphological changes are represented by partial-differential equations that are derived by minimizing certain energy functions; this model has been applied to both cellular migration [27] and proliferation [28]. Because the phase-field model can accommodate flexible morphological changes at the single cell level, it is suitable for simulations of population migration with proliferation.

A

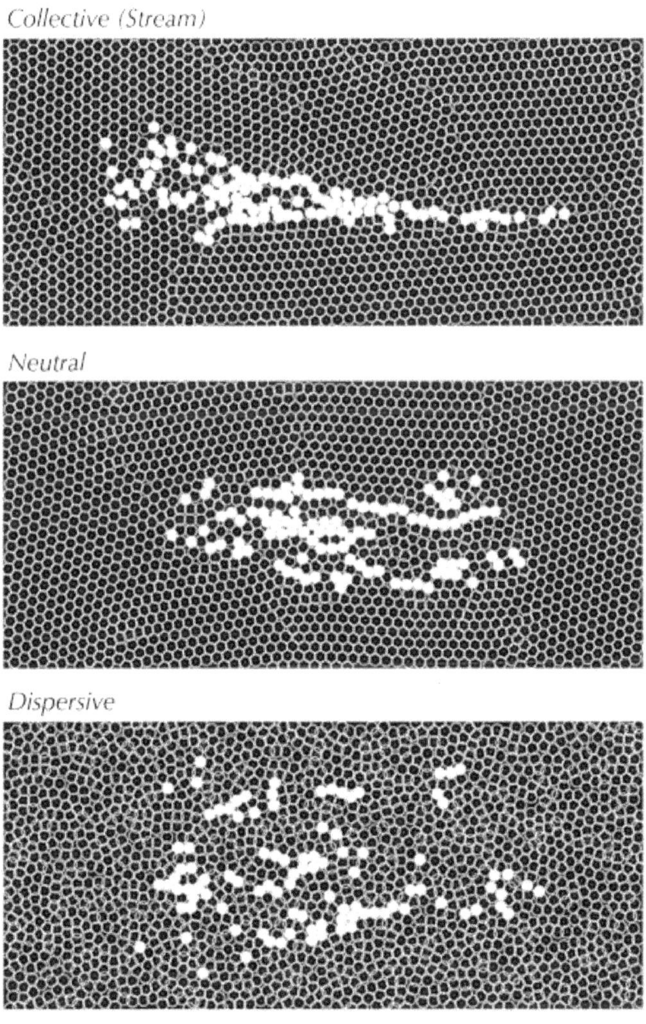

B

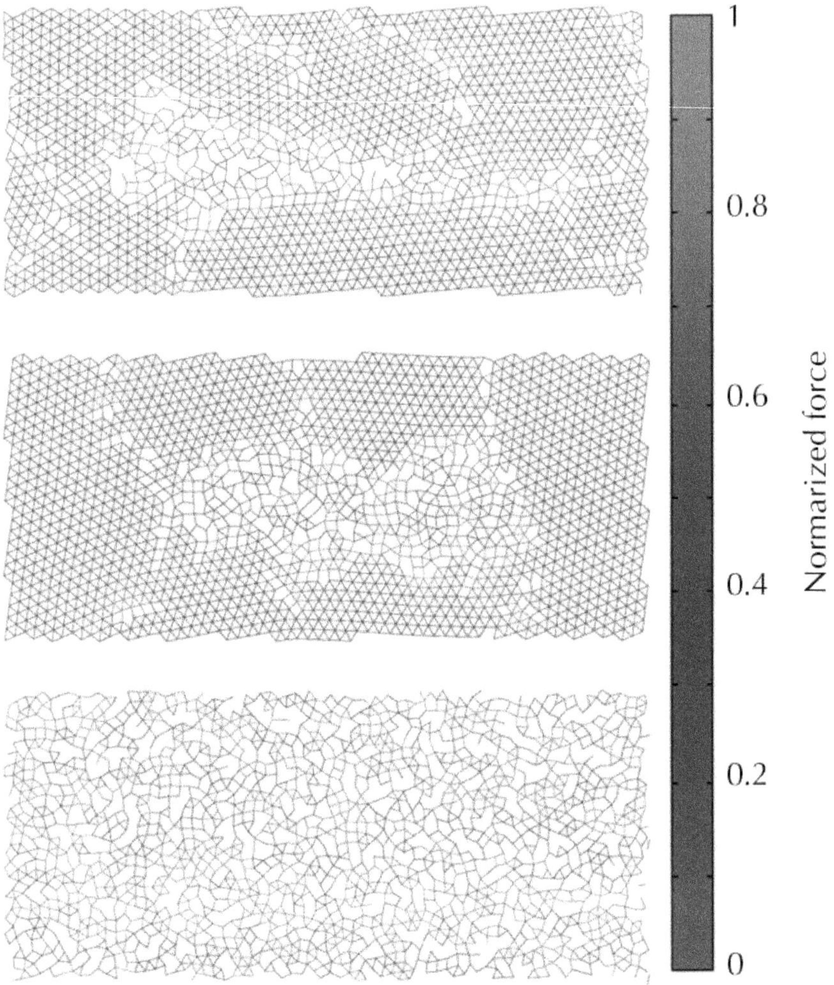

FIGURE 2 Snapshots of migration patterns, cell contacts and the migratory cell population. The upper, middle, and lower panels show the migratory patterns corresponding to the parameters indicated by α, β, and γ in Figure 3D. (A) The migration patterns at a specific point in time are shown. The white and black circles indicate migratory and non-migratory cells, respectively. (B) The cell contacts are shown at the same time point as in (A). The links depict contacts between cells that interact by repulsive elastic forces (Figure 4A), the strengths of which are indicated by their brightness (for red), or darkness (for blue).

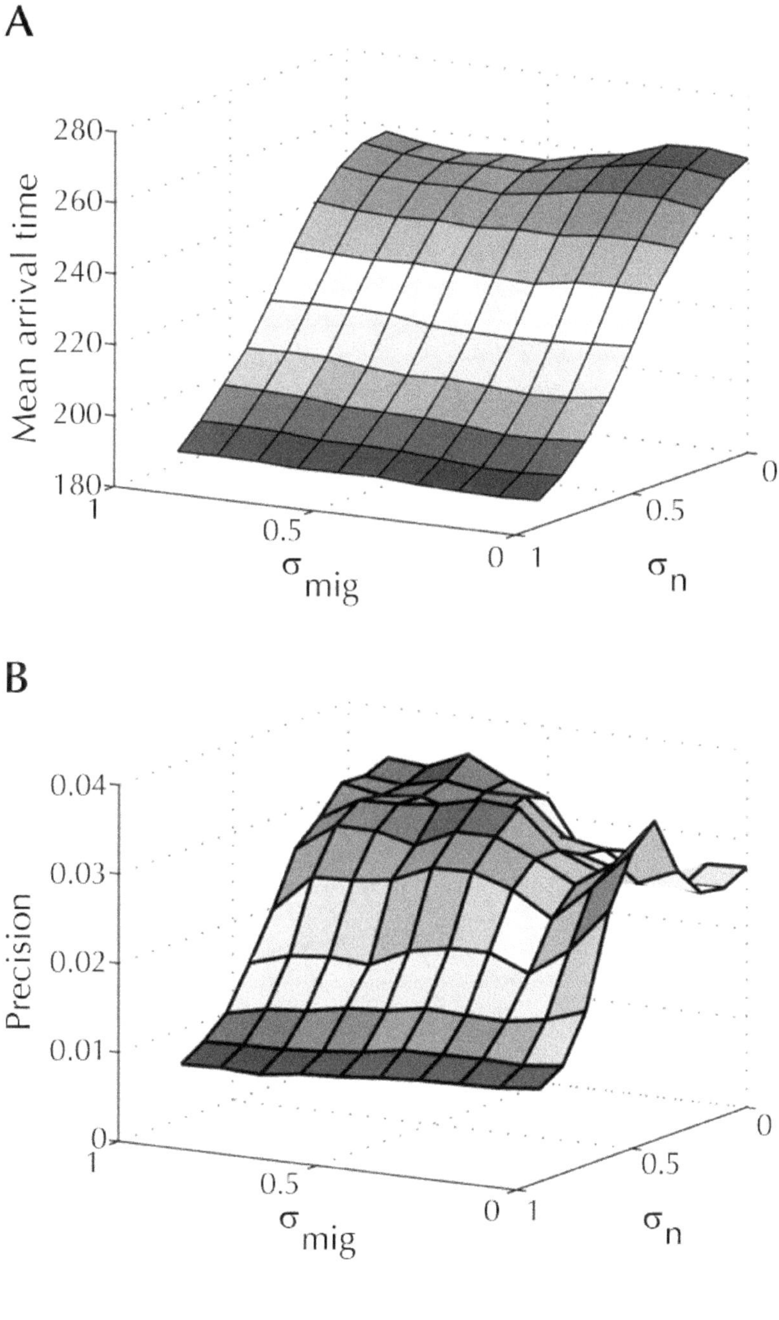

FIGURE 3 *(Continued)*

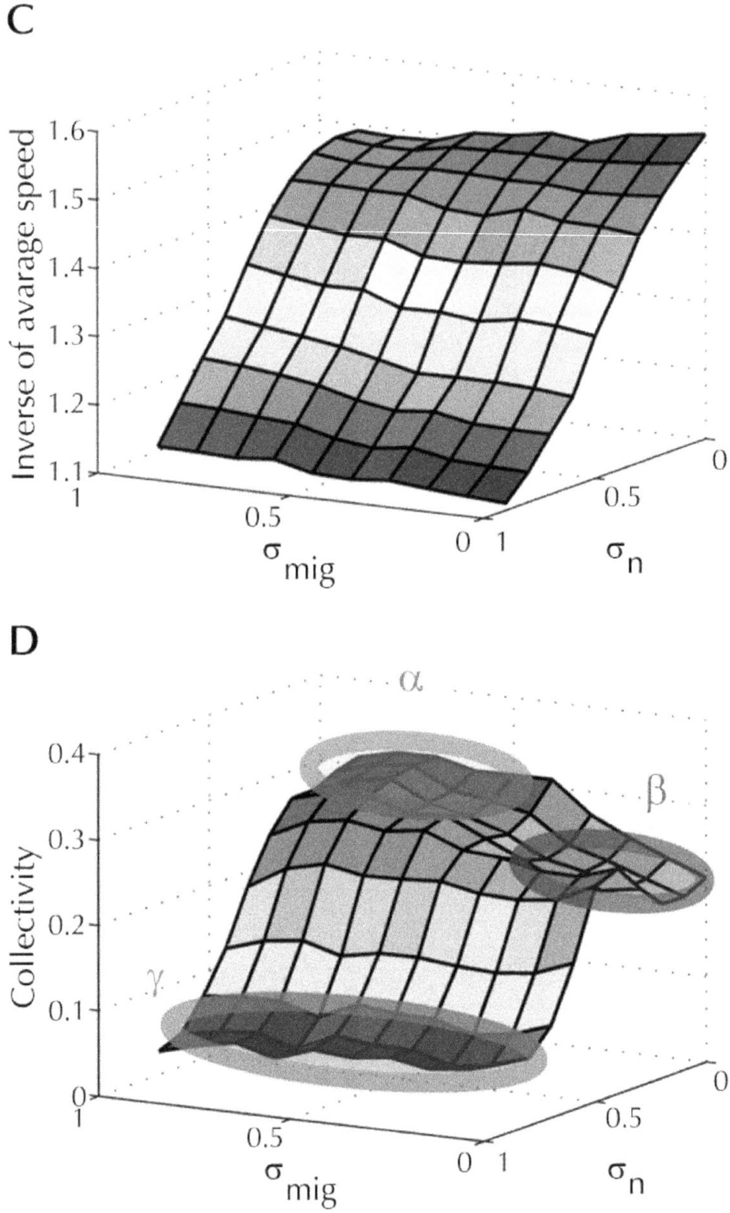

FIGURE 3 The cellular migration characteristics of multi-cellular systems depend on the relative fluctuation levels of the migratory (σ_{mig}) and non-migratory (σ_n) cells. The average time for a migratory cell to reach its target position x = 150 (A) and the inverse of the variance in the position of the migratory cell after arriving at the position (B) are plotted. The inverses of the mean velocity (C) and collectivity (D) of the migratory cells are plotted at a quasi-steady state after the initial transient phase. Here, collectivity is defined by Equation (18), with $\overline{R} = 5$ and $\theta = 4$. In (D), there are three typical collectivity patterns, signified by α, β, and γ.

A

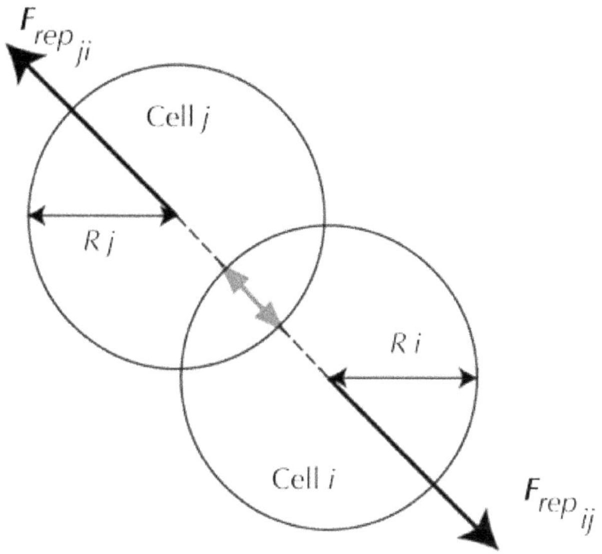
$F_{rep_{ji}}$
Cell j
$R\,j$
$R\,i$
Cell i
$F_{rep_{ij}}$

B

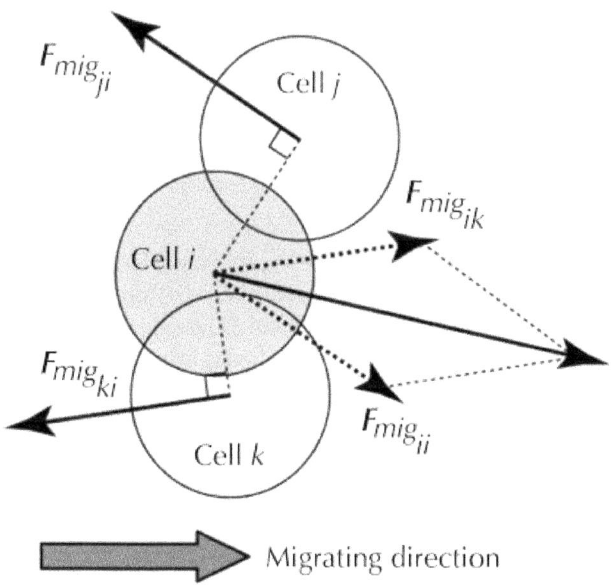
$F_{mig_{ji}}$
Cell j
$F_{mig_{ik}}$
Cell i
$F_{mig_{ki}}$
$F_{mig_{ij}}$
Cell k
Migrating direction

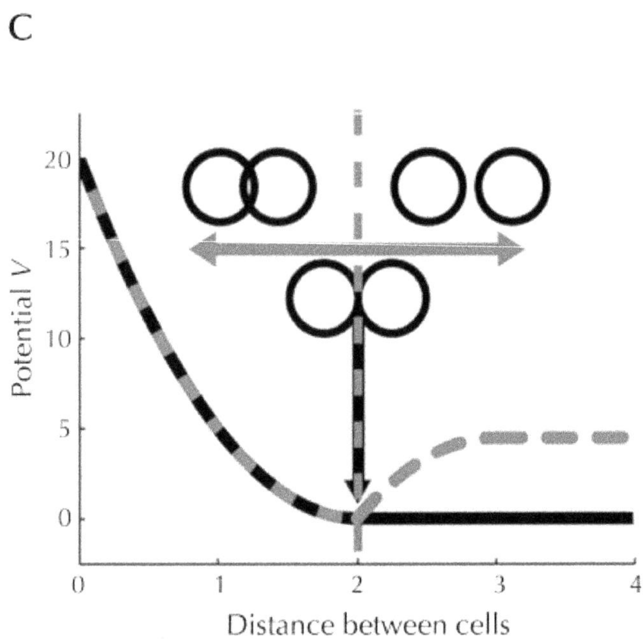

FIGURE 4 Model for simulation. (A) When two neighboring cells indicated by the white circles overlap, the repulsive force (F_{repij} and F_{repji}) is proportional to the degree of overlap, as indicated by the red arrow. (B) The migrating cell (indicated by a gray circle) is assumed to be attracted by a chemo-attractant gradient. Its driving force (the sum of F_{migij} and F_{migik}) is generated at points of contact with other cells, whereas reactive forces (F_{migji} and F_{migki}) are applied in the direction opposite to that of the attractant gradient regardless of the cell type (migratory or non-migratory). (C) The repulsive forces when the cells contact and the attractive adhesive forces when the cells are close are given by the gradient of the potential V. The black and dashed red lines indicate the potential V for Equations (13) and (20), respectively. The black arrow indicates a steady-state point at which the two cells just contact.

In model, multi-cellular migration was simply modeled using three forces: (1) the repulsive force between cells, (2) the driving force of migratory cells accompanied by the reactive forces of neighboring cells via adhesion, and (3) the stochastic forces involved in a random walk. These forces are biologically reasonable for the following reason. The repulsive force is induced by the compressive deformation of the cellular morphology and results from the elasticity of the cytoskeleton and plasma membrane [16]. The driving force is generated when a migratory cell extends a pseudopod that adheres to another cell [17]. The stochastic forces are autonomously generated by inherent intracellular dynamics [2], [3]. Although we ignored the complex rheological properties of such structures [8], the minimal model we adopted is still useful for understanding the system-level properties of multi-cellular migration. We propose that simulation and method of identifying cellular interactions can be applied to other simple developmental systems. Such systems include fibroblasts and neural precursor cells (which sometimes migrate one-dimensionally in a "chain"-like manner) [8], drosophila border cells during oogenesis, and the zebrafish neurons during the development of lateral lines that migrate as a "cluster".

Nevertheless, the *in vivo* mechanisms of cellular migration must be more complicated than those assumed in this chapter. In reality, a chemotactic cell shows morphological changes, such as extensions of special structures called filopodia and lamellipodia, through which the cytoskeletal network regulates cell motility [29]. When migratory neural crest cells collide, their migration transiently stops, and their morphological polarities are reorganized, a process known as "contact inhibition" [30]. Furthermore, proliferation and differentiation play important roles in the development of neural precursor cells and neural crest cells, and these behaviors are controlled by an extracellular Wnt signaling gradient. In the neural tube, proliferation and differentiation are induced by high and low levels, respectively, of the Wnt signal molecule, and the Wnt gradient thereby regulates pattern formation [31]. Neural crest cell requires the Wnt signal for their induction in the dorsal neural tube [32], their delamination from the dorsal neural tube [33], and to acquire motility [33]. Such effects are important for understanding the development of complete multi-cellular systems; because current chapter focuses on collective cellular migration, studying these effects remain as a future objective.

The collective behavior of populations of self-propelled particles has been studied in the context of many biological and social phenomena, such as schools of fish, flocks of birds [34], ant trails [35],[36], cars in traffic jams [37], and cellular slime molds [18]. In such systems, individual particles actively process external information provided by other particles, and this process induces collective behavior. By contrast, study obtained non-trivial simulation results in which cells collectively migrate solely in response to crowding effects and in the absence of active information processing. Therefore, work is the first to present a feasible model for the emergent collective behaviors displayed by multi-cellular systems in crowded situations. Moreover, model may provide general insight into the universal mechanisms underlying a large class of complex systems that consist of crowded self-propelled particles, such as pedestrian flow [37], solution of charged colloids in electric fields [38], and other multi-cellular developmental processes [39].

11.4 RESULTS

11.4.1 Model of Multi-Cellular Migration

To examine the general properties of multi-cellular migration, we developed a biophysical model that includes the essential characteristics of the mechanical nature of general multi-cellular systems. This model multi-cellular system consists of a number of mechanically interacting cells (Figure 4). Each cell is represented as a two-dimensional disk with a static body. The simple multi-cellular migration model consisted of three forces: (1) the repulsive force between cells, (2) the driving force of migratory cells accompanied by reaction forces of neighboring cells via adhesion, and (3) the stochastic forces involved in a random walk. Assuming that the viscosity of the cellular environment is sufficiently high, the inertia can be ignored, and the viscous drag force is exactly balanced between these forces. Thus, the dynamics of the cellular positions r_i are described as follows:

$$\mu \frac{d\mathbf{r}_i}{dt} = \sum_{j \in N_i} \mathbf{F}_{\text{rep}_{ij}} + \sum_{j \in M_i} \mathbf{F}_{\text{mig}_{ij}} + \mathbf{F}_{\text{flu}_i} \tag{13}$$

$$\mathbf{F}_{\text{rep}_{ij}} = k\{(R_i + R_j) - \|\mathbf{r}_i - \mathbf{r}_j\|\} \frac{\mathbf{r}_i - \mathbf{r}_j}{\|\mathbf{r}_i - \mathbf{r}_j\|} \tag{14}$$

$$\mathbf{F}_{\text{mig}_{ij}} = \begin{pmatrix} F_{\text{mig}_x} \\ F_{\text{mig}_y} \end{pmatrix} = s(i) F_{\text{m}} \begin{pmatrix} \dfrac{\|\Delta y_{ij}\|}{\|\mathbf{r}_i - \mathbf{r}_j\|} \\ -\dfrac{\|\Delta y_{ij}\|}{\Delta y_{ij}} \dfrac{\Delta x_{ij}}{\|\mathbf{r}_i - \mathbf{r}_j\|} \end{pmatrix} \tag{15}$$

$$\mathbf{F}_{\text{flu}_i} = \sigma_i \begin{pmatrix} \xi_{xi} \\ \xi_{yi} \end{pmatrix} \tag{16}$$

$$s(i) = \begin{pmatrix} 1 & \text{if cell i is migratory} \\ -1 & \text{if cell i is non-migratory} \end{pmatrix} \tag{17}$$

where k is the Young's modulus, R_i is the radius of cell i, σ_i is the fluctuation intensity of cell i, ξx_i and ξy_i are independent random functions of time with mean zero, $\langle \xi(t)\xi(t-\tau) \rangle = \delta(\tau)$ is the autocorrelation function, N_i is the index set of all cells contacting cell i, M_i is the index set of the other type of cells contacting cell i, $(\Delta x_{ij}, \Delta y_{ij})^{\text{T}} = \mathbf{r}_i - \mathbf{r}_j$, and μ is the viscous modulus. Equation (14) holds only when $\|\mathbf{r}_i - \mathbf{r}_j\| < R_i + R_j$. Otherwise $\mathbf{F}_{\text{rep}_{ij}} = 0$.

The repulsive force is induced by compressive deformation of the cells due to the elasticity of their cytoskeletons and plasma membranes [16]. Even though the cells are modeled as static bodies, we implicitly address this morphological compression by introducing a repulsive force; when two adjacent cells overlap (contact) each other, a repulsive force F_{rep} is generated between them (Figure 4A). This repulsive force is directed so as to separate the contacting cells, and its strength decreases in proportion to the distance between them.

The driving force is generated when a migratory cell adheres through a pseudopod, which is an actin rich peripheral structure that promotes cellular motility [17]. To model cell migration, we consider two types of cells: migratory and non-migratory. Migratory cells are assumed to have the chemotactic ability to be attracted by extra-cellular signals and thereby travel along their gradient. We do not focus on the molecular mechanisms sensing the gradient here and instead just set the migration direction. Because a migratory cell adheres to all contacting non-migratory cells so as to use them as footholds for migration, a driving force $\mathbf{F}_{\mathrm{mig}}$ is tangentially generated between the cells along the direction of the extra-cellular attractant gradient. Consistent with the principles of action and reaction, the non-migratory cells experience a force opposing the force driving the migration (Figure 4B), which causes the non-migratory cells to be pulled backward and the migratory cell to proceed forward. Note that when two migratory cells contact, the action (driving force) generated by one migratory cell is cancelled by the reaction from the other migratory cell›s driving force, and neither migratory cell is propelled forward. These assumptions are implemented by introducing simple geometrical rules: $\mathbf{F}_{\mathrm{mig}ij} \cdot \mathbf{F}_{\mathrm{rep}ij} = 0$, $\left\|\mathbf{F}_{mig}\right\| = F_m$, and $\mathrm{F}_{\mathrm{mig}x} > 0$. These rules lead to Equation (15).

The stochastic forces $\mathbf{F}_{\mathrm{flu}}$ are autonomously generated by the inherent intracellular dynamics [2], [3]. In the model, both the migratory and non-migratory cells are assumed to spontaneously display random movements even if they do not experience the above-mentioned forces. We modeled this randomness as a Gaussian random function.

To reduce the number of free parameters in the model equations of this chapter, we applied a non-dimensionalization technique to the original bio-physical Equations (13–17). We then have only three non-dimensionalized free parameters, which correspond to Young's modulus, the noise intensity of the migratory cells, and the noise intensity of the non-migratory cells (See the Materials and Methods section). Here, the typical value of the non-dimensionalized Young's modulus, $kR_i/F\mathrm{m}$, becomes approximately 10^{-1}~10^{1} by introducing typical values for Young's modulus, the cell radius, and the migration force into the original bio-physical model. These values are $k = 10^{-2}$ ~ 10^{-3}N/m [16], [18], $R = 10^{-5}$m, and $F_{\mathrm{m}} = 10^{-8}$ ~ 10^{-7}N [18], respectively. A typical value for the non-dimensionalized noise intensity, $\sigma/F_{\mathrm{m}} = 10^{-1}$ ~ 10^{1}, is also derived from the typical value of the noise intensity in the bio-physical model, $\sigma = 10^{-9}$~10^{-8}N√s [18 19]. Accordingly, we primarily use the parameter values $k = 10$, and $\sigma = 0$~1 as biologically plausible values in this chapter.

For the sake of ignoring boundary effects, all the cells are assumed to be packed into a two-dimensional rectangular space with boundaries that are connected to form a torus structure. At the beginning of each simulation run, 100 migratory cells were distributed around the center and then transferred rightward (Figure 5).

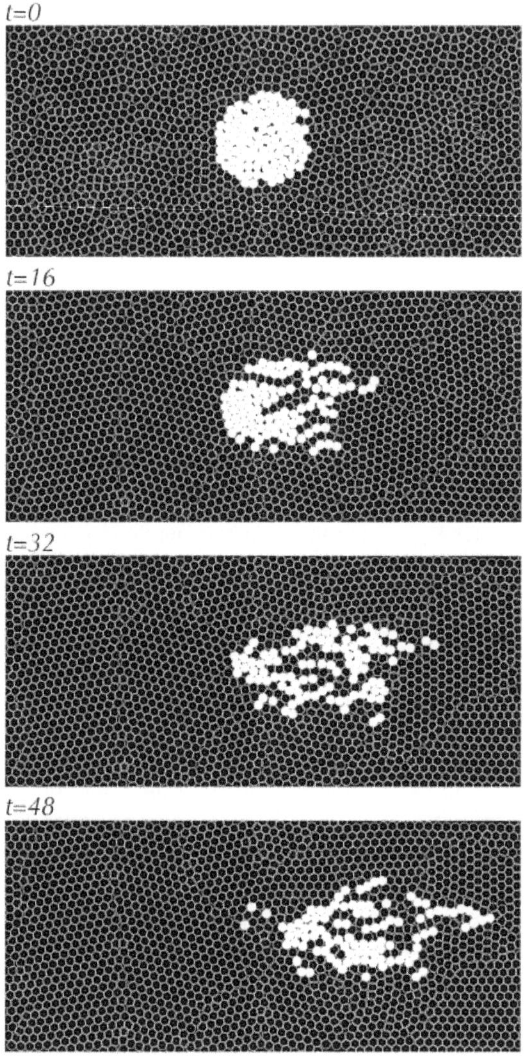

FIGURE 5 Snapshots of a single simulation series of multi-cellular migration. The white and black disks indicate migratory and non-migratory cells, respectively. The migratory cells are initially (at $t = 0$) distributed as a cluster (upper panel) and then migrate rightward progressively at $t = 16; 32; 48$ (the lower three panels). The fluctuation intensities for the migratory and non-migratory cells are set to $\sigma_{mig} = 0$ and $\sigma_n = 0$, respectively.

11.4.2 Effect of Single Cellular Stochasticity on Multi-Cellular Transportation

Because migratory cells in real developmental situations target a specific location and then differentiate within a specific developmental stage, their transportability and transportation accuracy are determinants of their eventual configuration. Therefore, these characteristics were examined in simulation. First, transportability was examined by varying the fluctuation intensities of the migratory and non-migratory cells, σ_{mig} and σ_n. We defined transportability here as the average time required for the migratory cells to reach a specific goal position ($x = 150$ of the rectangular space). The transportation speed (the inverse of the mean arrival time) was found to increase as the fluctuations of the non-migratory cells strengthened (Figure 3A). This noise-induced transportability can be understood on the basis of the following mesoscopic analysis (Figure 6). With relatively small fluctuations in the non-migratory cells, a migratory cell slowly migrates in a hopping manner by pushing other cells out of its way (Figure 6A, upper panel). When the non-migratory cells have large fluctuation, cell migration is smooth and rapid (Figure 6A, lower panel) because the large fluctuation cause the distances between adjacent non-migratory cells to vary significantly, which enables the migratory cell to move easily between them (Figure 6B).

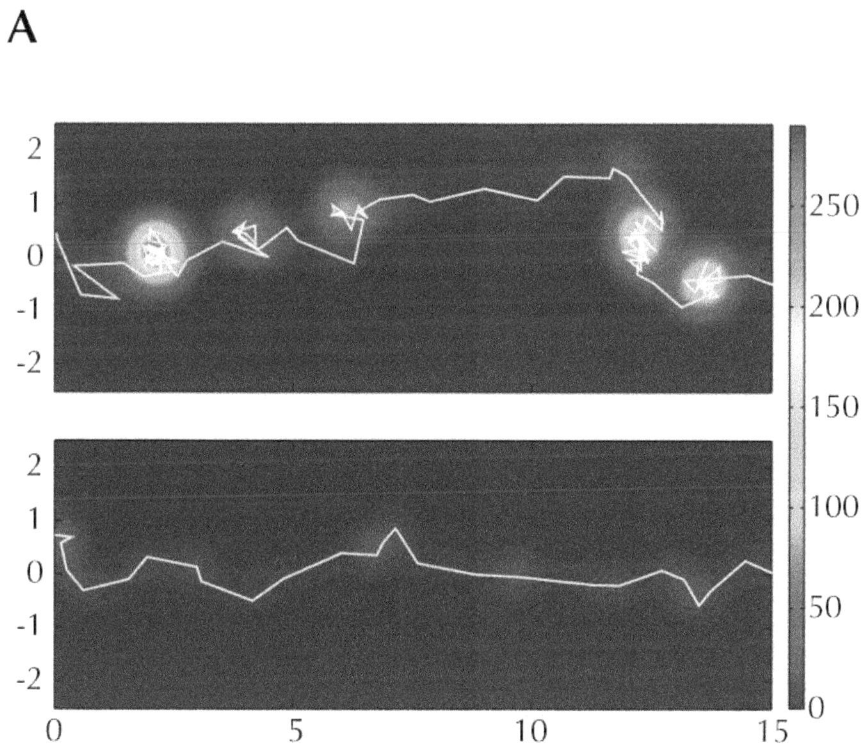

FIGURE 6 *(Continued)*

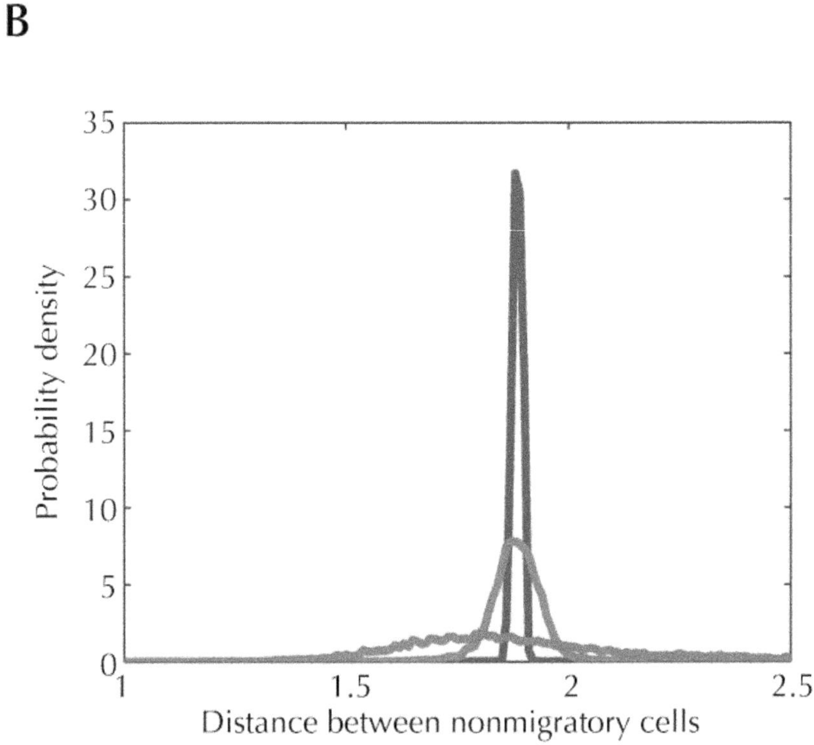

FIGURE 6 The mesoscopic behaviors of migratory and non-migratory cells. (A) Simulations were performed with a single migratory cell surrounded by non-migratory cells. The white lines indicate sample two-dimensional trajectories of the migratory cells with parameter values corresponding to the points indicated by α and γ in Figure 3D. The color contour indicates the density of the migratory cells as calculated by the kernel density estimation using Gaussian kernel functions. (B) Each line shows the distribution of distances between neighboring non-migratory cells when simulating a multi-cellular system that has only non-migratory cells and no migratory cell. The green, blue, and red lines correspond to the cases that have parameter values in the three regions, α, β, and γ in Figure 3D.

Second, we evaluated transportation accuracy, which is defined as the inverse of the variance in the position of a migratory cell that has arrived at the target position (Figure 3B). Transportation accuracy was found to be high, especially when the migratory cells fluctuated significantly and the non-migratory cells did not. It is interesting that significant fluctuation in the non-migratory cells naturally led to dispersed migration, whereas significant fluctuation in the migratory cells led to increase accuracy. Figures 3A and B suggest that transportation in the multi-cellular system exhibits diverse transportability and accuracy patterns depending on the fluctuation levels of the migratory and non-migratory cells and that there is tradeoff between transportability and accuracy.

Because the migratory cells are assumed to be initially aggregated (top panel in Figure 5) and then move to the target position, the cell-migration properties shown in Figures 3A and 3B include transient effects. To examine the population behavior that is independent of such transient effects, we performed additional long-term simulations in which the population behavior reaches a quasi-steady state. We then characterized the steady-state logistics in terms of the inverse of the mean velocity (Figure 3C) and the collectivity of the migratory cells (Figure 3D). Here, collectivity was quantified as

$$\phi = \frac{1}{\#C} \sum_{i,j \in C} \frac{1}{1 + \exp\{\theta(d_{ij} - \bar{R})\}} \tag{18}$$

where C, #, d_{ij}, θ, and $\bar{R}$ denote the index set of the migratory cells, the operator that counts the number of elements of a set, the distance between two migratory cells i and j, the steepness of the sigmoidal fitting, and the arbitrary radius of focused regions centered on each migratory cell, respectively. This equation approximates the average number of migratory cells around themselves within the radius of $\bar{R}$. Figures 3C and D are roughly comparable to Figures 3A and B, respectively. Figure 3D shows three characteristic parameter regions (indicated by α, β, and γ) wherein high, intermediate, and low collective patterns, respectively, are realized. Figure 2A shows the characteristic migratory patterns realized by different combinations of σ_{mig} and σ_n in each of the three parameter regions in Figure 3D. First, when the migratory cells fluctuate significantly and non-migratory cells do not (point α in Figure 3D), the migratory cells collectively converge into one large cellular stream (" *collective migration*") (upper panel in Figure 2A and Movie S1); this behavior is similar to that of neural crest cells migrating from rhombomeres to branchial arches [7]. Second, at the point β in Figure 3D where the fluctuations of all the cells are weak, the dispersion of the migratory cells does not change significantly during their migration (" *neutral migration*") (middle panel in Figure 2Aand Movie S2). Third, when the non-migratory cells fluctuate significantly (point γ in Figure 3D), the migratory cells disperse rapidly, and each migratory cell comes to migrate individually (" *dispersive migration*"), regardless of the fluctuation of the migratory cells (lower panels in Figure 2A and Movie S3). These simulation results show that even though the migratory cells are represented as mechanically passive disks lacking information processing by intra-cellular signal transduction, this multi-cellular system has the potential to exhibit cellular migration and to show various migration patterns that are induced by both intrinsic and environmental fluctuations.

11.4.3 Interaction Between Migratory Cells

How is collective migration (a macroscopic behavior) realized when cells mechanically interact with neighboring cells? In many migration patterns, migratory cells were found to follow other migratory cells (Movie S1). Migratory cells can easily invade non-migratory cells because a preceding migratory cell produces some null space in its wake, implying that a migratory cell affects the positional configuration of the surrounding cells. To visualize such configurations, we performed a simulation run with a single migratory cell surrounded by non-migratory cells (Figure 1A), and we evaluated the average density of the non-migratory cells around the single migratory cell (Figure 1B), which reflects the spatial profile of the pressure caused by repulsive interactions. With parameter values in the*collective* migration mode, the average density of non-migratory cells was much lower behind the migratory cell than in other location (Figure 7A). This low density region is similar to the null space and can be interpreted as a low-pressure region where cells easily invade due to their morphological deformation. The average density of non-migratory cells was slightly lower behind the migratory cell under *neutral* migration parameter values (Figure 7B), and it was almost constant with *dispersive*values (Figure 7C). These results revealed that the fluctuations of migratory cells help to form a null space that induces other migratory cells to follow in their wake, whereas the fluctuations of non-migratory cells erase the wakes of the migrating cells.

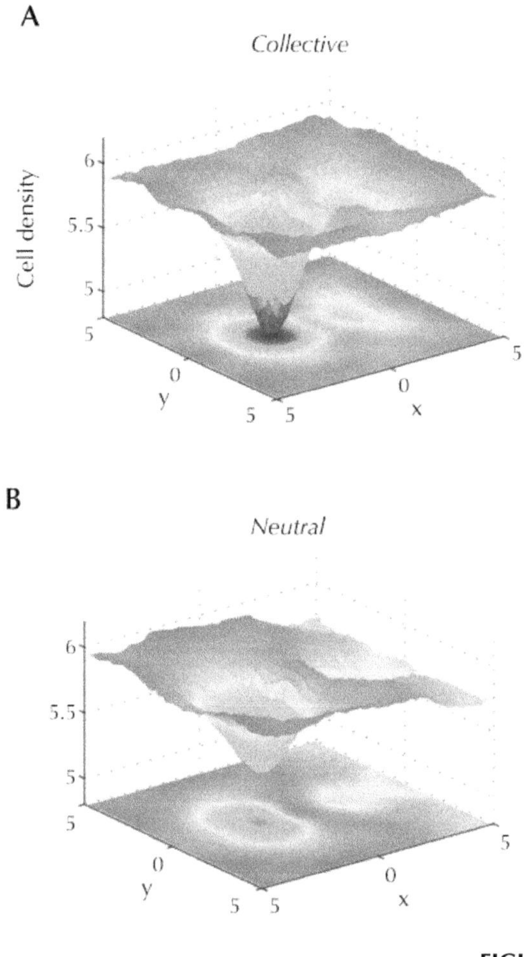

FIGURE 7 *(Continued)*

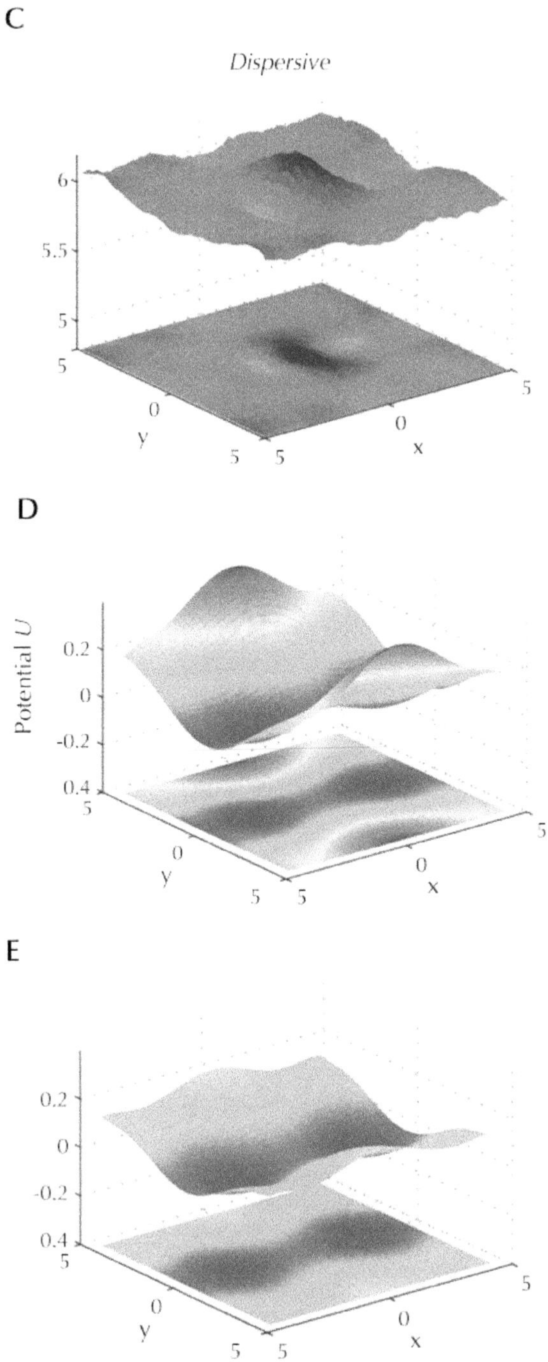

FIGURE 7 *(Continued)*

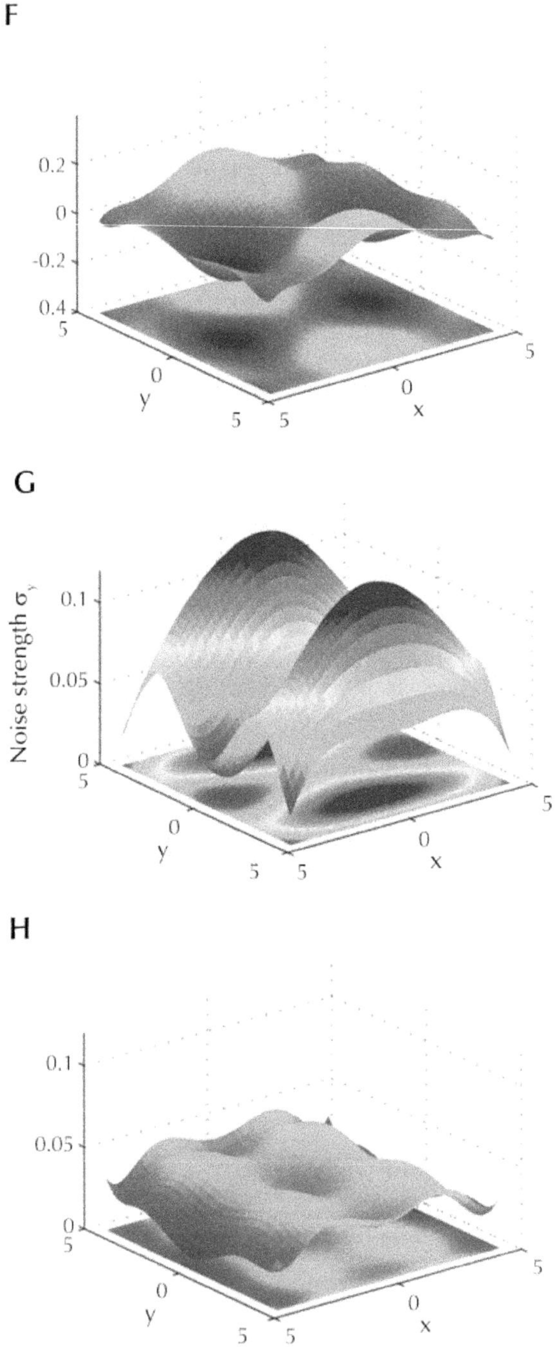

FIGURE 7 *(Continued)*

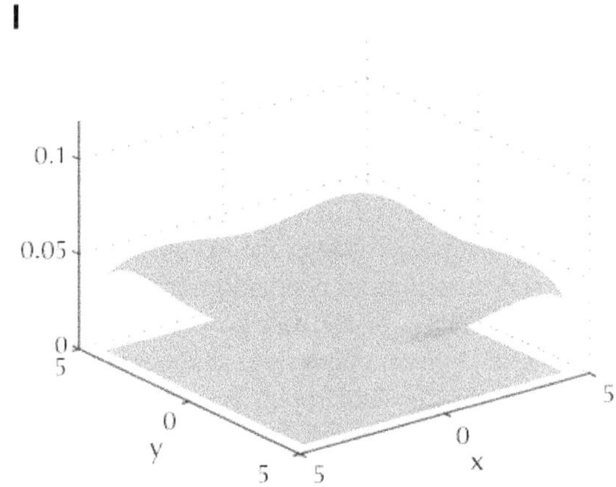

FIGURE 7 Simulation-based determination of effective cellular interaction. (A,D,G), (B,E,H), and (C, F, I) show the collective, neutral, and dispersive migration modes corresponding to the parameters indicated by α, β, and γ in Figure 3D, respectively. The *x* and *y* axes indicate spatial coordinates relative to a migratory cell in Figures 1A and C. (A-C) The average cellular density is estimated by the same method as in Figure 1B. The potential landscape (D-F) and effective noise intensity along the *y* axis (G-I) are estimated using a least-square regression for polynomial functions. Please see the Materials and Methods section.

Analysis above indicates that as long as migratory cells are close to one another, they stay close due to the effective attraction induced by the null space, and this effect contributes to the stability of the collective migration. However, there is still a missing link: how do separate migratory cells aggregate? Two possibilities are conceivable. First, they may randomly migrate, by chance encounter each other, and then follow one another. Alternatively, they may be actively attracted even in the absence of direct contact, through long-distance effects resulting from direct interactions with non-migratory cells. To examine which possibility is more plausible, we determined the degree of effective interaction between separated migratory cells.

For simplicity, we considered a two-body interaction between two migratory cells under the assumption that their movement follows Brownian motion under an effective potential field. If there is an effective potential field U, the Brownian dynamics of the migratory cells can be expressed by a stochastic differential equation:

$$\frac{d}{dt}\begin{pmatrix} r_x \\ r_y \end{pmatrix} = -\begin{pmatrix} \partial U/\partial x \\ \partial U/\partial y \end{pmatrix} + \begin{pmatrix} \sigma_x & 0 \\ 0 & \sigma_y \end{pmatrix}\begin{pmatrix} \xi_x \\ \xi_y \end{pmatrix} \tag{19}$$

where r_x and r_y denote relative location coordinates of the two migratory cells, σ_x and σ_y are the effective noise intensities along the horizontal and vertical axes of the rectangular space, ξ_x and ξ_y are independent random functions of time with mean zero, and $\langle \xi(t)\xi(t-\tau) \rangle = \delta(\tau)$ is the autocorrelation function. Because this dynamics is "effective", we identified the effective potential U and noise intensities, σ_x and σ_y, using simulation; we simulated a system that includes only two migratory cells (Figure 1C) and then sampled the velocity vector field $V = (v_x, v_y)^T$ as a function of the relative coordinates between the two cells (Figure 1D) (see the Materials and Methods section).

Figures 7D–F (and 7G–I) show the estimated effective potential U (and the noise intensities σ_y) for the collective, neutral, and dispersive migration modes, respectively. In all cases, the potential landscapes have one saddle node and two stable points. This equilibrium point structure implies a situation in which two migratory cells are effectively attracted to each other. In the case of collective migration, the potential gradient is steeper, and σ_y is higher on both sides of the migratory cell. The migratory cells each appear to search for the stable point by utilizing a higher σ_y and a steeper gradient; once they reach the stable point, it is difficult for them to escape from it because of the lower σ_y. In neutral migration, the potential gradient is gentle and the fluctuation intensity is low over the area around the migratory cells. In the case of dispersive migration, however, the potential gradient is gentle but disturbed by strong noise, suggesting that the migratory cells can easily escape from the stable points and move away from each other.

11.4.4 Other Properties of Multi-Cellular Logistics

We next investigated the parameter dependence of transportability (Figure 8). For the collective and neutral migration modes, we found that an increase in the number of migratory cells increased the speed of the collective migration in a saturating manner (Figure 8A, blue and green lines), which is consistent with experimental observations [20]. This population-based transportability likely occurred because migratory cells broke their contacts with the non-migratory cells (Figure 2B, upper and middle), leading to energetically efficient migration. By contrast, the velocity of dispersive migration was unaffected by population size (Figure 8A, red line) because the noisy environment broke the contacts between the non-migratory cells, enabling the migratory cells not for necessitate population-based migration (Figure 2B, lower).

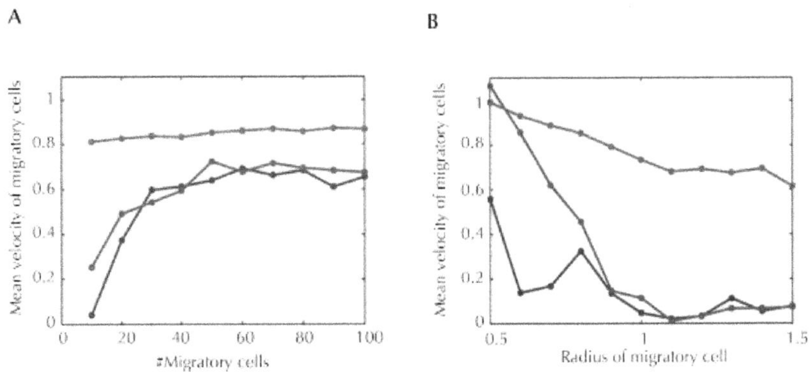

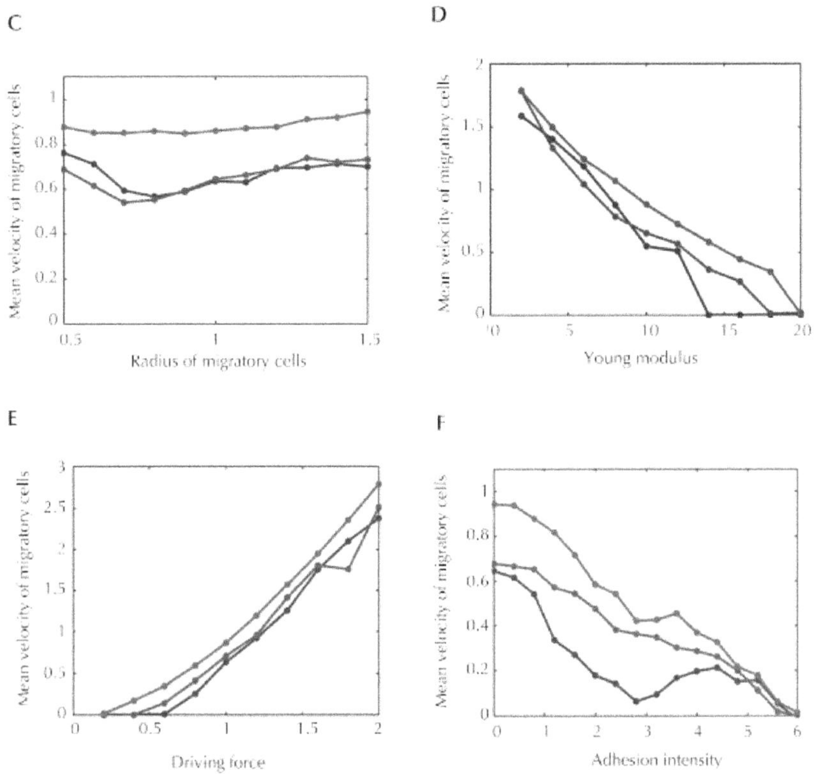

FIGURE 8 Dependence of transportability on the physical parameters of cells. The green, blue, and red lines represent collective, neutral, and dispersive migrations corresponding to the parameters indicated by α, β, and γ in Figure 3D, respectively. The average migratory cell speeds are plotted according to various values for population size (A), migratory cells radius (B, C), migration driving force (D), and Young's modulus for all cells (E). In (F), an additional attractive force from cell adhesion is included in the model by using Equation (20) (see also the text) instead of Equation (13), and its intensity *a* is varied. In (A), (C), (D), (E), and (F), the setting of the migratory and non-migratory cells is similar to that in Figure 5, with the addition of the attractive force in (F), whereas in (B), there is only a single migratory cell surrounded by non-migratory cells.

Because cell size changes drastically between different developmental stages, we next examined how the size of a migratory cell affects its transportability. When a cell migrated alone, its migration speed was found to decrease as its cellular radius increased, regardless of the migration mode (Figure 8B). Note that the migration speed is highest in the dispersive migration mode (Figure 2B, bottom). With population-based migration, however, cell size was found to affect migration in a complicated fashion (Figure 8C). For the collective and neutral migration modes, there are two characteristic phases; as the size increases, the migration speed first decrease and then increases when the size exceeds a certain threshold. The first phase exhibits behaviors similar to those observed when there is only a single migratory cell (Figure 8B). The second phase could be attributed to a population effect, through which a large migratory cell produces a large null space in its wake and is effectively followed by other migratory cells.

During cell migration, extra- and intra-cellular signaling actively control cellular stiffness, force generation, and adhesion via regulating cytoskeletal components such as actin filaments and microtubules. In simulation, the migration speed was found to decrease as the cells stiffen (Figure 8D), because stiff non-migratory cells require a larger driving force to allow the migratory cells to invade. When the cells are stiffer, i.e., when the Young's modulus exceeds 18, the multi-cellular system behaves completely differently; the migratory cells do not proceed any further.

A larger driving force was found to trivially increase the migration speed (Figure 8E); however, it also eliminates collective migration (Figure 9A) because a powerful migratory cell easily pushes the non-migratory cells out of its way and makes the effect of the null space less important in migration. This interpretation also suggests that there is a trade-off between migration velocity and collectivity, as seen in Figure 3.

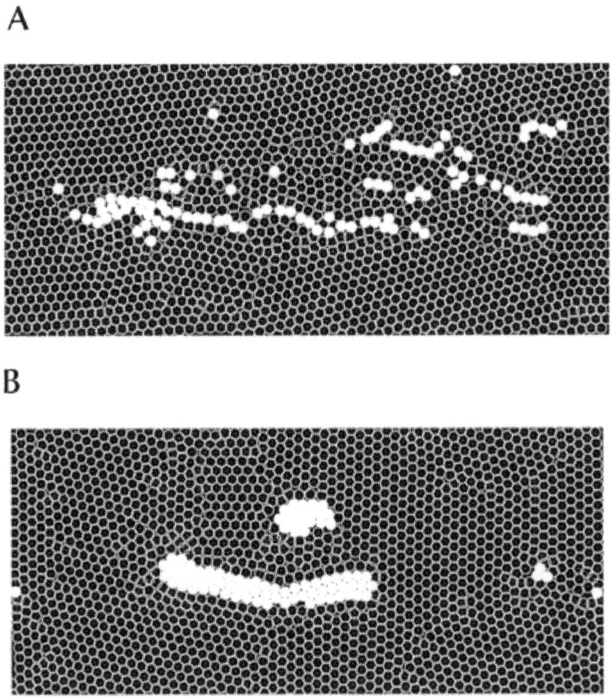

FIGURE 9 Dependence of migration mode on the driving force and cellular adhesion. (A) Collective migration collapsed when the driving force was too strong. The parameters values are identical to those in the top panel of Figure 2A, i.e., to the collective migration parameters with the driving force (F_m) doubled. (B) Chain- or cluster-like migration was induced by introducing an attractive force due to cellular adhesion, i.e., Equation (21). The attractive force intensity is a=4, the effective distance of the attractive force is L = 1, and the other parameter values are those of the characteristic point β in Figure 3D.

11.5 CONCLUSION

The effects of adhesion molecules are implicitly reflected in the model by the action-reaction force between contacting cells. However, adhesion molecules play other roles in attracting cells. To examine whether the adhesion force does or does not affect multi-cellular migration, we further extend the model to include an attractive adhesion force between neighboring migratory cells that are located within a distance of L:

$$\mu \frac{d\mathbf{r}_i}{dt} = \sum_{j \in N_i} \mathbf{F}_{\mathrm{rep}_{ij}} + \sum_{j \in M_i} \mathbf{F}_{\mathrm{mig}_{ij}} + \sum_{j \in L_i} \mathbf{F}_{\mathrm{adh}_{ij}} + \mathbf{F}_{\mathrm{flu}_i} \tag{20}$$

$$\mathbf{F}_{\mathrm{adh}_{ij}} = -a\{L + (R_i + R_j) - \|\mathbf{r}_i - \mathbf{r}_j\|\} \frac{\mathbf{r}_i - \mathbf{r}_j}{\|\mathbf{r}_i - \mathbf{r}_j\|} \tag{21}$$

where a is the intensity of the attractive force and $\mathcal{L}_i$ is the set of migratory cells that satisfy the condition, $R_i + R_j < \|r_i - r_j\| \leq R_i + R_j + L$. In this extension, the strength of the attractive force is designed to increase as the cells become closer (as in Equation (21)) because the number of binding adhesion molecules, which generates the attractive force, increases as the cells become closer. To see the difference between Equations (13) and (20), we depict the potential V for those equations; the repulsive and attractive forces depend on the potential gradient according to $\mathbf{F}_{rep} + \mathbf{F}_{adh} = \partial V / \partial \mathbf{r}$ (Figure 4C). In the previous equation, the migratory cells are only repelled when they come in contact (black line in Figure 4C), whereas in the extended model that employs this new equation, the migratory cells are attracted when they become close (red line in Figure 4C). The typical orders of a and L are approximately 10^{-3} N/m and 10^{-5}m [18], [21], respectively, and their non-dimensionalized values become 10°~10^{1} and 10°, respectively.

We found that the attractive force inhibits the migrating speed (Figure 8F). Interestingly, the additional attractive force changed the neutral migration mode to another mode characterized by forming a cluster or chain (Figure 9B), as experimentally observed in cancer cell migration [10]. This result implies that cellular adhesion can be involved in generating migration patterns accompanied by cluster and chain-like behaviors.

KEYWORDS

- **Cell Migration**
- **Cytoskeleton**
- **Multi-Cellular Transportation**
- **Neuronal Growth Cones**
- **Proliferation**

AUTHOR CONTRIBUTIONS

Conceived and designed the experiments: HN. Performed the experiments: MY. Analyzed the data: MY. Wrote the chapter: MY HN SI.

ACKNOWLEDGMENTS

We thank Y. Sakumura, S. Oba, Y. Igarashi, and M. Kon for their valuable comments on this chapter.

REFERENCES

1. Klambt, C. Modes and regulation of glial migration in vertebrates and invertebrates. *Nat Rev Neurosci* **10**: 769–79 (2009).
2. Naoki, H., Sakumura, Y., and Ishii, S. Stochastic control of spontaneous signal generation for gradient sensing in chemotaxis. *J Theor Biol* **255**: 259–66 (2008).
3. Ueda, M. and Shibata, T. Stochastic signal processing and transduction in chemotactic response of eukaryotic cells. *Biophys J* **93**: 11–20 (2007).
4. Naoki, H., Nakamuta, S., Kaibuchi, K., and Ishii, S. Flexible search for single-axon morphology during neuronal spontaneous polarization. *Plos One* **6**: e19034 (2011).
5. Friedl, P. and Wolf. K., Plasticity of cell migration: A multiscale tuning model. *J Cell Biol* **188**: 11–9 (2010).
6. Arrieumerlou, C. and Meyer, T., A local coupling model and compass parameter for eukaryotic chemotaxis. *Dev Cell* **8**: 215–27 (2005).
7. McLennan, R. and Kulesa, P. M., In vivo analysis reveals a critical role for neuropilin-1 in cranial neural crest cell migration in chick. *Dev Biol* **301**: 227–39 (2007).
8. Lois, C., Garcia-Verdugo, J. M. and Alvarez-Buylla, A. Chain migration of neuronal precursors. Science **271**: 978–81 (1996).
9. Thiery, J. P. Epithelial-mesenchymal transitions in tumour progression. *Nat Rev Cancer* **2**: 442–54 (2002).
10. Hegerfeldt, Y., Tusch, M., Brocker, E. B., and Friedl, P. Collective cell movement in primary melanoma explants: plasticity of cell-cell interaction, beta1-integrin function, and migration strategies. *Cancer Res* **62**: 2125–30 (2002).
11. Wolf, K., Wu, Y. I., Liu, Y., Geiger, J., and Tam, E. Multi-step pericellular proteolysis controls the transition from individual to collective cancer cell invasion. *Nat Cell Biol* **9**: 893–904 (2007).
12. Klambt, C. Modes and regulation of glial migration in vertebrates and invertebrates. *Nat Rev Neurosci* **10**: 769–79 (2009).
13. Naoki, H., Sakumura, Y., and Ishii, S. Stochastic control of spontaneous signal generation for gradient sensing in chemotaxis. *J Theor Biol* **255**: 259–66 (2008).
14. Ueda, M. and Shibata, T. Stochastic signal processing and transduction in chemotactic response of eukaryotic cells. *Biophys J* **93**: 11–20 (2007).
15. Naoki, H., Nakamuta, S., Kaibuchi, K., and Ishii, S. Flexible search for single-axon morphology during neuronal spontaneous polarization. *Plos One* **6**: e19034 (2011).
16. Friedl, P. and Wolf. K., Plasticity of cell migration: A multiscale tuning model. *J Cell Biol* **188**: 11–9 (2010).
17. Arrieumerlou, C. and Meyer, T., A local coupling model and compass parameter for eukaryotic chemotaxis. *Dev Cell* **8**: 215–27 (2005).
18. McLennan, R. and Kulesa, P. M., In vivo analysis reveals a critical role for neuropilin-1 in cranial neural crest cell migration in chick. *Dev Biol* **301**: 227–39 (2007).
19. Lois, C., Garcia-Verdugo, J. M. and Alvarez-Buylla, A. Chain migration of neuronal precursors. Science **271**: 978–81 (1996).
20. Thiery, J. P. Epithelial-mesenchymal transitions in tumour progression. *Nat Rev Cancer* **2**: 442–54 (2002).

21. Hegerfeldt, Y., Tusch, M., Brocker, E. B., and Friedl, P. Collective cell movement in primary melanoma explants: plasticity of cell-cell interaction, beta1-integrin function, and migration strategies. *Cancer Res* **62**: 2125–30 (2002).
22. Wolf, K., Wu, Y. I., Liu, Y., Geiger, J., and Tam, E. Multi-step pericellular proteolysis controls the transition from individual to collective cancer cell invasion. *Nat Cell Biol* **9**: 893–904 (2007).

12 Differential Adhesion and Cell Sorting

Ying Zhang, Gilberto L. Thomas, Maciej Swat, Abbas Shirinifard, and James A. Glazier

CONTENTS

12.1 Introduction 237
12.2 Methods 239
    12.2.1 Reaction-Kinetic Models of E-Cadherin Binding 239
    12.2.2 Glazier-Graner-Hogeweg Simulations of Cell Sorting 241
12.3 Simulation Implementation 245
12.4 Discussion 245
12.5 Results 246
Keywords 259
Author Contributions 259
References 259

12.1 INTRODUCTION

The actions of cell adhesion molecules, in particular, cadherins during embryonic development and morphogenesis more generally, regulate many aspects of cellular interactions, regulation and signaling. Often, a gradient of cadherin expression levels drives collective and relative cell motions generating macroscopic cell sorting. The computer simulations of cell sorting have focused on the interactions of cells with only a few discrete adhesion levels between cells, ignoring biologically observed continuous variations in expression levels and possible nonlinearities in molecular binding. In this chapter, it presented three models relating the surface density of cadherins to the net intercellular adhesion and interfacial tension for both discrete and continuous levels of cadherin expression. Then used the Glazier Graner Hogeweg (GGH) model to investigate how variations in the distribution of the number of cadherins per cell and in the choice of binding model affect cell sorting. It is find that an aggregate with a continuous variation in the level of a single type of cadherin molecule sorts more slowly than one with two levels. The rate of sorting increases strongly with the interfacial tension, which depends both on the maximum difference in number of cadherins per cell and on the binding model. The approach helps connect signaling at the molecular level to tissue-level morphogenesis.

The *cadherin* family of cell-adhesion membrane proteins plays a key role in both early and adult tissue morphogenesis [1–3]. Spatio-temporal variations in cadherin number and type help regulate many normal and pathological morphogenetic processes, including: neural-crest-cell migration [4], somite segmentation [5, 6], epithelial-to-mesenchymal transformations during tumor invasion and metastasis [7, 8], and wound healing [9, 10]. Many of these processes involve continuous variations in the expression level of a single type of adhesion moleculeduring proximo-distal limb growth [11] and rostro-caudal body-axis elongation [12], adhesion gradients resulting from variations in the number of a single type of adhesion molecule may maintain cells' relative positions. *In vitro* and in experiments *in vivo*, when cells from different domains of a limb are mixed together, they can sort out according to their original positions [11, 13]. In *Drosophila*, an adhesion gradient drives the oocyte towards the posterior follicle cell, which expresses the highest level of DE-cadherin [14]. A cellcell adhesion gradient along the dorso-ventral axis directs lateral cell migration during zebrafish gastrulation [15]. Thus, understanding the role of cadherins in creating and stabilizing tissue structures, especially the role of continuous variation in the level of a single cadherin, is crucial to understanding embryonic morphogenesis.

Steinberg's Differential Adhesion Hypothesis (DAH) originated the idea that cell sorting can result from variations in cellcell adhesivity [16–19]. Cell sorting depends on the effective molecular binding strength between opposing cadherins, which in turn depends on their types and expression levels in each cell and potentially the cells' internal biochemistry and cytoskeletal structures [20]. Both differences in expression levels of a single type of cadherin [18, 19] and differences in the types of cadherins expressed [19, 21] can lead to sorting.

The relation between forces at the molecular level (pairs of cadherins), cell level (cellcell adhesion), tissue level (surface tension) and cell sorting is more complicated than the simple physics suggested by the DAH. Experimental measurements of cadherin binding employing a variety of approaches have obtained widely differing estimates of the per-cadherin pair-binding force, cell-cell adhesion force and surface tension at the tissue level [19, 22–25]. In some experiments, the scaling between cadherin expression levels and surface tension, as given by equation (7), is quadratic (see equation (9)) [23]; in others, the scaling between cadherin expression levels and the cell-cell adhesion force is linear (see equation (10)) [19]. The cadherin organization within the cell membrane and the underlying cytoskeleton also change over a period of hours after two cells come into contact [3, 26–28]. Bindings between cadherin pairs differ for cadherins in different conformational states [3], *e.g.*, cadherin reorganization into adhesive patches on the cell membrane due to both passive diffusion and interaction with the actin cytoskeleton [26–28] can greatly increase the effective binding strength per cadherin pair between two cells. Cluster formation depends on the proper functioning of the actin cytoskeleton, so actin-disrupting drugs like cytochalasin-D and latrunculin greatly decrease cellcell adhesivity [29].

Multiple transcriptional and post-translational signaling cascades can regulate cadherin expression levels, localization and per-cadherin binding strengths [3, 30]. In turn, cadherin binding can modify gene expression [3]. This complexity obscures the role of the cadherin-binding force in cell sorting [25]. As a result, different classes of experiments on specific types of cadherin have led to at least four simplified cadherin-binding models: the *linear-zipper model* (*LZM*) based on experiments on N-cadherin [31, 34], the *cis-dimer model* (*CDM*) (equation (8)) based on experiments on E-cadherin [35], the *trans-homophilic-bond model* (*THBM*) (equation (9)) based on experiments on C-cadherin [36], and the *saturation model* (*SM*) (equation (10)), based on the observation that, for both the CDM and THBM models, when the cadherin binding between cells saturates, the number of bonds depends on the cell with the minimum cadherin concentration.

This chapter therefore proposes a simple framework to explore how homotypic cadherin binding at the molecular level could produce intercellular adhesion and eventually determine cell sorting at the tissue level. Neglect complex spatial and temporal changes in cadherin behavior, assuming that cadherin distributions are uniform and constant on the cell membrane and that adhesion-strength per molecular bond is also time-independent (*i.e.*, we assume no conformational changes in molecular structure during a simulation). Then explore how the sorting configuration and rate depend on a few essential parameters in our models. Compared to the rate of sorting for an aggregate with two levels of a single cadherin, simulations with more intermediate levels sort more slowly but the sorting rate is similar for aggregates with the same number of cadherin levels for all binding models. The speed of sorting increases strongly with the interfacial tension, which depends both on the maximum difference in number of cadherins per cell and on the binding model.

12.2 METHODS

12.2.1 Reaction-Kinetic Models of E-Cadherin Binding

The nature of cadherin binding determines the way the cellcell adhesion energy, depends on cells' cadherin surface densities, and thus the correct binding model to use in simulations of cell sorting. Since, more recent mutagenesis studies do not support the linear-zipper model [3], we use the cis-dimer (*CDM*), the trans-homophilic-bond (*THBM*), and the saturation (*SM*) models to relate the cells' cadherin surface densities to thc cell-cell adhesion energy.

The cis-dimer model (CDM) [35] assumes that cis-dimers first form on the surfaces of individual cells and that two dimers on apposing cells then bind together to form homophilic tetramers. Dimerization of monomers (*A* and *A* or *B* and *B*) on individual cells' surfaces to form dimers A_2 and B_2 has the form:

$$A + A \rightleftharpoons A2;\ B + B \rightleftharpoons B2 \tag{1}$$

Similarly, when the trans-tetramer *A*2*B*2 forms between dimers (*A*2 and *B*2) on two apposing cells, the reaction has the form:

$$A2 + B2 \rightleftharpoons A2B2 \tag{2}$$

It is assume that the cadherin concentrations on the cells' surfaces are constant and that one can apply the Law of Mass Action. Dimerization and tetramerization quickly equilibrate if K_D and K_T,'the *equilibrium dimerization* and *equilibrium tetramerization dissociation constants* are large and the cadherin concentrations, $C_A=N_A/(S_Ah)$ and $C_B=N_B/(S_Bh)$, are lower than the dissociation constants [37]. Here N_A and N_B are the number of cadherin molecules distributed on the cell surfaces S_A and S_B, respectively, and h is the amplitude of cadherin fluctuations normal to the cells' surfaces. In this case, the total number of tetramers is less than the number of dimers, which in turn is less than the number of monomers. Then, the equilibrium concentration of tetramers in the CDM is, approximately,

$$[A2B2]=C_A^2C_B^2/(K_D^2K_T)=k_TN_A^2N_B^2, \tag{3}$$

where $k_T=\left(K_D^2K_T\left(S_AS_Bh^2\right)^2\right)^{-1}$ is the tetramer effective equilibrium constant.

According to the trans-homophilic-bond model (THBM) [36], cadherins bind individually between cells, so the concentration of bound pairs is given by:

$$[A2B2]=C_AC_B/K_D=k_DN_AN_B, \tag{4}$$

where $k_D=(K_DS_AS_Bh^2)^{-1}$ is the *dimer effective equilibrium constant.*

Finally, for the saturation model (SM), which applies for strong clustering of cadherins, or large differences in the number of molecules per cell, the concentration of bound cadherin pairs is given by

$$[A2B2]=\min\{C_A,C_B\}=k_M\min\{N_A,N_B\} \tag{5}$$

where $k_M=\left(\left(S_A|S_B\right)h^2\right)^{-1}$ is the effective equilibrium constant and the surface $S=S_A|S_B$ corresponds to the smaller of C_A or C_B.

Relate the concentration of cadherin pairs to the cellcell intercellular adhesion energy density due to cadherin binding via the relation:

$$J(N_A,N_B)=[A2B2]\Delta g+c \tag{6}$$

where Δg is the cadherincadherin-binding free-energy per cadherin bond [37], which is negative, since bond formation releases energy, and where c is the energy density due to adhesion unrelated to cadherins [19].

The *interfacial-tension density* over the contact area between two cells expressing different numbers of a single type of cadherin is defined [38, 39] as:

$$\begin{aligned}\gamma_{A,B}&=[J(N_A,N_A)+J(N_B,N_B)]/2-J(N_A,N_B)\\&=(([A2]+[B2])/2-[AB])\Delta g\end{aligned} \tag{7}$$

For the three models just listed, equations (3–5), we have:

$$\gamma_{CDM} = -k_T(N_A^2 - N_B^2)^2 \Delta g/2, \tag{8}$$

$$\gamma_{THBM} = -k_D(N_A - N_B)^2 \Delta g/2 \tag{9}$$

$$\gamma_{SM} = -k_M(N_A - N_B)\Delta g/2, \text{ for } N_A > N_B \tag{10}$$

12.2.2 Glazier-Graner-Hogeweg Simulations of Cell Sorting

To simulate cell sorting due to cellcell adhesion, we used the *GlazierGranerHogeweg* model (*GGH*) [40] (also known as the *Cellular Potts Model* [38, 39]). The GGH is a multi-cell, lattice-based model, which uses an *effective energy*, *H*, to describe the behavior of cells, for instance, due to cellcell adhesion. GGH simulations agree quantitatively with simple cell-sorting and other experiments [41–49].

Cells in the GGH are extended domains of pixels (on a regular lattice, denoted $\vec{i}$), which share the same *cell index*, $\sigma(\vec{i})$. The effective energy governs how the lattice evolves as cells attempt to displace other cells by extending their pseudopodia [50]. At each step, we select a lattice site $\vec{i}$ and change its index into the index of a neighboring lattice site $\vec{i}$ with probability:

$$P(\sigma(\vec{i}') \to \sigma(\vec{i})) = \begin{cases} \exp(-\Delta H/T) & \text{if } \Delta H > 0; \\ 1 & \text{if } \Delta H \leq 0, \end{cases} \tag{11}$$

where ΔH is the energy gain from the change and T is the *intrinsic cell motility* corresponding to membrane fluctuations resulting from cytoskeleton fluctuations. If the lattice has Q pixels, we define one *Monte Carlo Step* (*MCS*) to be Q displacement attempts.

For a two-dimensional simulation of an aggregate containing cells expressing varying levels of a single type of cadherin, we assume that: (1) The effective energy between cells is due to cell-cell adhesion. (2) The cells have fixed and identical target volumes, membrane areas, and intrinsic motilities. (3) Cells do not grow, divide or die. (4) Cells are isotropic, so, cadherins are uniformly distributed on the cell membrane and the cadherin concentration is constant in time. With these assumptions, the effective energy is:

$$\begin{aligned} H = & \sum_{\vec{i},\vec{i}' \text{ neighbors}} \left\{ J_0 + J\left(N_{\sigma(\vec{i})}, N'_{\sigma'(\vec{i}')}\right)\left(1 - \delta_{\sigma(\vec{i}),\sigma'(\vec{i}')}\right) \right\} \\ & + \sum_{\sigma} \lambda (V(\sigma) - V_t)^2, \end{aligned} \tag{12}$$

where, J_0 is the energy per unit contact area between two cells in the absence of cadherin binding, which may be positive since such cells may not cohere. $J(N_{\sigma(\vec{i})},N_{\sigma(\vec{i}')})$ is the adhesion-energy per unit contact area between cells σ and σ' expressing N and N' adhesion molecules, respectively. This term is always negative, since forming cadherin bonds decreases the effective energy. Sums go up to fourth nearest neighbors on a square lattice. λ, $V(\sigma)$, and V_t are the volume elasticity, actual volume and target volume of cell σ, respectively. $\delta_{\sigma}(\vec{i}),\sigma(\vec{i})$ is the usual Kronecker delta function.

Each cell expresses a specific number of cadherins. The cellcell adhesion energy relates to N and N' according to equation (6) together with equations (3), (4) or (5). Since we can rescale the energy by the intrinsic cell motility, we are free to pick the energy scale and set $\Delta g = -1$.

The relative strengths of cellcell adhesions result in net forces which act on each cell. Depending on the relative hierarchy of cellcell adhesive interactions the generated forces can either drive or suppress cell sorting. Equation (13) is the condition for the sorting to occur.

Why does sorting occur for most of the conditions that we consider in this paper? For two cadherin levels with $N_A > N_B$, complete sorting requires that the less cohesive cell type wet the more cohesive cell type [39]:

$$\begin{aligned} &J(N_A,N_A)<[J(N_A,N_A)+J(N_B,N_B)]/\\ &2<J(N_A,N_B)<J(N_B,N_B). \end{aligned} \tag{13}$$

Since $-\frac{N_A^2+N_B^2}{2}<-N_AN_B$ for the THBM, $-\frac{N_A^4+N_B^4}{2}<-N_A^2N_B^2$ for the CDM, and $-\frac{N_A+N_B}{2}<-\min\{N_A,N_B\}$ for the SM, the binding energies all satisfy the sorting condition. Therefore, cells should sort for all three binding models. Even cells with a continuous distribution of cadherin levels satisfy the sorting inequality, so cells with fewer adhesion molecules envelop cells with more adhesion molecules, which sort towards the center of the aggregate, creating an adhesion gradient, decreasing from the center to the periphery (Figure 1E), with a small amount of local mixing due to intrinsic cell motility. As mentioned above, sorting is a simple mechanism for cells to reach and maintain their positions during morphogenesis, *e.g.*, during limb outgrowth, in which cells maintain both their antero-posterior and proximo-distal positions through differential adhesion.

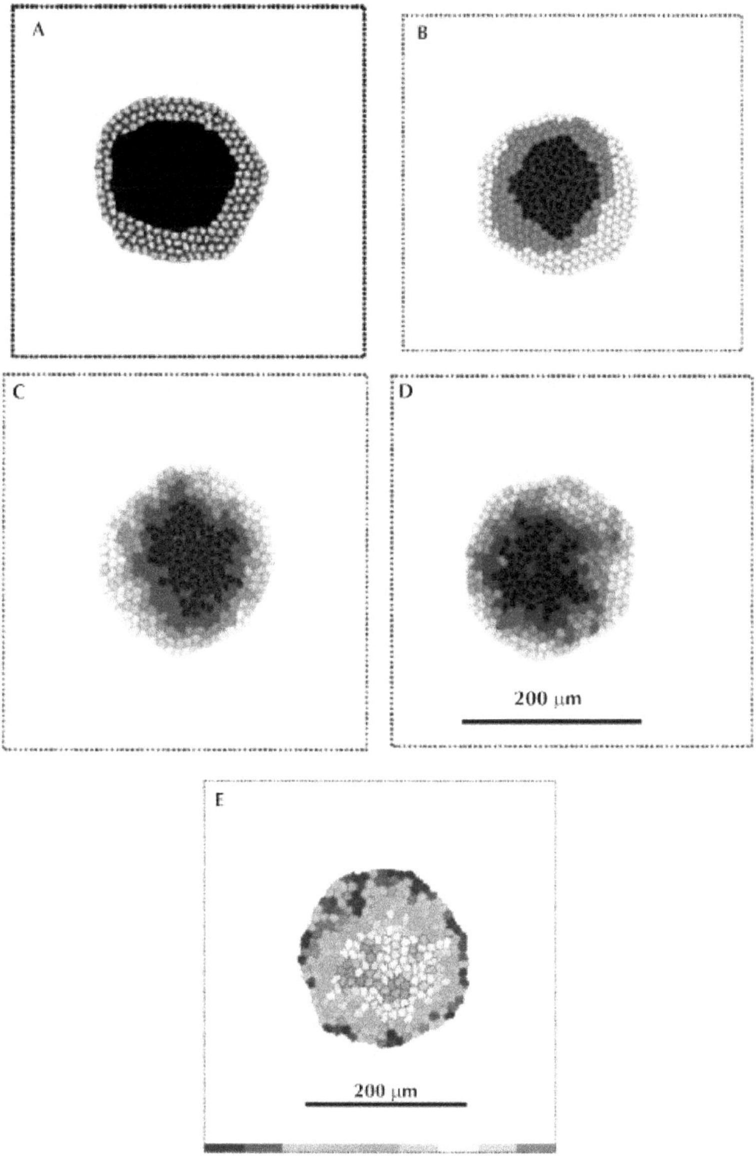

FIGURE 1 Typical simulated sorted configurations for aggregates of cells for the trans-homophilic-bond model (THBM). All images shown at time t = 999,000 MCS. In A–D, the gray-scale represents the cadherin-expression level. The darkest color (gray level = 0) represents the highest cadherin-expression level. The lightest color (gray level = 200) represents the lowest cadherin-expression level. The cell culture medium is white (gray level = 255). In (E), represent the expression levels, (H=[1–(N–N_{min})/(N_{max}–N_{min})] 255, S = 255, V = 255), where N is the cadherin-expression level, and N_{min} and N_{max} are the minimum and maximum cadherin-expression levels, respectively. Red (H=0) is the highest expression level, (H=255) the lowest expression level. The cell culture medium is white. Sorting for: (A) 2 levels. (B) 3 levels. (C) 5 levels. (D) 9 levels. (E) Continuous levels. Cadherin expression ranges from N_{min} = 1 to N_{max}=23. In all simulations, T = 20 and λ = 25.

- In an ideal, fully-sorted configuration, cells expressing the higher levels of cadherins will cluster together and round up into a solid sphere, surrounded by successive spherical shells of cells expressing successively lower levels of cadherins. To monitor the progress of cell sorting in our simulations, we define the heterotypic boundary length (*HBL*), the total contact length between cells with different cadherin levels, measured in pixels:

$$L_h = \sum_{\vec{i},\vec{i}'\ \text{neighbors}} \left(1 - \delta(N_{\sigma(\vec{i})}, N'_{\sigma'(\vec{i}')})\right). \tag{14}$$

- The simulations time evolution gradually minimizes L_h.
- If cells express multiple cadherin levels, L_W, the heterotypic boundary length weighted by the energy differences between neighboring cells is a better metric for cell sorting. This weighted heterotypic boundary length (WHBL) is simply the total interfacial tension (equations (8–10)) multiplied by the lengths:

$$L_{W_{CDM}} = -k_T \sum_{\vec{i},\vec{i}'\ \text{neighbors}} \left(N^2_{\sigma(\vec{i})} - N'^2_{\sigma'(\vec{i}')}\right)^2 \left(1 - \delta_{N_{\sigma(\vec{i})}, N'_{\sigma'(\vec{i}')}}\right) \Delta g/2, \tag{15}$$

$$L_{W_{THBM}} = -k_D \sum_{\vec{i},\vec{i}'\ \text{neighbors}} \left(N_{\sigma(\vec{i})} - N'_{\sigma'(\vec{i}')}\right)^2 \left(1 - \delta_{N_{\sigma(\vec{i})}, N'_{\sigma'(\vec{i}')}}\right) \Delta g/2, \text{ and} \tag{16}$$

$$L_{W_{SM}} = -k_M \sum_{\vec{i},\vec{i}'\ \text{neighbors}} \left(N_{\sigma(\vec{i})} - N'_{\sigma'(\vec{i}')}\right) \left(1 - \delta_{N_{\sigma(\vec{i})}, N'_{\sigma'(\vec{i}')}}\right) \Delta g/2. \tag{17}$$

Different aggregates may have different maximum (initial) and minimum heterotypic boundary lengths (HBL) or weighted heterotypic boundary lengths (WHBL). To compare sorting in different aggregates, we normalize these lengths using the transformation:

$$l_{norm} = \frac{L(t) - L_{Min}}{L_{Max} - L_{Min}}, \tag{18}$$

where $L_{Max} = \max\{L(t)\}$, $L_{Min} = L_k$ or L_{thea}, and $L(t)$ is the HBL or WHBL at time t. $L_k = \min\{L(t)\}$ is the minimum value of HBL or WHBL over the typical simulation duration of 10^6 MCS. L_{thea} is the theoretical minimum HBL or WHBL for the fully sorted and rounded aggregate, assuming that the cells form perfect concentric rings with perimeters equal to $2\pi R$ (R is the real radius of the ring of cells from the center of the aggregate). Experimentally, this value is easily calculated with digital imaging analysis, which gives us the total area of each type of cell. The *sorting relaxation time*, τ, is the time at which the aggregate reaches its typical, maximally-sorted configuration. τ is defined via the relation:

$$l_{norm}(\tau) = \frac{l_{norm}(0)}{e}, \tag{19}$$

The sorting rate, R_s, is the inverse of the sorting relaxation time:

$$R_S = \tau^{-1}. \tag{20}$$

12.3 SIMULATION IMPLEMENTATION

We first investigated sorting completeness for the trans-homophilic-bond model (THBM, equation (4)), with $k_D = 0.02$, as we moved from two levels of cadherin expression towards a continuous distribution of levels (two, three, five, nine and continuous levels) with the same range of cadherin numbers, $[N_{min} = 1, N_{max} = 23]$. The same range of cadherin expression numbers provides the same range of adhesion energies, independent of the number of levels.

It can be implemented simulations using the open-source software package CompuCell3D which allows rapid translation of biological models into simulations using a combination of CC3DML and Python scripting.

All simulations for cell sorting use aggregates of 305 cells, close to the size of a 2D section of the 3D aggregates experimentally studied by Armstrong, Steinberg and others [18, 41, 51], which are about 200 microns in diameter. Each cell has a 25-pixel target volume, which sets the lattice length scale to approximately 2 microns per pixel. We begin with a circular-disk aggregate with cells randomly assigned cadherin expression numbers, with each allowed number having equal probability. Each simulation uses $T = 20$ and runs for 10^6 MCS, to allow for complete sorting for continuous variations of cadherin expression over the range [1, 23]. We set $\lambda=25$, which allows patterns to evolve reasonably fast without large cell-volume or cell-surface-area fluctuations. Changing λ around this value does not greatly affect the relaxation of cells' shapes and positions. We further set $J_0 = 16$ (in equation (12)) for all simulations. For different cadherin binding models and for the cadherin expression range [1, 23], we choose the values of k_T, k_D and k_M (according to equation (3–5)), so that cells neither pin to the lattice nor dissociate.

12.4 DISCUSSION

At the beginning of a particular developmental phase, patterns of gene expression are often fuzzy initially, then gradually become distinct. Both changing cell identity and cell movement are possible mechanisms for refining initially fuzzy expression patterns or for fixing transient patterns of morphogens. Glazier *et al.* 2008 [49] and Watanabe *et al.* 2009 [55], found that, during somite segmentation, the fuzzy boundary formed by cells, disregarding positional cues and differentiating inappropriately, can reorganize to form a sharp boundary due to cell motility and differential adhesion. The sorting rate, and hence the rate of patterning, depend on the interfacial tensions, which in turn depend on the range of cadherin expression, equilibrium constants and free energies of cadherin bonds (see equations (8–10)). These mechanisms may act in parallel with, or coordinate with, other morphogenic mechanisms, such as Turing-type reaction-diffusion instabilities or Wolpertian threshold-based positional coding. Adhesion mechanisms act as an effective low-pass filter, reducing the effect of stochasticity in gene expression. During development, signaling cascades modulate cadherin expression. Because cell sorting is slow compared to fluctuations in gene-expression levels and because sorting rectifies noise into a stable gradient, transient fluctuations in cadherin expression will not change final morphology, increasing developmental robustness.

To provide better links/interplay between computer simulations and biological experiments, we would suggest carrying out measurements of the following key parameters [19, 22–25]: individual cell motilities, positions, contours and boundary lengths and tissue and single-cell level adhesion protein expression, elasticity and viscosity. While not always accessible, measurements of one or more adhesion-related parameters such as interfacial tension between cell aggregates, cell-cell adhesion forces or energies, molecular binding forces or energies and molecular binding and junction-formation kinetics would facilitate constructions of more realistic computer simulations. In particular, the ability to measure and then model temporal variation of adhesion related parameters is essential for simulations of complex developmental phenomena such as somitogenesis, limb growth, *etc*. Therefore future measurements should concentrate on dynamics of intra and inter-cellular mechanisms (*e.g.* intercellular signaling and regulatory networks) related to cellular adhesion [3, 26–28].

Our studies based on the GGH model, investigated how homotypic cadherin binding at the molecular level affects cellcell adhesion and determines cell sorting speeds at the tissue level. We have used three different microscopic models of cadherin-binding for discrete and continuous levels. The three binding mechanisms lead to similar cell-sorting behavior, although the saturation binding model is somewhat faster for larger aggregates with more cadherin levels. Sorting speed decreases with increasing numbers of cadherin levels. For classical sorting with two cadherin levels, sorting speed increases with the ratio between the two levels. Additionally, in each case a single optimum value for the cell motility results in the fastest sorting. Cell motilities above or below the optimum sort more slowly.

12.5 RESULTS

Figures 1A–E show final aggregates for cells expressing discrete or continuous levels of cadherins. Cells with higher expression (darker gray in Figures 1A–D, red in Figure 1E) assume more central positions, while cells with lower expression (lighter gray in Figures 1A–D, blue in Figure 1E) move to the periphery. For multiple discrete levels, cells follow a sorting hierarchy [17]; each layer of cells has a given expression number and surrounds the layer of cells with the next-higher level. For continuous levels, expression numbers decrease continuously from the center to the periphery of the aggregate (Figure 1E).

We investigated the evolution of the effective energy and the heterotypic boundary length (HBL)/weighted heterotypic boundary length (WHBL) for the THBM (equation (16)) in three cases:

1. Cells with different numbers of levels of cadherin expression, but the same range between maximum and minimum expression number.
2. Cells with different ranges between maximum and minimum expression number but with the same number of levels (two, for simplicity).
3. Cells with different motilities, but with the same cadherin levels.

We also investigated:

1. Cells with different cadherin binding models, but the same range between maximum and minimum expression number for two, five, nine and continuous levels.

Figure 2 shows sets of snapshots of simulations for cell aggregates with the THBM (equation (4) with $k_D = 0.02$, $T = 20$, and $\lambda = 25$) with cells expressing two [1, 23], three [1, 12, 23] , five [1, 6.5, 12, 17.5, 23], nine [1, 3.75, 6.5, 10.25, 12, 14.75, 17.5, 20.25, 23] cadherin levels.

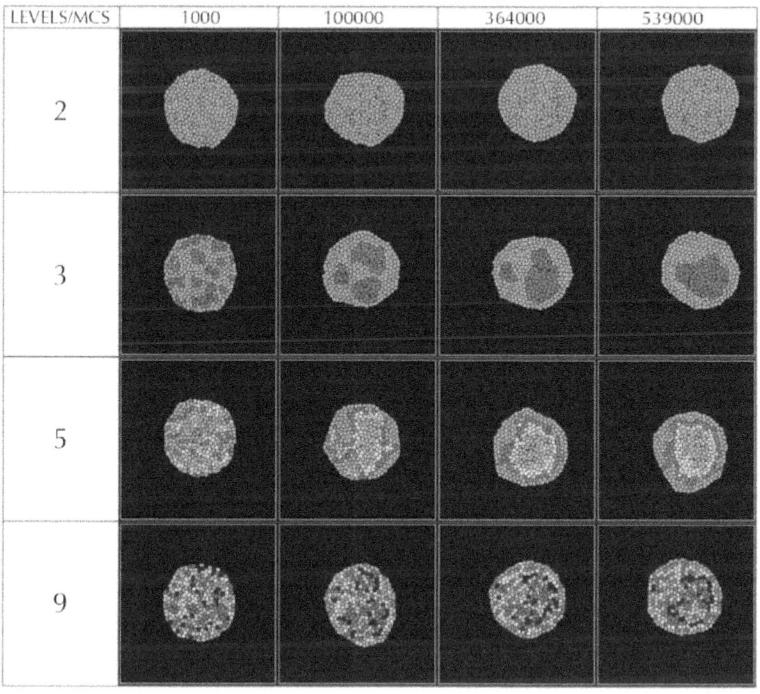

FIGURE 2 Simulation snapshots for aggregates with differing numbers of cadherin levels, with the same maximum to minimum expression range [1, 23], for the THBM.

Figure 3A shows the evolution of the effective energy H for the cell aggregates presented in Figure 2, and for cell aggregates with continuous cadherin levels in the range [1, 23] calculated using the THBM (equation (4) with k_D=0.02, T = 20, and λ = 25).Figures 3B and 3C illustrate the evolution of the normalized weighted heterotypic boundary length (NWHBL) for the cell aggregates in Figure 3A, setting $L_{min} = L_k$ and $L_{min} = L_{thea}$, respectively. Aggregates with two or three levels sort quickly, while those with more levels take more time to sort (Figure 2D).

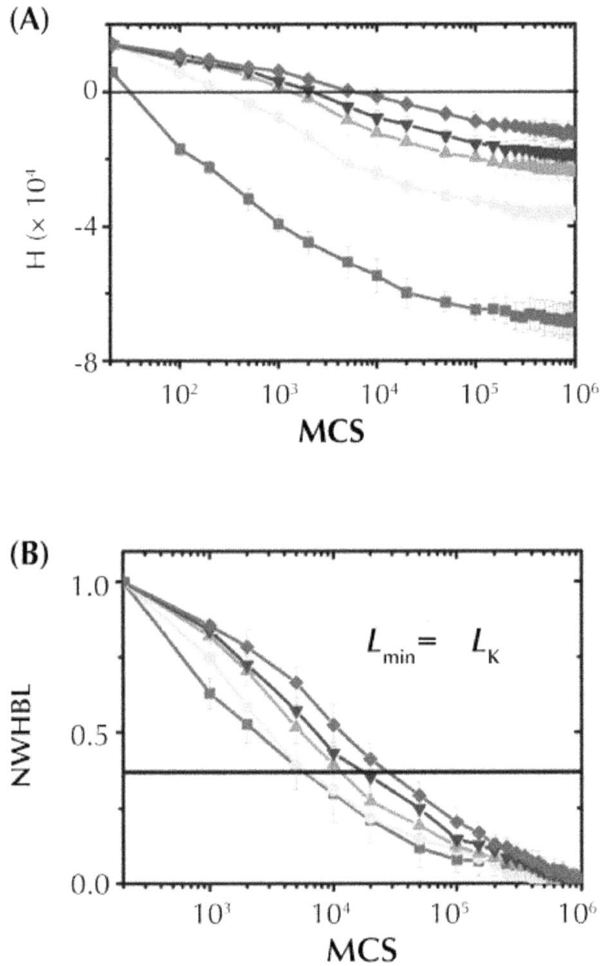

FIGURE 3 *(Continued)*

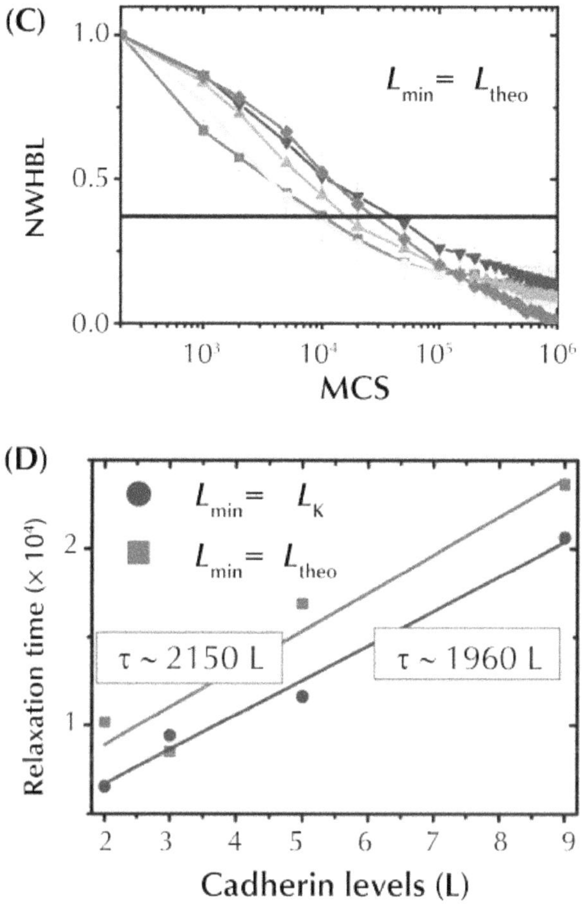

FIGURE 3 Evolution of the effective energies (H) and normalized weighted heterotypic boundary lengths (NWHBL) for aggregates with differing numbers of cadherin levels, with the same maximum to minimum expression range [1, 23], for the THBM. (A)–(C) ■ ▪– 2 levels; • – 3 levels; ▲ – 5 levels; ▼ – 9 levels; ● – continuous levels. The black horizontal lines mark 1/e. (A) Evolution of H. (B)–(C) Evolution of the NWHBL for the simulations in (A), with $L_{min}=L_k$ in (B) and $L_{min} = L_{thea}$ in (C). (D) Relaxation time vs. number of levels. ▪: $L_{min} = L_{thea}$. •: $L_{min} = L_k$. The graphs are calculated from ten simulation replicas.

Figure 4A shows the evolution of the effective energy H for aggregates with two cadherin levels, but different expression ranges: [1, 12], [1, 14.75], [1, 17.50], [1, 20.25], [1, 23], [12, 23], and [19.62, 23], also calculated using the THBM (equation (4) with k_D=0.02, T = 20, and λ=25). Figures 4B and 4C show the evolution of the NWHBL for the same aggregates, using $L_{min} = L_k$ and $L_{min} = L_{thea}$, respectively. Sorting is quickest ($\tau \simeq 14,000$ MCS) for aggregates with the widest cadherin expression range [1, 23], and is slowest (no complete sorting, τ = ∝) for aggregates with the smallest expression range [19.62, 23].

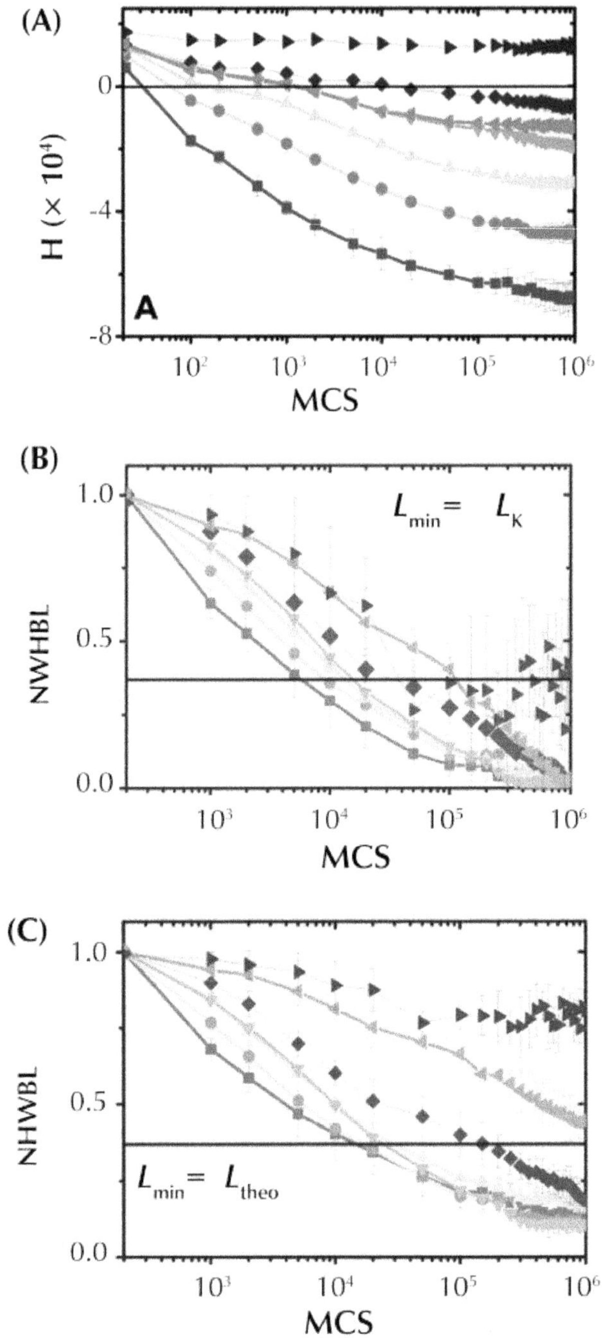
(A)
0
-4
-8
H (× 10⁴)
A
10²
10³
10⁴
10⁵
10⁶
MCS
(B)
1.0
0.5
0.0
NWHBL
L_min = L_K
10³
10⁴
10⁵
10⁶
MCS
(C)
1.0
0.5
0.0
NHWBL
L_min = L_theo
10³
10⁴
10⁵
10⁶
MCS

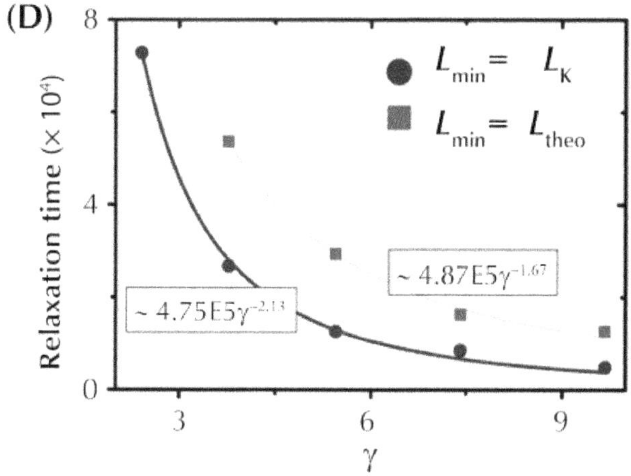

FIGURE 4 Evolution of the effective energies (H) and normalized weighted heterotypic boundary lengths (NWHBL) for aggregates expressing 2 cadherins levels modeled with the THBM. Expression ranges: ▪ – [1, 23]; • – [1, 20.25]; ▴ – [1, 17.50]; ▾ – [1,14.75]; ◂ – [12, 23]; ● – [1, 12]; ▸ – [19.62, 23]. Black solid horizontal lines mark 1/e. (A) Evolution of H. (B)–(C) Evolution of NWHBL for the simulations in (A) with: (B) $L_{min} = L_k$, and (C) $L_{min} = L_{thea}$. (D) Relaxation time vs. interfacial tension γ. Dots – simulation, and Lines – fitting curves ax^b. • – $L_{min} = L_k$; ▪ – $L_{min} = L_{thea}$. The error bars in the graphs are calculated from ten simulation replicas.

According to the theory of phase separation in liquids, the sorting rate for simple fluids is proportional to the interfacial tension divided by the viscosity' [52]. A similar relationship may hold for cell sorting [53]. Figure 4D plots the sorting relaxation time against the interfacial tension (equation 9) for the simulated aggregates in Figure 4A, and a power law (of form $\tau = a\gamma^b_{THBM}$, with a and b constants), fitting for both the cases $L_{min} = L_k$ and $L_{min} = L_{thea}$, respectively:

$$\tau_k = 4.75 \times 10^5 \gamma_{THBM}^{-2.13} \quad \text{and} \tag{21}$$

$$\tau_{theo} = 4.87 \times 10^5 \gamma_{THBM}^{-1.67}. \tag{22}$$

The fitting is reasonable, since for $L_{min} = L_{thea}$, the adjusted coefficient of determination $R^2 = 0.89$, and for $L_{min} = L_k$, $R^2 = 0.98$, suggesting that the sorting relaxation time and interfacial tension may obey an approximate power law with an exponent $b \simeq -2$.

In Figure 5 we compare the evolution of the effective energy H and of the NWHBL for the different cadherin binding models (CDM, THBM, and SM), with two, five, nine and continuous cadherin levels (the same levels as in Figure 3). We chose the effective equilibrium constants (see equations (8)–(10)), $k_D = 0.02$, $k_T = 0.000038$, and $k_M = 0.46$, so the cell-cell adhesion energies fell in the same range, excluding changes in cell sorting rates due to differences in these ranges. Figure 5D3 shows that as the number of expression levels increases from 2 to 5 to 9, the relaxation time increases for each model.

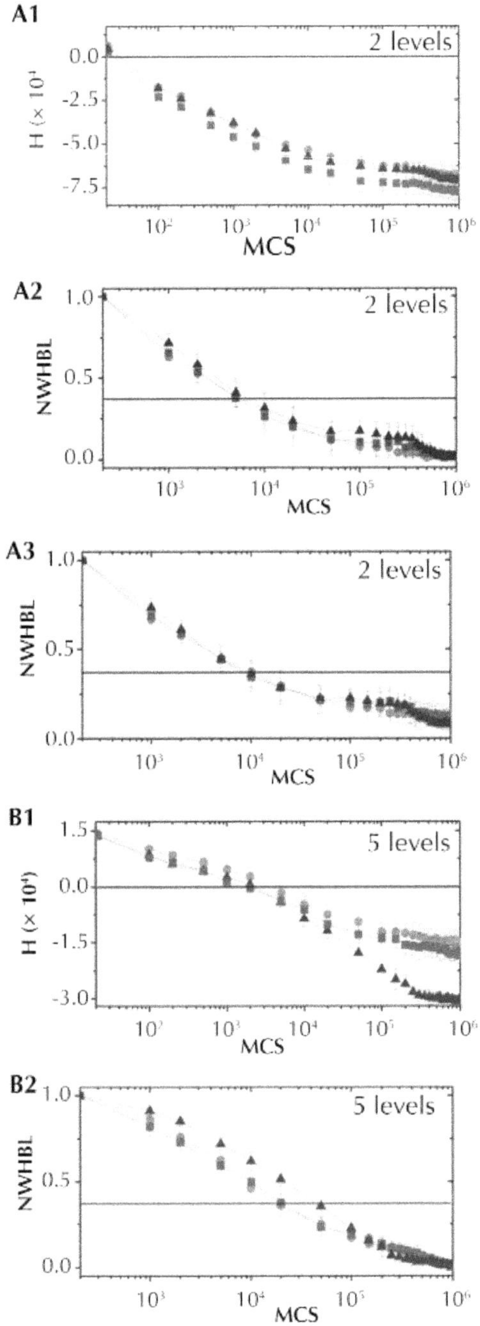

FIGURE 5 *(Continued)*

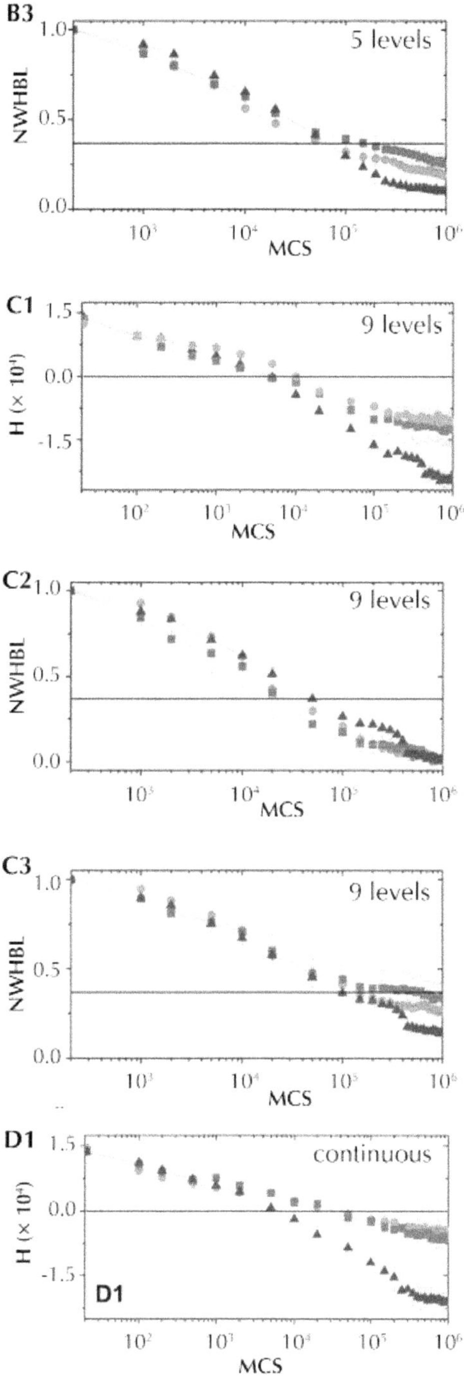

FIGURE 5 *(Continued)*

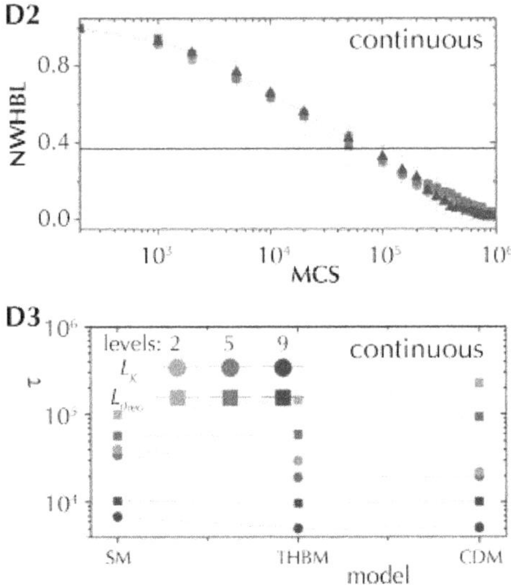

FIGURE 5 Evolution of the effective energies (H) and normalized weighted heterotypic boundary lengths (NWHBL) for aggregates with 2, 5, 9 or continuous cadherin levels using CDM, THBM or SM for the same expression range [1, 23]. In (A1)–(A3), (B1)–(B3), (C1)–(C3), and (D1)–(D2), ▪ – CDM; – THBM; SM. In (A2), (A3), (B2), (B3) and (C2)–(C4), the time at which the heterotypic boundary length of a given simulation crosses the horizontal black line is defined as its relaxation time. In (A2), (B2), (C2) and (D2) $L_{min} = L_k$. In (A3), (B3) and (C3) $L_{min} = L_{thea}$. (A1), (B1), (C1) and (D1) Evolution of the H for aggregates with cells expressing 2, 5, 9 and continuous cadherin levels respectively. (A2), (B2), (C2) and (D2) Evolution of NWHBL for the aggregates in (A1), (B1), (C1) and (D1), respectively, with $L_{min} = L_k$. (A3), (B3) and (C3) Evolution of the NWHBL for the aggregates in (A1), (B1) and (C1) respectively, with $L_{min} = L_{thea}$. (D3) Relaxation time vs. bond model for different cadherin expression levels. Circles– $L_{min} = L_{thea}$. Squares – $L_{min} = L_k$. The error bars in the graphs are calculated from ten simulation replicas.

For different models with the same cadherin expression levels, for two-level aggregates (Figure 5A1), sorting times are equal, as we expect because equations (8–10) give almost identical interfacial tensions. For aggregates with five and nine cadherin levels (Figures 5B1 and 5C1), sorting is more rapid for the saturation model (SM) and slowest for the trans-homophilic-bond model (THBM). The average minimum WHBLs are largest for the SM, but are the same for the cis-dimer model (CDM) and THBM. Since the weighted heterotypic boundary length (WHBL) is actually the interfacial tension, it is the main factor which determines the sorting rate.

Figure 6 shows sets of snapshots of simulations for cell aggregates with the THBM (equation (4) with $k_D = 0.02$, $T = 20$, and $\lambda = 25$) with five cadherin levels [1, 6.5, 12, 17.5, 23] and different cell motilities: 5, 10, 20, 40, 60, and 80.

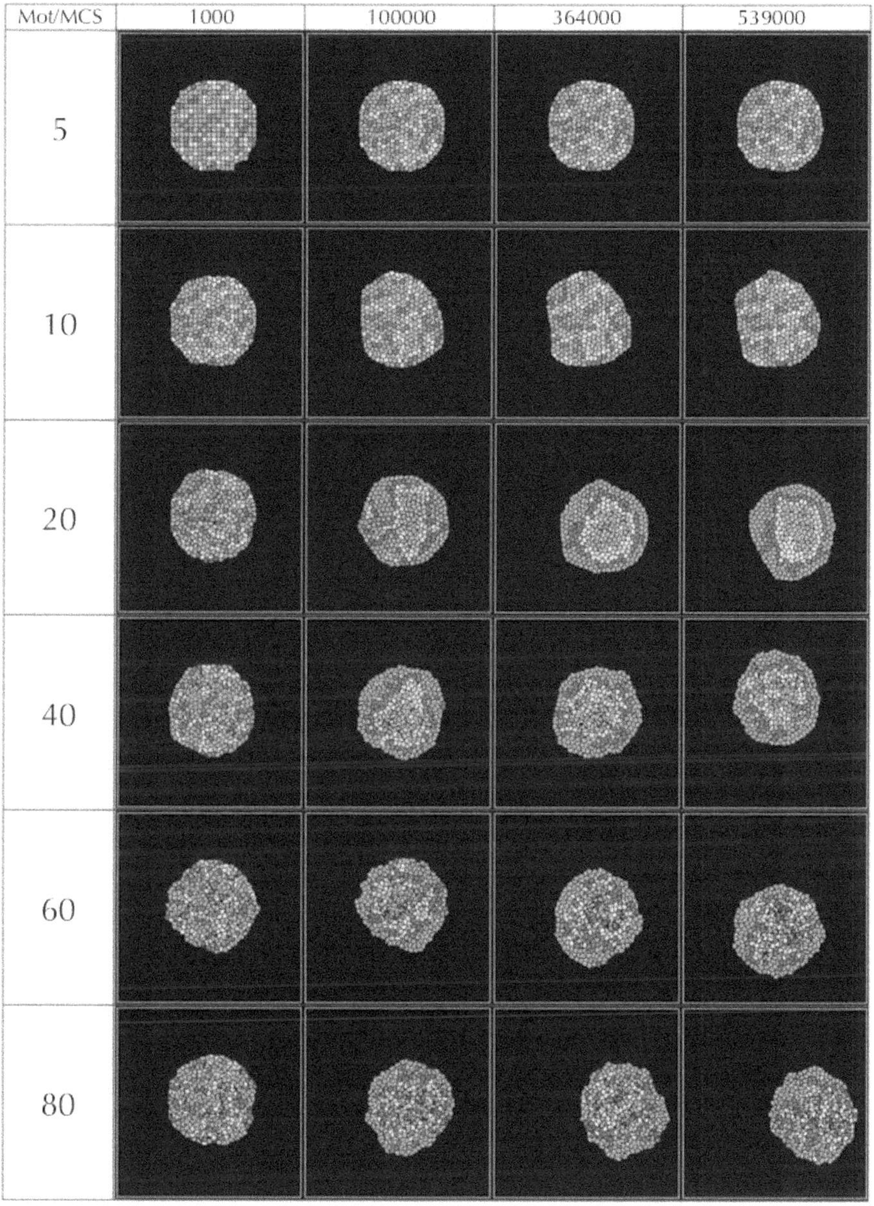

FIGURE 6 Simulation snapshots for aggregates with five levels [1, 6.5, 12, 17.5, 23] of cadherins and different cell motilities (5, 10, 20, 40, 60, 80), for the THBM.

Figure 7 shows the effect of cell motility on the evolution of the effective energy and normalized WHBL for aggregates with two cadherin levels using the THBM (with $k_D = 0.02$ and $\lambda = 25$). Figures 7A and 7B show the evolution of the effective energy for fixed λ. If the cell motility is very low ($T = 5$), cells pin before reaching their lowest-energy positions and sorting is slow. As the motility grows, the aggregates sort faster (Figure 5A). However, if the cell motility is too large ($T = 60$ and $T = 80$), sorting is rapid but remains incomplete (Figures 7B, 7C and 7D).

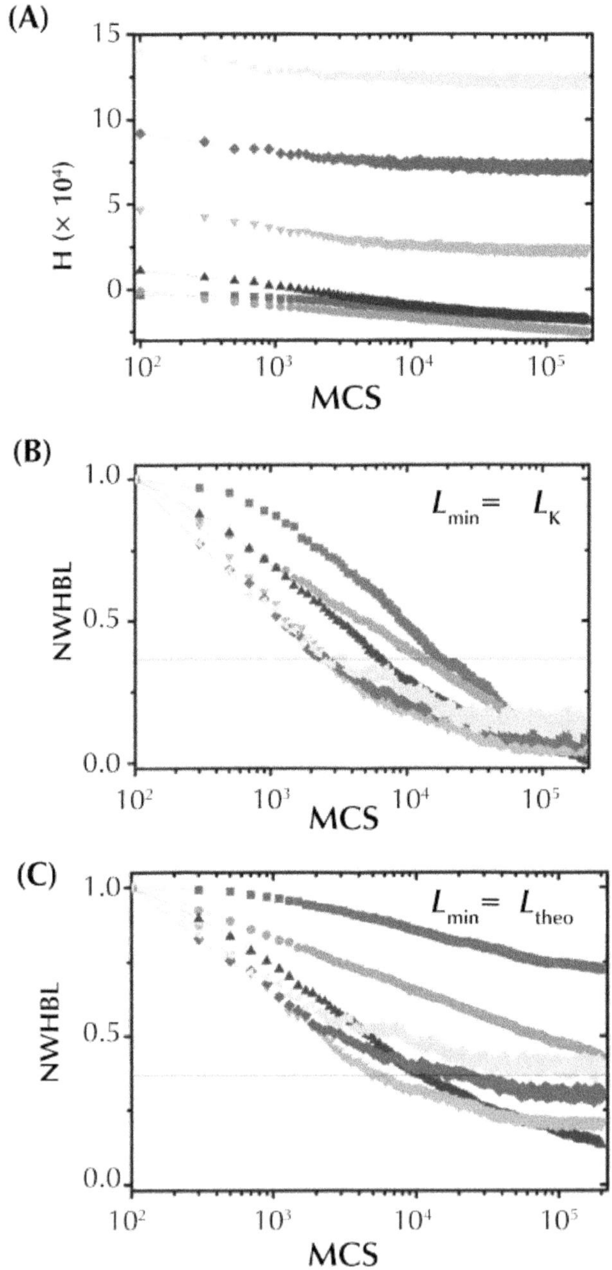

FIGURE 7 *(Continued)*

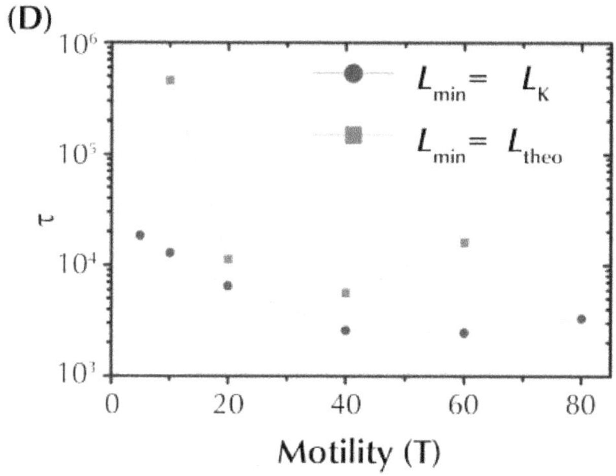

FIGURE 7 Evolution of the effective energies (H) and normalized weighted heterotypic boundary lengths (NWHBL) for aggregates with 5 cadherin levels and the same maximum to minimum expression range [1, 23] using the THBM with different motilities. ▪ – 5; • – 10; ▴ – 20; ▾ – 40; ● – 60; ◂ – 80. (A) Evolution of *H*. In (B) and (C) the time at which the heterotypic boundary length of a given simulation crosses the horizontal black line is defined as its relaxation time τ. In (B) $L_{min} = L_k$ and in (C) $L_{min} = L_{thea}$. (D) Relaxation time vs. relative cell motilities. ▪ – $L_{min} = L_{thea}$; • – $L_{min} = L_k$. The error bars in the graphs are calculated from ten simulation replicas.

When cells' expression of cadherin varies continuously, sorting still occurs, but more slowly than for discrete expression levels. The final configuration is imperfectly sorted since the intrinsic cell motility can overcome small differences in adhesion energy due to local missorting. The sorting rate depends on the interfacial tension rather than directly on the expression levels or the cadherin-binding model. Again, insufficient or excessive motility prevents complete sorting.

From the considerations above we can say that, although individually the sorting kinetics in aggregates with each binding model are sensitive to the number of cadherin levels and the energy range, all models have similar global behaviors. For each model, sorting is always faster for smaller numbers of cadherin levels, independent of the energy expression range. The dependence of sorting time and completeness on the number of cadherin levels is also similar for the three models, although the SM model seems to sort slightly faster and more completely for large numbers of cadherin levels. In the absence of experiments determining the model to use, the SM is computationally more efficient for larger aggregates.

The results could be checked by experiments controlling cadherin expression. *E.g.* we could transfect a GFP-cadherin plasmid construct into normally non-adherent CHO cells, so the amount of cadherin in each cell would be proportional to its fluorescence intensity. For discrete levels we could use multiple fluorescent tags. The cotransformation with a nuclear-targeted fluorescent protein of a different color would allow real-time cell tracking to determine cell motilities and positions.

Using the interfacial boundary length as a measure of sorting is experimentally inconvenient because current automated image segmentation cannot accurately extract the interfacial lengths from a stack of images. Instead, measuring the autocorrelation of the intensity in experimental and simulation image stacks would be much simpler. To represent a nuclear-targeted label in our simulations we could place a dot at each cell›s center of mass with an intensity proportional to the cell›s the number of cadherins. To represent cytoplasmic labeling, we could fill the entire cell volume with an intensity corresponding to the cadherin level and similarly for membrane labeling, we could label the cell›s contour.

An alternative measure of sorting would use a clustering algorithm to track the number and size of homotypic cell clusters. This approach is straightforward in CC3D and relatively easy to implement in experiments using K-Means or K-Median clustering algorithms, as described in [54]. Figure 8 shows an example of this procedure.

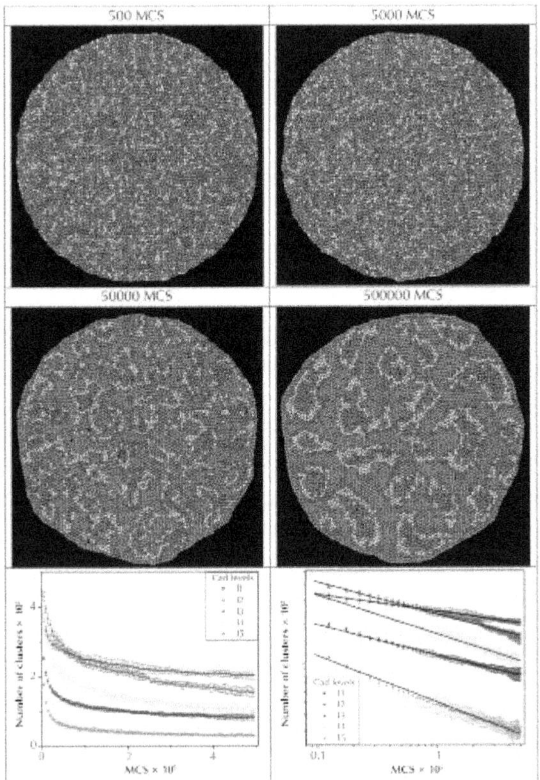

FIGURE 8 Clustering dynamics. First and second rows: snapshots taken from a 5,000 cell aggregate simulation with five levels of cadherins [1(11), 6.5(12), 12(13), 17.5(14), 23(15)] showing the dynamics of cluster formation. Bottom row: the left graph shows the evolution of the number of cluster for each cadherin level. The log-log graph (right) shows that the dynamics is adequately fitted by a power law of at^b, as indicated by the black lines. The error bars in the graphs are calculated from six simulation replicas.

We have used a bigger aggregate, with about 5000 cells in order to have a reasonable statistics. The cells have five levels of cadherins (as in Figure 6) and initially they are randomly distributed within the aggregate (top left snapshot). In our simple clustering algorithm, cells that express the same amount of cadherin and are in direct contact belong to the same cluster. The initial small clusters rapidly coalesce and form large clusters (top right and second row snapshots). The graphs at the bottom row show that the clustering rates decrease with time (left graph) and that they are adequately fitted by a power law of ab^t, as can be seen from black lines in the *log-log* graph at right. Mean and error bars for these graphs are calculated from six simulation replicas.

Comparing any of these bulk cell-sorting measures for experiments and simulations would allow us to infer the specific binding mechanism in a particular experiment, information otherwise difficult to obtain.

KEYWORDS

- **Adhesion Energy**
- **Cadherin**
- **Cell Clusters**
- **Cell-Sorting**
- **Molecular Level**

AUTHOR CONTRIBUTIONS

Conceived and designed the experiments: YZ JAG GLT AS. Performed the experiments: YZ GLT AS. Analyzed the data: GLT YZ JAG. Contributed reagents/materials/analysis tools: YZ GLT MS AS. Wrote the paper: GLT YZ JAG MS. Helped with coding: MS.

REFERENCES

1. Radice, G. L., Rayburn, H., Matsunami, H., Knudsen, K. A., Takeichi, K., Developmental defects in mouse embryos lacking N-cadherin. *Dev Bio* **181**: 64–78 (1997).
2. Price, S. R., de Marco Garcia, N. V., Ranscht, B., and Jessell, T. M. Regulation of motor neuron pool sorting by differential expression of type II cadherins. *Cell* **109**: 205–216. (2002).
3. Gumbiner, B. M. Regulation of cadherin-mediated adhesion in morphogenesis. *Nature Rev Mol Cell Bio* **6**: 622–634 (2005).
4. Xu, X., Li, W. E., Huang, G. Y., Meyer, R., and Chen, T., Modulation of mouse neural crest cell motility by N-cadherin and connexin 43 gap junctions. *J Cell Bio* **154**: 217–230 (2001).
5. Linask, K. K., Ludwig, C., Han, M. D., Liu, X., and Radice, G. L., Ncadherin/catenin-mediated morphoregulation of somite formation. *Dev Bio* **202**: 85–102 (1998).
6. Horikawa, K., Radice, G. L., Takeichi, M., and Chisaka, O. Adhesive subdivisions intrinsic to the epithelial somites. *Dev Bio* **215**: 182–189 (1999).
7. Takeichi, M. Cadherins in cancer: implications for invasion and metastasis. *Curr Op Cell Bio* **5**: 806–811 (1993).
8. Berx, G., and Van Roy, F. The E-cadherin/catenin complex: an important gatekeeper in breast cancer tumorigenesis and malignant progression. *Breast Cancer Res* **3**: 289–293 (2001).
9. Bement, W. M., Forscher, P., and Mooseker, M. S. A novel cytoskeletal structure involved in purse string wound closure and cell polarity maintenance. *J Cell Biol* **121**: 565–578 (1993).

10. Lorger, M. and Moelling, K. Regulation of epithelial wound closure and intercellular adhesion by interaction of AF6 with actin cytoskeleton. *J Cell Sci* **119**: 3385–3398 (2006).
11. Yajima, H., Yoneitamura, S., Watanabe, N, and Tamura, K., Ide, H. Role of N-cadherin in the sorting-out of mesenchymal cells and in the positional identity along the proximodistal axis of the chick limb bud. *Dev Dyn* **216**: 274–284 (1999).
12. Bitzur, S., Kam, Z., and Geiger, B. Structure and distribution of N-cadherin in developing zebrafish embryos: morphogenetic effects of ectopic over-expression. *Dev Dyn* **201**: 121–136 (1994).
13. Omi, M., Anderson, R., and Muneoka, K. Differential cell affinity and sorting of anterior and posterior cells during outgrowth of recombinant avian limb buds. *Dev Biol* **250**: 292–304 (2002).
14. Godt, D. and Tepass, U. Drosophila oocyte localization is mediated by differential cadherin-based adhesion. *Nature* **395**: 387–391 (1998).
15. Von der Hardt, S., Bakkers, J., Inbal, A., Carvalho, L., and Solnica-Krezel, L., The Bmp gradient of the zebrafish gastrula guides migrating lateral cells by regulating cell-cell adhesion. *Curr Bio* **17**: 475–487 (2007).
16. Steinberg, M. S. Reconstruction of tissues by dissociated cells. Some morphogenetic tissue movements and the sorting out of embryonic cells may have a commonexplanation. *Science* **141**: 401–408 (1963).
17. Steinberg, M. S. and Wiseman, L. L. Do morphogenetic tissue rearrangements require active cell movements? The reversible inhibition of cell sorting and tissue spreading by cytochalasin B. *J Cell Bio* **55**: 606–615 (1972).
18. Steinberg, M. S. and Takeichi, M. Experimental specification of cell sorting, tissue spreading, and specific spatial patterning by quantitative differences in cadherin expression. *Proc Nat Ac Sc USA* **91**: 206–209 (1994).
19. Foty, R. A. and Steinberg, M. S. The differential adhesion hypothesis: a direct evaluation. *Dev Bio* **278**: 255–263 (2005).
20. Friedlander, D. R., Mege, R. M., Cunningham, B. A., and Edelman, G. M. Cell sorting-out is modulated by both the specificity and amount of different cell adhesion molecules (CAMs) expressed on cell surfaces. *Proc Nat Ac Sc USA* **86**: 7043–7047 (1989).
21. Niessen, C. M., Gumbiner, B. M. Cadherin-mediated cell sorting not determined by binding or adhesion specificity. *J Cell Biol* **156**: 389–399 (2002).
22. Baumgartner, W., Hinterdorfer, P., Ness, W., Raab, A., and Vestweber, H., Cadherin interaction probed by atomic force microscopy. *Proc Nat Ac Sci USA* **97**: 4005–4010 (2002).
23. Chu, Y. S., Thomas, W. A., Eder, O., Pincet, E., and Perez, E., Force measurements in E-cadherin-mediated cell doublets reveal rapid adhesion strengthened by actin cytoskeleton remodeling through Rac and Cdc42. *J Cell Biol* **167**: 1183–1194 (2004).
24. Panorchan, P., Thompson, M. S., Davis, K. J., Tseng, Y., and Konstantopoulos, K., Single-molecule analysis of cadherin-mediated cell-cell adhesion. *J of Cell Sc* **119**: 66–74 (2006).
25. Prakasam, A. K., Maruthamuthu, V., and Leckband, D. Similarities between heterophilic and homophilic cadherin adhesion. *Proc Nat Ac Sc USA* **103**: 15434–15439 (2006).
26. Angres, B., Barth, A., Nelson, W. J. Mechanism for transition from initial to stable cell-cell adhesion: kinetic analysis of E-cadherin-mediated adhesion using a quantitative adhesion assay. *J Cell Biol* **134**: 549–557 (1996).
27. Adams, C. L. and Nelson, W. J. Cytomechanics of cadherin-mediated cell-cell adhesion. *Curr Op Cell Biol* **10**: 572–577 (1998).
28. Adams, C. L., Chen, Y. T., Smith, S. J., and Nelson, W. J. Mechanisms of epithelial cell-cell adhesion and cell compaction revealed by high-resolution tracking of E-cadheringreen fluorescent protein. *J Cell Biol* **142**: 1105–1119 (1998).
29. Behrens, J., Birchmeier, W., Goodman, S. L., Imhof, B. A. Dissociation of Madin-Darby canine kidney epithelial cells by the monoclonal antibody anti-arc-1: mechanistic aspects and identification of the antigen as a component related to uvomorulin. *J Cell Biol* **101**: 1307–1315 (1985).
30. Halbleib, J. M. and Nelson, W. J. Cadherins in development: cell adhesion, sorting, and tissue morphogenesis. *Genes & Development* **20**: 3199–3214 (2006).

31. Shapiro, L., Fannon, A. M., Kwong, P. D., Thompson, A., and Lehmann, M. S., Structural basis of cell-cell adhesion by cadherins. *Nature* **374**: 327–337 (1995).
32. Sivasankar, S., Brieher, W., Lavrik, N., Gumbiner, B. M., and Leckband, D. Direct molecular force measurements of multiple adhesive interactions between cadherin ectodomains. *Proc Nat Ac Sc USA* **96**: 11820–11824 (1999).
33. Chappuis-Flament, S., Wong, E., Hicks, L. D., Kay, C. M., and Gumbiner, B. M. Multiple cadherin extracellular repeats mediate homophilic binding and adhesion. *J Cell Biol* **154**: 231–243 (2001).
34. Zhu, B., Chappuis-Flament, S., Wong, E., Jensen, I E., and Gumbiner, B. M., Functional analysis of the structural basis of homophilic cadherin adhesion. *Biophys J* **84**: 4033–4042 (2003).
35. Pertz, O., Bozic, D., Koch, A. W., Fauser, C.,and Brancaccio, A., A new crystal structure, Ca^{2+} dependence and mutational analysis reveal molecular details of E-cadherin homoassociation. *EMBO J* **18**: 1738–1747 (1999).
36. Boggon, T. J., Murray, J., Chappuis-Flament, S., Wong, E., and Gumbiner, B. M., C-cadherin ectodomain structure and implications for cell adhesion mechanisms. *Science 296*: 1308–1313 (2002).
37. Chen, C. P., Posy, S., Ben-Shaul, A., Shapiro, L., and Honig, B. H. Specificity of cell-cell adhesion by classical cadherins: Critical role for low-affinity dimerization through beta-strand swapping. *Proc Nat Ac Sc USA* **102**: 8531–8536 (2005).
38. Graner, F. and Glazier, J. A. Simulation of biological cell sorting using a twodimensional extended Potts model. *Phys Rev Lett* **69**: 2013–2016 (1992).
39. Glazier, J. A., and Graner, F. Simulation of the differential adhesion driven rearrangement of biological cells. *Phys Rev E* **47**: 2128–2154 (1993).
40. Glazier, J. A., Balter A, Poplawski, N. J. *Single-Cell-Based Models in Biology and Medicine*, Birkhauser-Verlag, Basel, Switzerland, chapter Magnetization to morphogenesis: A brief history of the Glazier-Graner-Hogeweg model. pp. 79–106 (2007).
41. Mombach, J. C. and Glazier, J. A. Single cell motion in aggregates of embryonic cells. *Phys Rev Lett* **76**: 3032–3035 (1996).
42. Rieu, J. P., Upadhyaya, A., Glazier, J. A., Ouchi, N. B., and Sawada, Y. Diffusion and deformations of single hydra cells in cellular aggregates. *Biophys J* **79**: 1903–1914 (2000).
43. Zajac, M., Jones, G. L., and Glazier, J. A. Model of convergent extension in animal morphogenesis. *Phys Rev Lett* **85**: 2022–2025 (2000).
44. Maree, A. F. and Hogeweg, P. How amoeboids self-organize into a fruiting body: multicellular coordination in Dictyostelium discoideum. *Proc Nat Ac Sc USA* **98**: 3879–3883 (2001).
45. Zeng, W., Thomas, G. L., and Glazier, J. A. Non-Turing stripes and spots: a novel mechanism for biological cell clustering. *Physica A* **341**: 482–494 (2004).
46. Dan, D., Mueller, C., Chen, K., and Glazier, J. A. Solving the advection-diffusion equations in biological contexts using the cellular Potts model. *Phys Rev E* **72**. 041909 (2005).
47. Merks, R. H. M., Glazier, J. A. A cell-centered approach to developmental biology. *Physica A* **352**: 113–130 (2005).
48. Poplawski, N. J., Swat, M., Gens, J. S., Glazier, J. A. Adhesion between cells, diffusion of growth factors, and elasticity of the AER produce the paddle shape of the chick limb. *Physica A* **373C**: 521–532 (2007).
49. Glazier, J. A., Zhang, Y., Swat, M., Zaitlen, B., and Schnell, S. Coordinated action of N-CAM, N-cadherin, EphA4, and ephrinB2 translates genetic pre-pattern into structure during somitogenesis in chick. *Curr Top Dev Biol* **81**: 205–247 (2008).
50. Metropolis, N., Rosenbluth, A. W., Rosenbluth, M. N., Teller, A. H., and Teller, E. Equations of State Calculations by Fast Computing Machines. *J ChemPhys* **21**: 1087–1092 (1953).
51. Armstrong, P. B. Light and electron microscope studies of cell sorting in combinations of chick embryo neural retina and retinal pigment epithelium. *Wilhelm Roux'Archiv* **168**: 125–141 (1971).
52. Frenkel, J. Viscous flow of crystalline bodies under the action of surface tension. *J Phys* **4**: 385–431 (1945).

53. Beysens and D. A., Forgacs, G., Glazier, J. A. Cell sorting is analogous to phase ordering in fluids. *Proc Nat Ac Sc USA* **97**: 9467–9471 (2000).
54. Steinhaus, H. Sur la division des corps materiels en parties. *Bull Acad Polon Sci* **4**: 801–804 (1956).
55. Watanabe, T., Sato, Y., Saito, D., Tadokoro, R., and Takahashi, Y. EphrinB2 coordinates the formation of a morphological boundary and cell epithelialization during somite segmentation. *Proc Natl Acad Sci* USA **106(**18): 7467–7472 (2009).

Author Notes

CHAPTER 1

Funding

This work was supported by NIH Subaward No. 5710002511 to J.J.T. through the Integrative Cancer Biology Program at MIT (Prime Award No. U54-CA112967); by NSF award DMS-0342283 to J.J.T.; by NIH award R01-CA73413 to J.W.J., and P30-CA43703 for Core Facilities at CWRU. The funders had no role in study design, data collection and analysis, decision to publish, or preparation of the manuscript.

Competing Interests

The authors have declared that no competing interests exist.

Acknowledgments

We thank Keith E. Schultz for collecting the data in Fig. 4, as well as Kathy Chen and Tongli Zhang for helpful suggestions. The cyclin A2 antibody was a gift from Vincent Shankey, Beckman-Coulter (Miami). Mammalian Cell Cycle Regulation was originally published as "A Hybrid Model of Mammalian Cell Cycle Regulation" in PLoS Computational Biology 2011, 7(2): e1001077. Used with permission of the authors.

Author Contributions

Conceived and designed the experiments: RMS JWJ. Performed the experiments: RMS. Analyzed the data: RS JWJ JJT. Wrote the paper: RS RMS JWJ JJT. Conceived modeling approach: JJT. Implemented the model and ran all simulations: RS. Designed the model: JWJ.

CHAPTER 2

Competing Interests

The authors declare that they have no competing interests.

Author Contributions

DAH, WJB, and GG designed research; DAH performed simulations and analyzed data; DAH, WJB, and GG wrote the paper. All authors read and approved the final manuscript.

Acknowledgments

Financial support of this project by the European Network of Excellence "SoftComp" through a joint postdoctoral fellowship for DAH is gratefully acknowledged. Spindles and Active Vortices was originally published as "Spindles and Active Vortices in a Model of Confined Filament-Motor Mixtures" in BMC Biophysics 2011; 4:18. Used with permission of the authors.

CHAPTER 3

Funding

This study was funded by National Institutes of Health grants GM 55507 to Jonathan Scholey, GM068952 to Alex Mogilner, and GM 068952 to Alex Mogilner and Jonathan Scholey. The funders had no role in study design, data collection and analysis, decision to publish, or preparation of the manuscript.

Competing Interests

The authors have declared that no competing interests exist.

Acknowledgments

We thank Dr. Jonathan Scholey for hosting all experimental work in his laboratory and for fruitful discussions. Actomyosin-Dependent Cortical Dynamics was originally published as "Actomyosin-Dependent Cortical Dynamics Contributes to the Prophase Force-Balance in the Early *Drosophila* Embryo" in PLoS ONE 2011, 6(3): e18366. Used with permission of the authors.

Author Contributions

Conceived and designed the experiments: PS DC IB AM. Performed the experiments: PS DC IB. Analyzed the data: PS DC IB AM. Wrote the paper: PS IB AM.

CHAPTER 4

Funding

We gratefully acknowledge the support of the NSF (http://www.nsf.gov) through DMR-0605044 (EJB and AJL), DMR-0520020 (AJL), and PHY-0957573 (NSW). ZG is supported by a HFSP (http://www.hfsp.org/) Young Investigator Award and NIH (http://www.nih.gov/) grant 1DP2OD004389. The funders had no role in study design, data collection and analysis, decision to publish, or preparation of the manuscript.

Competing Interests

The authors have declared that no competing interests exist.

Acknowledgments

We thank Rob Phillips for helpful discussions. Filament Depolymerization During Bacterial Mitosis was originally published as "Filament Depolymerization Can Explain Chromosome Pulling During Bacterial Mitosis" in PLoS Computational Biology 2011, 7(9): e1002145. Used with permission of the authors.

Author Contributions

Conceived and designed the experiments: EJB ZG NSW AJL. Performed the experiments: EJB MAG. Analyzed the data: EJB. Wrote the paper: EJB NSW AJL.

CHAPTER 5

Funding

This work was funded by the Max Planck Society.

Competing Interests

The authors have declared that no competing interests exist.

Author Contributions

Conceived and designed the experiments: AH HB. Performed the experiments: HB. Analyzed the data: AZ. Wrote the paper: FJ AZ KK. Other: Developed the theory: FJ AZ KK. Actin Ring Constriction was originally published as "Stress Generation and Filament Turnover During Actin Ring Constriction" in PLoS ONE 2007, 2(8): e696. Used with permission of the authors.

CHAPTER 6

Funding

N.G. thanks the Alvin and Gertrude Levine Career Development Chair, and the BSF grant No. 2006285 for their support. This research is made possible in part by the historic generosity of the Harold Perlman Family. G.S. and A.D. thank the AIRC (Associazione Italiana Ricerca sul Cancro) grants No. 4874 and 8678 for their support. G.S. thanks the PRIN2007 (progetti di ricerca di interesse nazionale) and the International Association for Cancer Research (AICR 09-0582) for their support. The funders had no role in study design, data collection and analysis, decision to publish, or preparation of the manuscript.

Competing Interests

The authors have declared that no competing interests exist.

Author Contributions

Conceived and designed the experiments: BP GS AD NG. Performed the experiments: BP GS AD NG. Analyzed the data: BP GS AD NG. Contributed reagents/materials/analysis tools: BP GS AD NG. Wrote the paper: BP GS AD NG. Curved Activators and Cell-Membrane Waves was originally published as "Propagating Cell-Membrane Waves Driven by Curved Activators of Actin Polymerization" in PLoS ONE 2011, 6(4): e18635. Used with permission of the authors.

CHAPTER 7

Acknowledgments

The work reported here has been supported by SPP 1128 of the Deutsche Forschungsgemeinschaft and the Max-Planck-Gesellschaft. It was performed in collaboration with the Light Microscope Facility of the MPI-CBG in Dresden, The Oklahoma Medical Research Foundation, and the Nikon Imaging Center of the University of Heidelberg. I thank Carsten Beta, University of Potsdam for discussions on bistability, Mary Ecke and Jana Prassler for expert assistance, and Terry O'Halloran, University of Texas, Austin, for the GFP-clathrin light-chain vector. Actin Waves and Phagocytic Cup Structures was originally published as "Self-Organizing Actin Waves That Simulate Phagocytic Cup Structures" in PMC Biophysics 2010, 3:7. Used with permission of the author.

CHAPTER 8

Funding

CNRS. The funders had no role in study design, data collection and analysis, decision to publish, or preparation of the manuscript.

Competing Interests

The author has declared that no competing interests exist.

Acknowledgments

I would like to thank Paul Nurse for insightful discussions and for his reading of the manuscript; Albert Libchaber for fruitful discussions; and Axel Buguin, Felice Kelly, Mark Lekarew, James Moseley, Frank Neumann, and the Nurse lab for stimulating discussions and comments. Yeast and Scaling was originally published as "Explaining Lengths and Shapes of Yeast by Scaling Arguments" in PLoS ONE 2009, 4(7): e6205. Used with permission of the author.

Author Contributions

Wrote the paper: DR. Conceived and developed the approach: DR.

CHAPTER 9

Funding

This study was supported by the U.S.-Israel Binational Science Foundation (grant 2006285 to NSG), and the Ministry of Absorption of Israel (DK). The funders had no role in study design, data collection and analysis, decision to publish, or preparation of the manuscript.

Competing Interests

The authors have declared that no competing interests exist.

Acknowledgments

We thank useful discussions with Kinneret Keren. Cellular Shapes from Protrusive and Adhesive Forces was originally published as "Theoretical Model for Cellular Shapes Driven by Protrusive and Adhesive Forces" in PLoS Computational Biology 2011, 7(5): e1001127. Used with permission of the authors.

Author Contributions

Conceived and designed the experiments: DK RS KS TS NSG. Performed the experiments: KS TS. Analyzed the data: DK RS NSG. Contributed reagents/materials/analysis tools: KS TS. Wrote the paper: DK RS NSG.

CHAPTER 10

Competing Interests

The author has declared that no competing interests exist.

Author Contributions

Conceived and designed the experiments: PVH. Performed the experiments: PVH. Analyzed the data: PVH. Contributed reagents/materials/analysis tools: PVH. Wrote the paper: PVH. Amoeboid Cells' Protrusions was originally published as "Amoeboid Cells Use Protrusions for Walking, Gliding and Swimming" in PLoS ONE 2011, 6(11): e27532. Used with permission of the author.

CHAPTER 11

Funding

Masataka Yamao was supported by Global COE Program in NAIST (Frontier Biosciences: Strategies for survival and adaptation in a changing global environment), MEXT (Ministry of Education, Culture, Sports, Science, and Technology), Japan. Honda Naoki was supported by a research fellowship from the Japan Society for the Promotion of Science from the MEXT (Ministry of Education, Culture, Sports, Science, and Technology) of Japan. Grant number: 19-11235, http://www.jsps.go.jp/j-pd/index.html. Shin Ishii and Honda Naoki were supported by the Next Generation Supercomputing Project (Shin Ishii and Honda Naoki) from the MEXT (Ministry of Education, Culture, Sports, Science, and Technology) of Japan. Grant number: 100081400012, http://www.nsc.riken.jp/index-eng.html. The funders had no role in study design, data collection and analysis, decision to publish, or preparation of the manuscript.

Competing Interests

The authors have declared that no competing interests exist. Collective Cell Migration was originally published as "Multi-Cellular Logistics of Collective Cell Migration" in PLoS ONE 2011, 6(12): e27950. Used with permission of the authors.

CHAPTER 12

Funding

The authors acknowledge support from Indiana University and The Biocomplexity Institute at Indiana University and National Institutes of Health grants NIGMS R01-GM76692 and R01-GM077138 and The EPA grant "Texas-Indiana Virtual STAR Center." GLT acknowledges support from the Brazilian agencies Conselho Nacional de Pesquisa e Desenvolvimento (CNPq) and Fundação de Amparo à Pesquisa do Estado do Rio Grande do Sul (FAPERGS) under the grant PRONEX-10/0008-0. The funders had no role in study design, data collection and analysis, decision to publish, or preparation of the manuscript.

Competing Interests

The authors have declared that no competing interests exist. Differential Adhesion and Cell Sorting was originally published as "Computer Simulations of Cell Sorting Due to Differential Adhesion" in PLoS ONE 2011, 6(10): e24999. Used with permission of the authors.

Author Contributions

Conceived and designed the experiments: YZ JAG GLT AS. Performed the experiments: YZ GLT AS. Analyzed the data: GLT YZ JAG. Contributed reagents/materials/analysis tools: YZ GLT MS AS. Wrote the paper: GLT YZ JAG MS. Helped with coding: MS.

Index

A

Actin ring constriction
  C. Elegans
    in Embryo, 98–99
    end-on imaging, 97
  continuum description, 105–106
  cross-links keep filaments, 97–98
  cytokinesis, 96
  and depolymerization, force generation in filament bundles, 104
  Dictyostelium cells, 96–97
  experimental observations, comparison, 102
  filament polymerization, force generation in filament bundles, 104
  force balance and material exchange, 100–101
  minimal model, 98
  molecular motors, force generation in filament bundles, 102–104
  Schizosaccharomyces pombe, 96
Actin waves and phagocytic cup structures
  in bistable systems, comparable to trigger waves, 137
  coronin, 133
  Dictyostelium cells, 133–134
  Dictyostelium discoideum, cells, 134–135
  LatA, 134
  phosphoinositide composition, organization and membrane areas, 139–142
  and PIP3 in phagocytic cups, 142
  PIP3 synthesis, 134
  propagation in *Dictyostelium* to other actin-based cell functions, relation, 135–137
  protein patterns in phagocytic cups, 142
  shape dynamics, 138
  spatial organization, 138
  structure and dynamics
    myosin-IB and coronin, 137
  TIRF microscopy, 142
Actomyosin-dependent cortical dynamics
  actin cap area in myosin-inhibited embryos, astral MTs, 62
  actin cap size, quantification, 56–57
  actin distribution at cortex, myosin II inhibition changes, 59–62
  additive effect, 62–67
  Drosophila embryo, 49–51
    injection, 53
  fly stocks, 52–53
  force balance in prophase, 52
  image acquisition, 53
  microtubule (MTs) and actin-myosin arrays, 50
  mosin inhibition in a *Drosophila* syncytial embryo, 58
  NEB, 50
  principal findings, 50
  Scrambled (Sced) mutants, 50, 51
  spindle-cortex interaction, 51
  synergistic actions, 54
  WT embryo, 55
Amoeboid cells' protrusions
  adhesion-defective mutant cells, 198
  amoeboid cell, protrusions, 193–194
  amoeboid movement, 193
  cAMP gradient, 194
  cell migration and chemotaxis, 194
  cells, pseudopods, 201
  chemoattractant, shallow gradient, 194
  convexities in tail-attached cells, traveling waves, 197
  cytoplasm, 201
  Dictyostelium cells, 194, 201–202
  gliding *gbpD*-null cells, 200
  GTP-binding protein, 197
  methods, 195
  pseudopods, 197
  swimming cells, 199
  swimming cells use pseudopods, 197–198, 200
  walking cells
    side bumps, 195

on solid support, 195
Anaphase Promoting Complex-Cyclosome (APC-C), 14–16
Anaphase Promoting Complex—Cyclosome (APC-C), 14
AndersonDarling statistics, 39
AndersonDarling test, 45
Andrea J. Liu, 71–94
APC-C. *see* Anaphase Promoting Complex-Cyclosome (APC-C)

B

Banigan, E. J., 71–94
Bin/Amphiphysin/Rvs (BAR), 159
Boolean networks (BNs), 3–4
Briels, W. J., 25–47
Bringmann, H., 95–108
Brust-Mascher, I., 48–69
Budding yeast, 26

C

Cadherin, 238
CAM. *see* Cell adhesion molecule (CAM)
Caulobacter crescentus, 72
Cdk's. *see* Cyclin-dependent kinases (Cdk's)
CDR. *see* Circular Dorsal Ruffles (CDR)
Cell adhesion molecule (CAM), 207
Cellular shapes from protrusive and adhesive forces
actin polymerization and adhesion, effects, 185
actin polymerization, shape evolution driven by, 186–187
adhesion, shape evolution driven by, 186
BAR, 159
cell motility, 184
cells, 158, 170
coalescence, evolution of cellular protrusions, 179–181
coarse-grained model, 162
crescent-cell shape transition, 178
cytoskeleton activity, 162
cytoskeleton-membrane structures, 184
ECM, 158
Ena/VASP, localization, 179
equations of motion, 165–168
experimental methods, 163
flat geometry, 168–169
fluorescence signal, 174
IMD, 159
linear stability analysis, 168
model details, 163–165
myosin, 161–162, 177
numerical simulations, 169–170
observed and calculated cell shapes, qualitative comparison, 174
protrusion coalescence, dynamics, 172
protrusive force, 159, 187
round cell geometry, 169
steady-state shape, coalescence dynamics, 187–188
steady-state shapes, 173
Cheerambathur, D., 48–69
Circular Dorsal Ruffles (CDR), 109–110, 127
Cis-dimer model (CDM), 239
Collective cell migration
CAM, 207
cellular transportation, system-level control, 206
collective migration, 206
contact inhibition, 219
effective interaction identification, 208–212
filopodia and lamellipodia, 219
interaction between, 225–230
logistics, 206
migration patterns cell
snapshots of, 214
multi-cellular, 206
multi-cellular logistics, properties, 230–232
multi-cellular migration
model of, 219–221
multi-cellular systems
cellular migration characteristics, 216
multi-cellular transportation, single cellular stochasticity effect, 222–225
non-dimensionalization, 207–208
proliferative activity, 212–213
stochastic forces, 218

stream formation, 212
theoretical models, 212
Wnt signal molecule, 219
Collective migration, 206
Cultured cells, contact inhibition, 20
HUVEC, 19, 21
Curved activators and cell-membrane waves
actin polymerization activity, 123
bifurcation analysis, 116
Boltzmann factor, 114
CDRs, 109–110
counting, 112
time lapse, 112
cell culture and reagents, 112
drug treatment and staining, 110–111
experimental results, 127
group velocity, 124
immunofluorescence microscopy, 112
linear stability analysis, 114–115
living cells, 110
lnear stability and simulation results, 122
model details, 112–114
non-linear simulations, 117
strong concave activator approximation, 118
theoretical results, 127–130
wave instability phase diagram, 116
Cyclin-dependent kinases (Cdk's), 3
Cyclin-dependent kinases (Cdk's), 3
Cyclin distributions in asynchronous culture
data, 18–19
test for, 18

D

Dictyostelium cells, 96–97, 137–138, 194, 201
Differential adhesion and cell sorting
actin-disrupting drugs, 238
binding model, 239
cadherin molecule, 237
cell-cell adhesion force, 238
cell membrane, 239
cell motilities, 246
cell sorting, GGH simulation
adhesion molecules, 242
antero-posterior position, 242
cadherin expression range, 243
cellular potts model, 241
cohesive cell, 242
effective energy, 241
heterotypic boundary length (HBL), 244
Monte Carlo step (MCS), 241
proximo-distal positions, 242
weighted heterotypic boundary length (WHBL), 244
clustering dynamics, 258
conformational changes, 239
discrete levels, 257
E-cadherin binding (*See* reaction-kinetic models)
effective energies, 249
excluding changes, 251
gene expression pattern, 245
GGH model, 237
intercellular adhesion, 239
interfacial boundary, 258
interfacial tensions, 246, 254
junction-formation kinetics, 246
molecular level morphogenesis, 238
multiple transcriptional signaling, 238
normalized weighted heterotypic boundary length *(NWHBL),* 248
post-translational signaling, 238
simulation implementation
biological models, 245
cadherin expression, 245
cell-volume fluctuations, 245
python scripting, 245
simulation replicas, 259
somite segmentation, 245
sorting speed, 246, 249
tissue-level morphogenesis, 238
Differential Adhesion Hypothesis (DAH), 238
Disanza, A., 109–132
Drosophila, 238
Dynamics and active vortices
AndersonDarling statistics, 39
fillaments, collective rotation, 36
microtubule-kinesin assays, 36

vortex, snapshot, 37

E

ECM. *see* Extra-cellular matrix (ECM)
Euler formula, 153
Extra-cellular matrix (ECM), 158

F

Filament depolymerization during bacterial mitosis
biochemistry method, 75
brownian dynamics simulations, 72
Caulobacter crescentus, chromosome, 72
equations of motion, 77–78
fast hydrolysis and depolymerization, 84
interactions method, 75–77
ParB-binding, 73
ParB-*parS-ori* complex, 73
ParB polymer
detachment force for, 89–91
para bundle, 91
ParB translocation
simulation, 83
velocity, rate of disassembly controls, 84–88
Par-mediated chromosome pulling, comparisons, 80–81
phenomena, implications for, 81–83
reliable chromosome segregation, 72
robust for side-binding of ParB to ParA with disassembly, translocation, 80
self-diffusiophoresis, 78–80
side-binding with, 84
tip-only
binding, 84
disassembly, side-binding with, 84
translocation velocity, 83
dependence of, 89
three regimes, 88–89
units method, 75
Vibrio cholerae, chromosome, 72
weak binding, 84
Fission yeast cell, 146
shape, two laws, 146–147
Fission yeast cells, 25–26

G

Gelbart, M. A., 71–94
Gerisch, G., 133–143
Gitai, Z., 71–94
Glazier Graner Hogeweg (GGH) model, 237
Glazier, J. A., 237–262
Gompper, G., 25–47
Gov, N., 109–132
Gov, N. S., 157–192

H

Head, D. A., 25–47
Human Umbilical Vein Endothelial cell line (HUVEC), 19
HUVEC. *see* Human Umbilical Vein Endothelial cell line (HUVEC)
Hybrid modeling approach
APC-C, 14
Boolean variables, 14, 16
cyclin, 16, 18
Gaussian-distributed, 18
of mammalian cell cycle control, 17
proliferation signals, 16
SCF labels cyclin, 14
TFE and TFB drives, 14
transcription factors, 16
G2—M transition, 16
Hyman, A. A., 95–108

I

IRSp53-Missing-In-Metastasis (IMD), 159
Ishii, S., 205–235

J

Jacobberger, J. W., 1–24
Jülicher, F., 95–108

K

Kruse, K., 95–108

L

Linear-zipper model (LZM), 239
Logistics, 206

M

Mammalian cell cycle regulation
- antibodies, 7–8
- asynchronous culture, cyclin distributions, 18–19
- Boolean networks (BNs), 3–4
- Boolean switching networks, 1–2
- Cdk's, 3
- cell-cycle location, 6
- cell division cycle, 1, 2
- culture and fixation, 7
- cultured cells, contact inhibition, 19, 21
- cyclin
  - dynamics, model predictions, 13
  - synthesis, 2
- cyclin abundances, 2
- cyclin synthesis, 2
- data pre-processing, 8
- discrete variables, 2
- DNA synthesis, 1
- flow cytometry, 7–8
- Gaussian distribution, 7
- G1 phase, 2–3
- human tumor cell line use, 8
- hybrid modeling approach, 14–16, 18
- ideal probability density, asynchronous population, 6
- immunofluorescence staining, 7–8
- methods
  - daughter cell, mass, 6
  - each cell in asynchronous sample, DNA and cyclin levels, 6–7
  - 'life histories' for thousands of cells, creation, 4–6
- mitosis, 1
- physiological behaviors of cells, 2
- scatter plots, 12
- switching network, Boolean representation, 8
- temporal dynamics, 1
- transformation method, 6

MEFs. *see* Mouse embryo fibroblasts (MEFs)

Mogilner, A., 48–69

Mosin inhibition in a *Drosophila* syncytial embryo
- results in greater centrosome separation during prophase, 58–59

Mouse embryo fibroblasts (MEFs), 112

Multi-cellular migration, 206

N

Naoki, H., 205–235

NEB. *See* Nuclear envelope breakdown (NEB)

NETO length, New–End Take Off (NETO) length

New-End Take Off (NETO) length, 150–152

Nuclear envelope breakdown (NEB), 50

P

Peleg, B., 109–132

R

Reaction-kinetic models
- cadherin binding, 239
- cadherin concentrations, 240
- equilibrium concentration, 240
- individual cells, 239
- interfacial-tension density, 240–241
- saturation model (SM), 240
- THBM, 239

Riveline, D., 145–156

S

Saccharomyces cerevisiae, 135

Schizosaccharomyces pombe, 96, 145

Schloen, K., 157–192

Scita, G., 109–132

Shirinifard, A., 237–262

Shlomovitz, R., 157–192

Singhania, R., 1–24

Sommi, P., 48–69

Spindles and active vortices
- AndersonDarling test, 45
- *budding yeast,* 26
- controlled volume, 43–45
- discrete numerical model, 27
- and dynamics, 36–41
- end-detachment rate, 30
- ends, enhanced detachment, 42–43

filament-motor protein models, 26
in *fission* yeast cells, 25–26
fixed volume vortices, 44
high-dimensional parameter space, 30
mean-squared angular changes, 41
mesoscopic models, 27
methods, 27
3D filament gels, 29
motors, 28
Monte Carlo Metropolis, 29
results
stationary states, 31–33, 35
static vortex, 30
steady-state structure, variation, 30
vortex-like motion, 27
Xenopus, 26
Sramkoski, R. M., 1–24
Stationary states of spindles and active vortices, 31
mode amplitudes *Qm,* 33
motor density, 35
motors per filament, mean number, 33
plus-end motors, 36
snapshots of steady-states, 32
state diagrams, 34
Stradal, T., 157–192
Swat, M., 237–262

T

Thomas, G. L., 237–262
Trans-homophilic-bond model (THBM), 239
Traveling waves, convexities in tail-attached cells
Dictyostelium cells, 197
Tyson, J. J., 1–24

V

Van Haastert, P. J. M., 193–204
Vorticity
Binomial test, 41
and contraction time
reciprocal relationship, 39
statistics, 40
defined, 38
diffusive limit, 41
mean-squared angular changes and deviation, 41
onset of, 41

W

Walking cells
side bumps, 195
on solid support, 195
Wingreen, N. S., 71–94

X

Xenopus, 26

Y

Yamao, M., 205–235
Yeast and scaling
cells, 145
C-shape, 152–154
Euler formula, 153
fission yeast cell, 146
fission yeast shape, two laws, 146–147
membrane surface tension as motor for recruiting growth machinery, low value, 147–148
molecular mechanisms, role, 147
NETO length, 150–152
pressure difference with use of cell wall tension, estimate, 148–149
Schizosaccharomyces pombe, 145
septum location, length at mitosis, 149–150
Young-Laplace equation, 146
Ying Zhang, 237–262
Young-Laplace equation, 146

Z

Zumdieck, A., 95–108

For Product Safety Concerns and Information please contact our EU
representative GPSR@taylorandfrancis.com
Taylor & Francis Verlag GmbH, Kaufingerstraße 24, 80331 München, Germany

www.ingramcontent.com/pod-product-compliance
Ingram Content Group UK Ltd.
Pitfield, Milton Keynes, MK11 3LW, UK
UKHW021505240926
475702UK00035B/450
* 9 7 8 1 9 2 6 8 9 5 3 6 9 *